THE
MYSTERIOUS ORIGINS
OF HYBRID MAN

Crossbreeding *and the* Unexpected Family Tree *of* Humanity

Susan B. Martinez, Ph.D.

Bear & Company
Rochester, Vermont • Toronto, Canada

Bear & Company
One Park Street
Rochester, Vermont 05767
www.BearandCompanyBooks.com

Bear & Company is a division of Inner Traditions International

Library of Congress Cataloging-in-Publication Data
Martinez, Susan B.
  The mysterious origins of hybrid man : crossbreeding and the unexpected family tree of humanity / Susan B. Martinez, Ph.D.
      pages cm
  Includes bibliographical references and index.
  Summary: "A provocative challenge to Darwin's theory of evolution"—Provided by publisher.
  ISBN 978-1-59143-176-3 — ISBN 978-1-59143-754-3 (e-book)
  1. Human evolution. 2. Breeding. 3. Heredity. I. Title.
  GN281.M377 2013
  599.93'8—dc23
                                                                    2013011642

Printed and bound in the United States by McNaughton & Gunn

10 9 8 7 6 5 4 3 2 1

Text design by Brian Boynton and layout by Virginia Scott Bowman
This book was typeset in Garamond Premier Pro with Gill Sans used as the display typeface

To send correspondence to the author of this book, mail a first-class letter to the author c/o Inner Traditions • Bear & Company, One Park Street, Rochester, VT 05767, and we will forward the communication, or contact the author directly at **poosh8@gmail.com.**

Thank you, John White, my "jungle guide," for midwifing this project with such enthusiasm and depth of understanding.
Much gratitude to Stephanie Howard for expert scans.
And many thanks to Dr. Marvin E. Herring ("meh") for his wonderful cartoons, and to Meghan MacLean for saving the (editorial) day.

CONTENTS

Beware the science of man.

EDWARD BULWER-LYTTON, *THE COMING RACE*

I think I shall avoid the whole subject.

CHARLES DARWIN, IN 1857 WHEN ASKED IF HIS
FORTHCOMING BOOK, *THE ORIGIN OF SPECIES,*
WOULD TREAT OF MAN'S EVOLUTION

A Note on Evolution, Chronology, and Abbreviations

When I use the word *evolution* it does not mean I endorse the concept. The word appears many times in references or inside quotes or as part of an argument. Also, when I cite certain dates, this does not mean an endorsement either, but quoted for the record and sometimes in a relative, not absolute, sense. When I mention such dates for man's existence on Earth, particularly any date earlier than eighty thousand years ago (80 kya), it is only according to the evolutionists' time scale—not mine! I do not take it as a true calendar date. There are many places in this book where I cite an age in the millions of years; while I don't believe it for a second, I use it in the context of the argument presented. Please also note that the word *man* is used as a convenient shorthand for humanity (feminists, please forgive).

In addition to being able to recognize the frequently mentioned fossil men throughout this book, it may be useful to be familiar with the following abbreviations:

Abbreviation	Meaning
AMH	anatomically modern human
Ar.	Ardipithecus (genus)
Au.	Australopithecus (genus)
Au	australopithecine, australopith
BP	before present
[e.a.]	emphasis added (to a quote)
H.	Homo (genus)
kya	thousand years ago
kyr	thousand years old
mod	morphologically modern
mya	million years ago
myr	million years old

Below is a table of the fossil types most frequently referenced throughout the text, with the corresponding date or range of dates attributed to them, along with a brief description of relevant features.

HOMINID GALLERY

Most Frequently Mentioned Fossil Men

NAMES	WHERE/WHEN	DESCRIPTION
Ardipithecus ramidus (Ardi, Asu)	Ethiopia/4.4 myr	Curved fingers, 350 cc
Australopithecus/ Praeanthropus anamensis (Kanapoi)	Kenya/4.2 myr	No chin but mod tibia

HOMINID GALLERY

Most Frequently Mentioned Fossil Men

NAMES	WHERE/WHEN	DESCRIPTION
Australopithecus afarensis (Lucy)	Africa/3 myr (date controversial)	Pygmy sized, modern footprint
Australopithecus africans (Mrs. Ples)	South Africa/1 to 3 myr	Au but mod pelvis
Australopithecus africanus (Taung Child, Dart's Child)	South Africa/2 myr	Brain and teeth more mod than face
Australopithecus/ Paranthropus robustus	Southern Africa/2 myr	Stocky, rugged, huge molars
Australopithecus sedipa	South Africa/2 myr	Upgraded Au
Homo habilis (handy man)	East Africa/2 myr	Short, gracile, long arms; apparently an improved Au
Homo rudolfensis (Skull 1470)	East Africa/2 myr	Pre–*H. erectus* with mod-shaped brain
Australopithecus/ Paranthropus boisei (Zinj)	East Africa/1.7 myr	Robust Au with huge cheek teeth
Homo ergaster (ER 3883, 3733)	Koobi Fora, Africa/1.5 to 1.75 myr	Tall, early *H. erectus?*; archaic features mixed with modern ones
KNM-WT 15000 (Turkana Boy)	Kenya/1.6 myr	Mosaic: primitive head on mod body
Homo erectus (Druk, ground people)	Various locations/1.8 myr to 300 kyr	Some very large
Homo heidelbergensis (Mauer Man, Heidelberg man)	Europe, China, Israel/130 to 750 kyr	Pre-Neanderthal but modern dentition
Pithecanthropus erectus (Java Man)	Java/500 kyr	First *H. erectus* discovered
Homo rhodesiensis/ heidelbergensis (Kabwe Man, Rhodesian Man, Broken Hill Man)	Zambia/40 to 400 kyr	Tall, strong *H. erectus–* Neanderthal mix
Sinanthropus/Homo erectus pekinensis (Peking Man)	China/300 kyr	Classic burly *H. erectus*

HOMINID GALLERY

Most Frequently Mentioned Fossil Men

NAMES	WHERE/WHEN	DESCRIPTION
Qafzeh hominid	Israel/30 to 100 kyr	Proto-Cro-Magnons contemporary with Neanderthals
Homo floresiensis (hobbit, Flo)	Flores, Indonesia/12 to 95 kyr	Tiny, very mixed morphology
Skhul hominid	Israel/40 to 80 kyr	AMH crania and vocal tract mixed with Neanderthal traits
Fontéchevade Man	France/70 kyr?	Pre-Neanderthals with mod traits
LM1, LM2, LM3 (Mungo Man)	Australia/up to 60 kyr	Delicate AMHs *before* erectoids
Denisova hominid	Altai, Siberia/40 kyr	Confusing hybrid of mod and primitive

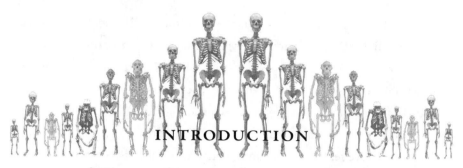

THE UNEXPECTED FAMILY TREE

How was the beginning of man? How was it with the first of the living?

OAHSPE, BOOK OF JEHOVIH, 6:8

IN THE DOCKET

As the world looks on, America the beautiful has a nasty squabble on its hands that won't go away. We just can't seem to agree on the right way to teach human origins to our schoolchildren. More than once, the U.S. Supreme Court has stepped in, slapping the wrist of Bible belt states that saw fit to introduce creation science to the curriculum.* Emotional, confrontational, and guileful, too, the schism between God-free science and intelligent design (ID) has laid bare a great divide in our country—and in our world.

---

*As of 2012 Tennessee and Louisiana are the only states that allow teachers to challenge the theory of evolution. Curiously, the 1925 Scopes Monkey Trial, which upheld the law against teaching Darwinism in schools, was in the state of Tennessee.

1

The U.S. Constitution's First Amendment—ironically?—favors the nonbelievers, for religious doctrines may not be imposed (by law) on Americans. Yet religious Americans themselves feel imposed upon by school boards that have swung to the science side, not allowing God or design into the discussion. Science and religion, argued one of Darwinism's more fluent advocates in 1999, Stephen Jay Gould, are "totally separate realms."[1] Yet today, with the theory of everything (ToE) looming large on the science horizon, Gould may be proven quite wrong about this intellectual apartheid. The single axiom, the great unity, that final theory that has not yet been ferreted from nature, promises to be sweeping enough to include amoeba, galaxy—and God.

Gould was not above using condescending terms like *ignorance, absurdity,* and *anti-intellectualism* in his (1999) polemic against America's "religious fundamentalists." He frankly thought God "an illusion" and the concept of soul "antiquated" (*Ever Since Darwin*). Like-minded citizens (most recently in Kansas, Pennsylvania, Georgia, Michigan, and Ohio) resent the creationist agenda for "posing a threat to the minds of our children." Could well-educated people "really believe that ID [intelligent design] can rival evolution?" asks *New Scientist,* the influential British magazine that is today's leading anticreationist gladiator; they swear that this is one "battle we cannot afford to lose."[2]

They are right. Intelligent design is an awesome rival that could put them all out of business. ID has been around a lot longer than "AD," Darwin's accidental design (I jump into the debate in chapter 8). Although ID is labeled a creation myth by opponents, evolution itself is beset with so many indefinites as to qualify as our own state mythology, an unimpeachable sacred cow. It is ironic that the word *indoctrination,* which was once applied to the overbearing dogmas of *religious* teaching, can now apply to the tenets of atheism and its pet theory—evolution.

Gould's remarks in that 1999 article make it abundantly clear that evolution brooks no heretics, for it is "one of the greatest triumphs of human discovery . . . as well documented as any phenomenon in science, as strongly as the earth's revolution around the sun"; therefore "no one ignorant of evolution can understand science." At the end of this book you are holding, the Earth will still be revolving around the sun, but will inquiring minds be revolving around Darwin's pedigree for man? I hope not.

THE ONLY GAME IN TOWN

If Darwin's evolution was the subversive underdog in the mid-nineteenth century, to *oppose* it today is the subversive, even unthinkable thing. Freedom of thought and speech? Not really. The new religion of the intelligentsia is nonreligion. This is a one-party system with its own secret policing, a kind of oligarchy, poised to keep adversaries at bay. Few among the educated, commented American anthropologist C. Loring Brace, could doubt evolution. What was once a nineteenth-century struggling theory is now an *idée fixe*. Human descent with modification (from apes) is now "an established truth . . . [like] the existence of atoms."[3] Darwinism, declared evolutionary biologist Ernst Mayr, is "the greatest intellectual revolution experienced by mankind."[4] But grandstanding doesn't make it so.

Politicized, the functionaries of evolution deliver explanations in standardized terminology that serves as the straitjacket of evolution; for example, the term *Java ape-man* clearly prejudices things in favor of a missing link. Anthropoid, which means "manlike," includes apes, humans, and monkeys! Even the agreement on a superfamily called Hominoidea is biased, for it lumps together the great apes and ourselves. Australopiths (Au) are called bipedal apes,[5] though, as we'll see, they were humans, not apes. Very recently indeed (2010) Douglas Palmer, like most other British anthropologists, still refers to *Australopithecus afarensis* as "these apes." (*Au. afarensis,* we will see in chapter 1, were the earliest humans—not apes at all.) But the deck is loaded: Definition of hominid? All primates more like man than ape.

Biased terminology sets the stage, the new word *hominin* means any descendant of the last common ancestor of humans and chimpanzees, including humans and their ancestors, as well as apes (chimps and gorillas). But the common ancestor dogma is pure assumption, which I confront in chapter 10.

PALEOPOLITICS

The subtext of all this confidence, this bravado, is the tumbling house of cards called evolution, laced with self-congratulation and softened with false modesty: When older Darwinian ideas are trumped by new evidence, why, this only shows that evolution is a marvelously flexible, healthy, and self-correcting science. But this is really built-in deniability, clothed as flexibility. John Feliks, writing about this, has said: "There is one far-reaching and influential field in which the quality of self-correction is notably absent. This is the field of human origins." Feliks goes on to comment that "the system of peer review in paleoanthropology is devoted entirely to belief in Darwinian evolution. . . . Conflicting evidence must be kept from publication."[6]

Dean Falk calls it paleopolitics: The VIPs of evolution disagree about almost everything except evolution itself. To my mind, the extent of quibbling is proportionate to its ill-founded premise. Almost every important find is contested amid a maelstrom of conflicting fossil information. One new fossil find, and everything gets reclassified. Referring to *Homo erectus, Homo sapiens,* and Neanderthal, paleoanthropologist Marvin L. Lubenow writes: "I cannot think of any scenario that more clearly demonstrates the meaninglessness of these categories."[7] What this means is that there exist no clearly defined taxa in the whole of human evolution. And for a good reason: All the overlapping among these types of fossil men is due not to evolution but to hybridization. Taxonomy doesn't stand a chance in the face of the ongoing amalgamations that this book is about. Again and again, the fossil record is an unsolvable puzzle of mosaics—each hominid type possessing hopelessly mixed racial traits.

Despite the disagreements and conflicts in this field, "disputation," argues the professional skeptic Michael Shermer, "is at the heart of a robust science."[8] But is nonstop infighting and sub rosa backstabbing really a sign of health and openness? Such debates, moreover, are all *within* evolutionary theory not between evolutionary theory and something else. Of course, the only disputation allowed is *among* insiders who may disagree on their competing models, triggering vicious debates, but they close rank and circle their wagons with the approach of "the empty logic of Darwin's doubters."[9]

The case of *H. sapiens* remains open and unsolved. Detective work is at its best when you seek whys and wherefores, not when you approach with forgone conclusions. Oh, there might be suspects, but even persons of interest are supposedly innocent until proven otherwise. Today, studies couched in evolution's special jargon are designed to *exemplify* human evolution, not to test it, not to put it on trial. All alternatives, goes the Darwinian cant, have been "thoroughly refuted."[10] So why do the refutations continue without stint? Helena Cronin calls critics of evolution "not even in the same league."[11] Censorship, I am afraid, is built into this bureaucracy of knowledge, which now basks in a near-perfect monopoly. When evolutionists refer to an alternate theory as discredited—know that this means it was shot down summarily by the evolution police: examples might include the lost race theory (see chapter 2) or high civilization in the Mesolithic.

Finds, no matter which science we are talking about, are interpreted to satisfy the leading theories of the day. As in any industry, career is at stake—grants, promotions, tenure, publications, professional reputation, respectability. One candid professor of paleoanthropology had this to say about the state of the art: "Human evolution is a big deal these days. Leakey's world known, Johanson is like a movie star. . . . Lecture circuit. National Science Foundation: big bucks. Everything is debatable, especially where money is involved. Sometimes people deliberately manipulate data to suit what they're saying."[12] Another frank observer noted that "as essential funding is brought more and more under centralized government control, researchers have no alternative but to concentrate upon the agenda set by the paradigm."[13]

Play by the rules, or you're out of the game. You cannot go out and dig without money; and it is harder and harder to be funded without a promise of groundbreaking work. Breakthroughs in evolution are the cash cow.

> *Scientists are in exactly the same position as Renaissance painters, commissioned to make the portrait the patron wants done. . . . [T]he system works against problem-solving. Because if you solve a problem, your funding ends. . . . Scientists are only too aware whom they are working for. Those who fund research . . . always have a particular outcome in mind.*
>
> MICHAEL CRICHTON, *STATE OF FEAR*

THE TRUTH WILL OUT

We live in an age when the truth will out, perhaps even more than we bargained for—bubbling to the surface in unexpected ways. As for myself, for forty years I have sat by and just kept my mouth shut, knowing my confreres in anthropology were not doing justice to the story of man's extraordinary origin. I had to go off the reservation to find out for myself if we are really apes who got culture and a great brain. Recently, in the course of writing about the little people (*Homo sapiens pygmaeus*),* the house of cards called human evolution came tumbling down on my head. I knew I could not turn back: I had to write about man, the hybrid. This then is his checkered history.

This book is not concerned with *biological* evolution, only the ascent of *man,* who is, of course, a zoological entity (corporeal) but at the same time something more than an animal (something incorporeal), in a class by himself: a separate kingdom.

---

*My book *The Lost History of the Little People: Their Spiritually Advanced Traditions around the World* and this one are really a set.

The origin of our genus . . . continues to inspire fiery debate.

DONALD JOHANSON AND ENRICO FERORELLI,
"FACE TO FACE WITH LUCY'S FAMILY,"
NATIONAL GEOGRAPHIC

I start this book on the premise that people (the races of man) don't, didn't, can't, and couldn't change or evolve (physically) to become something else. Notwithstanding hypothetical family trees and even more fanciful common ancestors, the question of origins today remains completely unsettled and open to interpretation. What evolutionists call change is, in my book, simply the result of commingling, different races mixing genes. Evolution's transmutation of species reminds me of nothing so much as the medieval transmutation of lead into gold, an alchemy fed by legend, ambition, and illusion.

It is man's ideas that have evolved.

LOREN EISELEY, *THE IMMENSE JOURNEY*

I don't believe in physical evolution; our real evolution is in mettle, ethos, character. George Morley of the Kosmon Church in England put it this way: "There have been times when Man was more advanced than today. . . . In the civilisations of the ancient Egyptians and Persians, and earlier still on the grand continent which sank beneath the Pacific (Pan), we see great knowledge and wisdom, when men had understanding of the gods. . . . Then what is it that has advanced? It is the conditions of life . . . the ordinary comforts. . . . But it must ultimately come about that there will be a spiritual apotheosis to which Man will attain."[14]

Human history, wrote author John W. White, my friend and publishing mentor, is a "process of ascent to godhood. . . . Cro-Magnon are distinguished from Neanderthal not so much by physical body design as by their greater intelligence which resulted in world's first art, statuary, engravings, music, personal ornamentation, star charts . . . [and]

more highly developed social systems. Altogether they showed a superior degree of consciousness."[15]

THE NOOKIE FACTOR

Yes, many have already rebutted Darwinian evolution, but "so far, they have failed to find . . . a better theory to put in its place."[16] The disproving of evolution has already been done many times and done well by biologists, mathematicians, lawyers, Christians, prehistorians, science writers, and even paleontologists. But a real meaty *alternative* remains to be offered. If anything, Darwinism has succeeded by default. An alternative comes to life in these chapters, which present our *two* common ancestors, Asu and Ihin, whose races mingled to produce a third, the Druks (*H. erectus*). (See "Cast of Characters" below.)

We have a lot of ground to cover and missteps to unravel: To begin with, it took many decades for scholars to finally admit that supposed ancestors were actually *contemporaries* (chapter 3 is devoted to this). Then, more decades elapsed before it was conceded that those contemporaries (living side by side) actually *interbred* (the subject of chapter 5), producing the entire pantheon of hybrids known to us as the fossil record.

Read any scientific book about the races in today's world and hardly a page goes by without mention of admixtures, crossings, half-breeds, and so on. It was the same in the ages of the *past*. Hiding in plain sight is the all-too-human factor of these mergers, what archaeologist Christopher Hardaker whimsically calls the nookie factor. So let's cut to the chase:

> *We could have babies together.*
>
> CHRISTOPHER HARDAKER

The nookie factor actually stands Darwinism on its head. Whereas evolution has species lines branching out and *separating* (splitting) at some time in the distant past, crossbreeding entails quite the opposite:

the different stocks *came together,* cohabited, to form new races. And we are all hybrids. Here's the nub of this book: Fossils taken as representing stages of evolution or changes (by mutations) are herein shown to represent nothing more than the unstoppable intermixing of the Paleolithic races. Man the mixer is our constant theme. The peopling of the world, as drawn in these chapters, is about the mingling and merging of disparate types. No evolution there, just the continual confection of half-breeds and quarter-breeds, an exchange of genes since day one.

THAT WAS THEN, THIS IS NOW

The present is not the key to the past.

MARVIN L. LUBENOW, *BONES OF CONTENTION: A CREATIONIST ASSESSMENT OF HUMAN FOSSILS*

Early man lived "amidst physical conditions far different from those which prevailed during Neolithic times," observed paleogeologist George Frederick Wright. As we will again have occasion to note, past processes cannot be judged wholly by present ones. We find, however, that cannibalism in the protohistorical world, as an example, has been judged by modern standards, sanitized and interpreted as "the result of mortuary practices. . . . In view of the extreme scarcity of cannibalism in historic times, its very existence in prehistory is becoming hard to swallow."[17] We will, however, see this is quite wrong (chapter 1).

The uniformitarian fallacy: Should we assume the past is comparable to the present? Immanuel Velikovsky, for one, protested this notion, "the shortsighted belief that no forces could have shaped the world in the past that are not at work also at the present time, a belief that is the very foundation of modern geology and of the theory of evolution."[18] Neither can such specious comparisons reveal the lost continent of Pan, which calamity changed the history of the planet (see prologue). Earth was different in earlier times with forces no longer operative after the Semuan age (discussed in chapters 6, 7, and 10).

The terrestrial atmosphere was not the same in the remote past as it is now.

FRED HOYLE, *EVOLUTION FROM SPACE*

Today, no fossiliferous rocks are forming anywhere; nowhere are bones and shells lithified; (fossilization requires rapid burial in sediments). The scale of mountain building, volcanic activity, and even nonterrestrial bombardments was much greater in early periods of Earth history (see chapter 6). The thickness of limestone sediments point to precipitation on a gigantic scale, though this is not taking place anymore. Today, no sea or lake is forming evaporate beds comparable to ancient deposits of immense thickness.

Darwin's theory of sexual selection in the distant past was based on the behavior of *living* creatures, working backward. Thus do his followers today make statements like: "Systematists can write a history of life on earth looking only at living species."[19] Paleontologists freely extrapolate from the present back to the past, called "back-reckoning"—a top-down approach.

Some have worked out *Homo erectus* demographics, assuming that their population dynamics were much like those of hunter-gatherer people today. They use population trends from historical time (along with other assumptions) to judge the extent of prehistorical populations (I challenge this in chapter 6).

In addition, our genetic history is supposedly revealed by a DNA tree that uses chromosomes from *present-day humans*. "One starts from . . . the present structure . . . and then looks at the fossil record . . . to derive filiation of these forms."[20] For example the out-of-Africa theory (explored in chapter 11) works back from present populations, using these mtDNA trees. The well-known 1987 study, based on 136 women from different parts of the world, purportedly leads us back to mitochondrial Eve, our presumed African ancestress. Yet it is pure assumption that "evolutionary" history can be read off our genes. And the idea of "colonizing" the world out of Africa is more congruent with today's poli-economic Zeitgeist than with anything that happened in the Paleolithic. Imposing a Western expansionist framework on early man is just the sort of Eurocentrism that anthropology supposedly beats out of its acolytes. Variation (heterogeneity, diversity), we will see, was once much greater in the human family. This former diversity, as these chapters unfold, is actually the key that unlocks the door to the past.

CAST OF CHARACTERS

The unfamiliar names of the five races of men used in this book have been borrowed from the historical portions of *Oahspe: A New Bible in the Words of Jehovih and His Angel Embassadors,* the dazzling, encyclopedic, lost records that open up many obscure aspects of Paleolithic culture to our understanding.*

First race of man, Asu: bedrock mortal, the first "adam," though insapient; a close precursor of *Australopithecus,* represented best by Africa's *Ardipithecus,* dwelling on land and in trees and without speech and meager in consciousness. Since *su* means "spirit," Asu means "man without spirit," which is to say, before upgraded to true human status.

Second race of man, Ihin: the sacred little people (only three feet tall), the gracile AMHs (anatomically modern humans) of the very early record, with Europoid features—the root stock of *Homo sapiens pygmaeus;* like biblical Abel, in the sense of being *able* to think, understand, and commune spiritually. Appeared on Earth only six thousand years after the first generation of Asu. (An extensive history of the Ihins can be found in my prequel *The Lost History of the Little People.*)

---

*Oahspe, first published in 1882, the year of Charles Darwin's death, was produced by automatic writing.

Third race of man, Druk: a cross (though forbidden) between Asu man and Ihin; equivalent in many cases to *H. erectus* and also to the biblical Cain; also known as the ground people (pit dwellers); omnivorous, long-armed, curved back, often quite large ("giants"). They were the third race, just as *H. erectus* has been called "the third identifiable hominid."[21] Known also as the barbarian hordes. In this book I often use the name *Druk* interchangeably with *H. erectus*.

Fourth race of man, Ihuan: has mostly AMH features. Three times produced (72, 39, and 20 kya) by crossing Ihin and Druk. Best known as the Aurignacian and Solutrean Cro-Magnon in Europe: tall, strong, and copper-colored "mighty men"; the last Ihuans to survive into the modern era were the American Paleo-Indians.

Fifth race of man, Ghan: a cross between Ihin and Ihuan; fully modern in type, *Homo sapiens sapiens,* beginning around the time of Apollo, eighteen thousand years ago in both the Old and New Worlds (see appendix B). Stately but willful souls with all the arts and sciences of man. Kings and born conquerors. Masters of the sun kingdoms in the Mesolithic. They tamed the Earth.

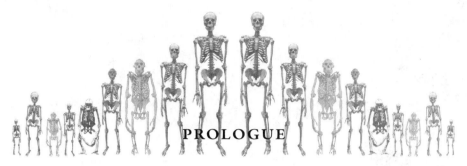

DIASPORA

The Story of Pan

The rules of science [allow] no appeals to divine inspiration and no references to nonexistent lost continents.

KENNETH L. FEDER, *FRAUDS, MYTHS AND MYSTERIES*

Orthodoxy has a thousand champions.

KEITH THOMSON, *THE WATCH ON THE HEATH*

Breaking the rules in this departure from orthodoxy and the insular (even dehumanizing) rule of science, we will not only appeal to divine inspiration but will, out of sheer necessity, begin our travel back in time to the lost continent in the Pacific Ocean. Atlantis has a thousand champions, but now it is time to lift the veil on a different lost continent, the forgotten land of Pacifica (aka Pan or Mu).

THE LIGHT BEARERS FROM PAN

Ever wonder why the Pacific Ocean is so big and so empty? Several names are associated with a missing continent therein: Mu, Tien-mu,

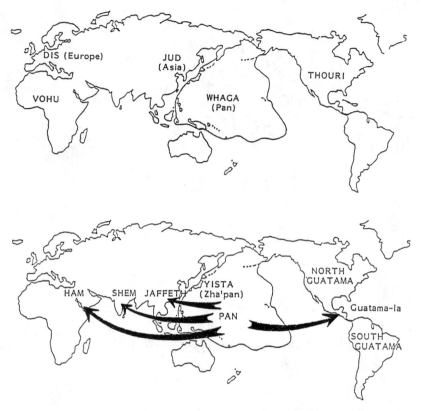

Figure P.1. Top: The continents were called by these names for 48,000 years until the sinking of Pan or Whaga. Bottom: Direcetion of each tribe's dispersal out of Pan. Maps by Zia Zymanksy, Oahspe supplement, appendix, 140.

Whaga (also spelled Wagga), Lemuria, Lumania, Hiva, Haiviki, Helani, Hoahomaitu, Fenua Nui, Patulan-Pa-Civan, Bolutu, Fu-Sang, Pacifica, Rutas, Adoma, Kalu'a, Kumari* Nadu, Olepanti, Pantiya—and Pan. Approximately 24,000 years ago a great destruction sank this continent; the only survivors were the Ihin, the little people.

Nothing could be older yet more permanent in the memory of man than the great deluge. Yet, contrary to old belief, it was not worldwide, just as Charles Darwin himself maintained, along with Charles Lyell, the father of modern geology (and Darwin's dear "chancellor"), who said

---

*Kumari and other *ku*-derived names are gathered together in appendix A.

this flood was not a universal event. The usual debate is whether the great flood was universal or local (in the Fertile Crescent). But it was neither. The Gilgamesh Flood Epic "leads us back to the very cradle of the human race."[1] And where did that cradle rock? Not in the Old World, which was seeded from elsewhere. Neither was the greatest diaspora of mankind out of Africa. It was out of Pan.* From Pan, the Ihins escaped in separate fleets to the five divisions of Earth—Japan, America, Asia, India, and Africa—and proceeded to mix with the indigenous races, bringing about a renaissance of civilization in each of those places.

As founders of the present races of men, these refugees "sapienized" the Near East, Europe, South America, Japan—bringing industry, tools, astronomy, temples, and the arts of peace. In each country, they founded colleges, teaching the priestly† arts and sciences. Their sudden arrival helps explain why the Semitic, Hamitic, and Indo-European genotype of "pure families . . . appeared almost overnight" (after the flood), for these are all of the Ihin family, whose legacy produced the centers of postcataclysmic civilization.

Panology (as I term this new study) entails numerous surprises. Noah, for one, did not hail from anywhere in the Old World but from the land now lost in the Pacific Ocean. Mesopotamian (Chaldean) annals point to such a homeland: "The lands we live in are surrounded by the ocean, but beyond that ocean there is another land . . . and in that land Man . . . lived in paradise. During the Deluge, Noah was carried in his Ark into the land his posterity now inhabit."[2]

> *The occurrence of an ark in the traditions of a deluge, found in so many distant times and places favors the opinion of these being derived from a single source.*
>
> EDWARD BURNET TYLOR, *RESEARCHES INTO THE EARLY HISTORY OF MANKIND AND THE DEVELOPMENT OF CIVILIZATION*

---

*See *The Lost History of the Little People,* where I found the name *Pan* retained to this day in scores of languages and locations (but especially in Mexico).

†The priestly arts are traced back to the sages of Khu (on Pan) as suggested in appendix A.

The Holy Bible refers to events of this sort as restocking the world with a more advanced people. The protohistorian's version predicates that "man's first civilization arose on Lemuria," whose émigrés became the initiators who went on to "stimulate the development of civilizations in Egypt, Peru, Tiahuanaco and the lands of the Celts."[3] And this is why the Egyptian god Ptah, who "re-created" the world after the flood, is pictured as a dwarf. For these émigrés, the sons of Noah, were of the Ihin race, a short, pale, and bearded people. These Egyptian-Lemurians, in turn, seem to correspond with the ancient people of Libya called the Garamantes, ancestors of the Egyptians who are painted on walls and cliffs as fair-skinned, blond charioteers living in the Sahara Desert.

This period, ca 24 kya, is generally recognized as the time of dramatic demographic dispersal of genetically modern peoples (AMHs). The deluge brought about a new configuration of the globe and was the true year one of the present human race.[4] It is the deluge, really,

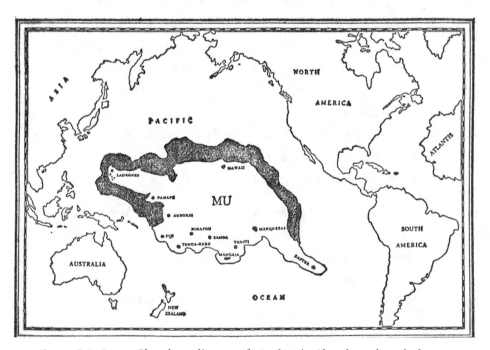

Figure P.2. James Churchward's map of Mu (Pan). Churchward made famous the phrase "the motherland of Mu."

that divides the Upper from the Lower Paleolithic. At that time, 24 kya, the only advanced culture left in the world was on the continent of Pan.

The |hin hath been destroyed off all the divisions of the earth save Wagga [aka Pan].

OAHSPE, SYNOPSIS OF SIXTEEN CYCLES 2:2

As we will be seeing, retrobreeding and genocide in the Old World had curbed the march of civilization. Indeed, the chosen people (Ihin) had everywhere been killed off and despoiled, and by 24 kya, even the land of Pan had become peopled over with men raised up to deeds of blood; yet a remnant of the peaceful Ihins remained, hiding from their evil pursuers.[5]

Preflood people are an older stratum, often with unclassifiable languages and unique customs and physiognomy. Today there seem to be a few surviving languages that were spoken by the little people before their near-universal extermination by barbarians (those "wicked" giants), which took place before the flood: The Filipino Abenlens, for instance, speak a language of their own that no one ever heard of. The Ainu language of Japan is also unlike any other. The (Bay of Bengal) Andamanese speak a completely unclassifiable language. The little people of Australia have a language different from the full-size Aborigines. The tiny Tapiro of New Guinea also retain a unique language. The Malagasy Mikea people have a vocabulary and way of life completely unlike their neighbors, their language thought to be a remnant of the Vazimba tongue. All of the above-mentioned are little people.

Centuries ago, when two African pygmies were brought to the king of Italy, it was soon learned that they spoke a dialect of their own, different from any known African tongue. Why? Because they go back to preflood days, *predating* the influx of Pan people and their subsequent mixes in Africa (explained in chapter 11). Schweinfurth believed the little Akka and Bushmen were the ultimate aboriginals of the continent. Today, the Mbenga pygmies are extremely divergent from all other human populations; they are among the unclassifiable groups and language speakers, which may well be anterior to the great postflood amalgamations.

There is reason to believe the "chosen" people were chosen to survive the devastation of Pan, for they were the light bearers (the "immortals") of deepest antiquity. Records kept by the Babylonian priesthood traced the first dynasty after the flood to the year 24,500 BP. Studying the Babylonians and their Gilgamesh Flood Epic, we find the hero setting out in search of immortality; a man named Utnapishtim (the Babylonian Noah) and his family had been the only survivors of the debacle and afterward *the gods made them immortal.* This immortality (as we touch on again in chapters 2 and 7) was the force of soul, the capacity for everlasting life—made possible through the influence and genes of the Ihins, sole survivors of the deluge, "seed" and spark of future humanity.

FIVE BROTHERS, FIVE RACES

Five great divisions of the earth there are . . . [for] the Lord established the five peoples who were saved from Pan.

OAHSPE, BOOK OF SETHANTES 2:4 AND
THE LORDS' FIFTH BOOK 5:1

While the doctrine called monogenism (one origin) has mankind arising in a single place, the school of polygenism posits separate development of the races in several different locations. A leading polygenist and anthropologist, Carleton Coon, thought *Homo sapiens* arose not in one place or cradle, but in five different areas of the world.

The ancient tradition of five divisions of Earth (iconically five men) seems to reinforce Coon's (much derided, as we will see) scheme of mankind established into five different races: Caucasoid, Congoid, Capoid, Mongoloid, and Australoid. Johann Friedrich Blumenbach, the father of physical anthropology, also divided the human races into five, a bit more accurately than Coon: Caucasian (Shem), Mongolian (Jaffeth), Ethiopian (Ham), American (Guatama), and Malayan (Yista). As we will see, these five races, Shem, Jaffeth, Ham, Guatama, and Yista, correspond to the five fleets out of Pan (see figure P.3 for depictions from all over the world of five men in a boat, representing the five races leaving Pan).

Five figures appear on the Sumatran "soul ship," a ceremonial motif; the frequent appearance of boats of the dead in Sundaland (Southeast Asia) may be understood as a memorial to the land of their ancestors to the east—Oceania. I think it is significant that the Oceanic Malekulans attribute their megalithic-building tradition to five culture-bearing brothers, white men with aquiline noses. In India as well, the Pan-davas are the five heroes/brothers of the *Mahabharata*, their name presumably based on Sanskrit *pan-du*, meaning "white, yellow-white pale." A tradition of five brothers is also recalled by the Haida Indians of British Columbia, who say the first people came out of clamshells (read dugouts, arks), and they were "five little bodies," which is to say the Ihins, *Homo sapiens pygmaeus*.

> And J. blew His breath upon the ships of His sons and daughters; blew them about upon the ocean; blew them to the east and west and north and south . . . the ships were congregated into four fleets; *thirty-four ships* [e.a.] into each fleet, except two ships, which were carried together in a fleet by themselves. The Lord said: I will name the fleets of my chosen, and their [five] names shall be everlasting on the earth: Guatama, Shem, Jaffeth, Ham and Yista. The Lord said: From these, my seed, I will people the earth over in all its divisions.
>
> OAHSPE, THE LORDS' FIRST BOOK 1:46–49

Hopi petroglyph carved
on a rock near Oraibi

Sign of Pan, from Oahspe;
the five men signify the five
fleets escaping Pan

Mayan glyph
from Bishop
Diego DeLanda's
alphabet

Undeciphered
Maori writing

Carved bone from
Tequixquiac, Mexico

From Colombia's Chibcha hoard, five
figures on a raft of gold logs
(two figures in rear are out of focus).
In their thanksgiving rite of offering
the chief-king sails out on Lake
Guatavita on a raft of rushes

Assyrian bas-relief at Koyunjik,
resembling an Egyptian wall painting of
reed vessels, which also shows five men

The only sense historians can make
of this Cro-Magnon cave drawing
is that it represents some sort
of "seagoing vessel"

Inuit drawing

*Figure P.3. Five men in a boat. The image of five men in a raft
has been portrayed in almost every part of the world.*

The Irish *Book of Invasions* records that the Nemhedh were five brothers who, bringing order, justice, and prosperity, established the five provinces of Ireland. These Nemedians "had *thirty-four* [e.a.] ships that traveled the open sea in search of new lands." The name *Nemedian* seems to be an anglicization of Neimhidh, being a "noble sacrosanct, worthy" people who arrived in Ireland some time (six hundred years) after the deluge. As Nemed, they are the king's poets and the forefathers of Ireland's diminutive Tuatha de Danaan, a "divine" race.

WHERE DID WHITE PEOPLE COME FROM?

Caucasoid origins remain . . . shrouded in mystery.
WILLIAM HOWELLS, *GETTING HERE*

Mystery, really, means that something, a piece of the puzzle, is missing (perhaps ignored or even suppressed). Too often mystery is the throwaway term for issues that today's paradigm enforcers would rather not deal with. But where *did* Caucasians come from? Cross-cultural comparisons may tell a tale: in many places, the sudden injection of culture (along with a different breed of people, a Europoid type) 24,000 to 25,000 years ago brought with it a raft of inventions that are *uncannily similar* in widely separated lands.

Every case is different. It's the similarities that intrigue.
JOSEPH KANON, *THE PRODIGAL SPY*

In the last two centuries, new light has begun to shine on these parallel works, which include: pyramids, mounds, arts, engineering, agriculture, medicine, astronomy, navigation, writing—the entire gestalt hinting at a mutual heritage, a common origin. But this wellspring of humanity, the key to all that followed, the key to the Upper Paleolithic and all its "mysteries," has not been a welcome subject on academic turf for a long time. It remains in limbo.

But it is not a mystery. Only the straitjacket rules of science have made it so. Europe's Cro-Magnon man, it is understood, came from elsewhere, since all that was before in Europe were brutish Neanderthals. The oldest rock painting in France is 25 kyr. French cave art at Lascaux, Ain, and Montespan, dated perhaps 20 kya, was the work of little people, for here the Cro-Magnon artist (usually over six feet) was less than five feet tall. France's Lascaux paintings are reminiscent of South African cave art dated 22 kya. "It would seem that the masters of Lascaux came from Asia or even from the legendary Mu."[6] Perhaps so, for the first Paleo-Caucasoids also turn up in China, Australia, and America at this time; and it is probably not a coincidence that Mexico's (Valsequillo) artifacts of the Cro-Magnon type are given this date: 24,000 BP.

Russian art may hold some answers: The earliest artists in the region of Lake Baikal are dated 23,000 years old; a 24,000-year-old Sunghir burial (near Moscow) contained thousands of ivory beads— one of the earliest examples of jewelry. Most intriguing is the tradition in Vladimir, Russia, that their village is 24,000 years old. Indeed, prehistorians say moderns first spread through central Siberia and Russia at that time. The earliest Venus figurines are also part of the mix, dated in Russia to 24,000 BP. In fact, the distribution of Venuses is so extensive in the Old World as to suggest "a large population movement [in] the final . . . period of the Ice Age."[7]

We can track this movement in cranial morphology, Russia appearing as the hub of a widespread round headed race; this brachycephaly, as it is called, is the distinctive head shape of the Ihins, who made their way from Pan to almost every continent. As one approaches "Hiva [Pan aka Mu], the more the people become brachycephalic."[8]

CAUCASIANS IN THE PACIFIC RIM

Italian prehistorian Egisto Roggero in *Il Mare* observed that "the great Oceanic race, an ancient people of whose story we know nothing . . . [were striking in] their resemblance to the white races of the west." Polynesians are neither Australoid nor Negroid nor Mongoloid, but rather a combination of these races, also possessing "a strong white element,"[9] their tall redheads so like the stately Inca nobility with their wheat-colored hair. Remarkably pale people abide at Baie des Francais and Mangea, as well as on other islands of the Pacific,[10] such as New Hebrides (the Ambat people), the Solomons, Vanuatu, Fiji, Rotumah, and Suuna Rii. The Polynesian race founder Kon-tiki is said to be bearded and white, as were the little people of Hawaii. Bearded people (most typical of the Caucasian stock) are also known among South America's Mapuche Indians, Arawaks, Tiahuanacans, and Incans. Builders of Brazil's Trans-Amazon Highway came across a white-skinned, red-bearded tribe called the Assurinis, their language different from the regional dialects. Though the ancient Mexicans were themselves beardless, most of their deities were bearded (like the Olmec statuettes), just as the Alux, a relict race of little people in the Yucatan, sport long jet-black beards. James Churchward in his classic, *The Lost Continent of Mu,* explained this by supposing that "far back there was a white race dominating Mexico and Central America. . . . All the kings and queens of Mayax during the twelve dynasties were of the white race. . . . The forefathers of the white Polynesians of today, the forefathers of the white Mayas of Yucatan and the forefathers of all our white races were one and the same."

In North America the Pomo and Hupa Indians (California) have remarkably abundant beards, as did the legendary little Nunnehi of Cherokee country as well as the white Eskimos. The 10-kyr Great Stone Face, found in Chicago and not Indian made, is bearded.

In the Far East the only heavily bearded and Caucasoid people are the Ainu, their name a variant of Ihin and Innu-it. Blond Inuit were seen by Sir John Franklin in the nineteenth century: an Alaskan people with blue eyes and remarkably abundant beards, possibly descendants of the "mummy people" (so called because they mummified their dead), a seven-thousand-year-old Caucasian-like race who lived on Alaska's Aleutian Islands. Early in the twentieth century, explorers again encountered

Figure P.4. Ainu with beard.

several tribes of white Eskimos, some with red hair, living on the shores of Davis Straits, their facial features "so well made."

The Inuit themselves assert "the men first made were white."[11] The Inuit's ancestors, perhaps ten thousand years ago, enjoyed an advanced culture and possessed a complexion as fair as the Cro-Magnons. American anthropologist Roland Dixon adduced a Neolithic population of blond, fair people extending from Asia to Siberia to the Bering Strait. They may have descended from the people who built an Alaskan "arctic metropolis."[12] Excavations at Ipiutak gave up astonishing ruins—a town a mile long with a population of probably four thousand, remains of eight hundred houses, long avenues of living sites, and thoroughfares. The town's inhabitants had produced sophisticated ivory carvings with motifs quite similar to the Ainu of Japan. Delicately made and engraved implements also bear an uncanny resemblance to items from North China. The intricate carvings as well as the detachable harpoon head of the Ainus and Alaskans have been carefully compared and deemed almost identical.

OCEANIA ITSELF

The peoples of Oceania have kept records of a great deluge: "All the legends of Hawaii, the New Hebrides and New Zealand talk about a white-skinned race with fair hair"[13] who survived that calamity. New Zealand natives called that long-sunken land Hawaiki, peopled over by "an ancient Caucasian race [from] . . . across the ocean."[14] The highland people on Flores, Indonesia, have their own white Noah, the forefather of their tribe who was "saved from a flood in a ship." J. Macmillan Brown, geologist and one-time chancellor of the University of New Zealand, found it impossible to account for the remains on Easter Island (in the South Pacific) without positing a submerged archipelago. Since his time, others have asked: Why are Easter Island pictograms so similar to those of India (Mohenjo Daro) and China? (See figure P.5.) Easter Island's *rongorongo* inscriptions have also been compared to the pictographs of Panama's Kuna Indians (see appendix A) and to the figures in Peru's Nazca lines (geoglyphs), as well as to Egyptian hieroglyphics and decorations at Turkey's Çatalhüyük.

Figure P.5. Rongorongo comparisons. On Easter Island this figure was the symbol of knowledge. In my own meanderings I have found variations on the theme of the quipu (an accounting device) on Easter Island, and in China, Babylonia, Peru, and Hawaii.

The Mu folk of Hawaii's Kauai forest are described as people with Caucasian features (ancestors, some say, of present-day Aryans[15]). They were civilized. A numerous and unrivaled race, they were, according to the Hawaiians, shipbuilders and skilled mariners, and one Polynesian legend records exactly how the ark was built. As long as 40 kya, the Ihins were navigating the globe, the regular commerce between Old and New Worlds extensively probed in the work of Barry Fell, Harvard marine biologist and epigrapher, and also in the books of Frank Joseph, Philip Coppens, and others.

Students of prehistory are taught that Australia was peopled by the migrations from Southeast Asia. In this view (which I do not share) the little people of Australia, as anthropologist J. B. Birdsell argued, can be traced to Negrito migrants moving across South Asia *to* the southerly islands. On the same wavelength, Joseph Campbell saw the peopling of Melanesia as a late diffusion also coming from the mainland, perhaps China. And in the same way, the ancestors of the Polynesians are thought to have come *from* Asia around 1500 BCE. I don't think the Polynesians "came from" anywhere. Nevertheless, genetic studies have found that the great migrations of the Paleolithic moved from the islands of Sundaland (insular Southeast Asia) to the north and west of the mainland, thus populating Asia and beyond, instead of the other way around.

The most far-fetched theories are offered to account for (explain away) the white genes of Polynesians; we will have to argue, though, that they (like the Australians, Papuans, Filipinos, etc.) are all indigenous to their locale and not "imported" from anywhere. If Oceania owes its history to migrants from China or Singapore or Southeast Asia, why are there no antecedents to Austronesian languages on the mainland of Asia? Because they "could have arisen in Oceania."[16] *Circling* the empty Pacific Ocean are languages that: (1) use numerical classifiers; (2) put the verb first in a sentence; and (3) have pronouns beginning with *m* and *n*. This geolinguistic mapping suggests a wave of people spreading out, from some lost Oceanic center, in all directions along the Pacific rim. With vocabulary, numerals, and grammar of Sundaland, Asia, and Native America matching up with Polynesian languages, there is good reason to ask if "at some remote period . . . they were originally one people."[17]

JAFFETH, SON OF NOAH, IN CHINA

In the Chinese deluge story "the pillars of heaven were broken . . . the heavens sank lower toward the earth . . . and the earth fell to pieces."[*] On the Asian mainland, "from the end of the flood onwards, we begin to see settlements all along the . . . coast of China."[18] The Uighurs of China are remembered as the first civilizers in the Old World after the flood. Chapter 10 of Patrick Chouinard's recent book *Forgotten Worlds* regales us with hundreds of blond and red-haired mummies found in China with Europoid features. "Asia," he wrote, "was once home to a lost tribe . . . a Caucasian population."

The sons of Noah (Jaffeth branch) who fled to China were, like most of the Ihins, white and yellow men: "The original inhabitants of the region, far from being Mongoloid, were actually Caucasian."[19] Neither are the blue-eyed Uighur (of Western China) and the "Paleo-caucasoid" Tocharians (who entered China 25 kya) of a Mongoloid caste. There must be some truth to the Chinese legend recalling that the Gobi Desert was once a great sea and there was an island in this sea inhabited "by white men with blue eyes and fair hair . . . [originally] inhabitants of Mu . . . who imparted civilization."[20]

GUATAMA, SON OF NOAH, IN THE AMERICAS

The Ihin fleet that reached America was named Guatama: Stephen Oppenheimer has people entering the New World 22,000 to 24,000 years ago, and they are Caucasian, as seen in the California specimen LA Man who is dated 24,000 BP.[21] In one AmerInd tradition, the only flood survivors were "little men of the mountains" (mounds), a lost race,

---

[*]Creator-god Pan-ku kept the sky from falling; (both *pan* and *ku* are terms reminiscent of the lost continent).

but indeed they were groups of Ihins—the little people, with white and yellow hair.

The same date, 24,000 BP, obtains for the fine engravings and AMH (anatomically modern human) skulls found at Flagstaff, Arizona, as well as for the Basket Makers, in their small-sized burial vaults, uncovered in southwestern Pennsylvania. These engravers and artisans were also in Peru—made famous by the Ica stones, some of which depict the submerged lands from which their forefathers had been saved.

There were made images of stone and copper, and engravings thereon of the children of Noe, and of the flood, and of the sacred tribes.

OAHSPE, THE LORDS' FOURTH BOOK 2:20

The striking resemblance between China's Choukoutien Man and America's Plains Indians, especially in cranial morphology, does not necessarily mean a migration of Siberians across that old saw, the Bering Strait. Rather, let us consider that they *both* are from a common source—Pan. DNA analysis tells us that the ancestors of Asians and Americans separated *earlier* than 15,000 years ago (a fairly standard date for the Bering crossing). And if that earlier time be 24,000 years ago, it marks the great diaspora, after the sinking of the continent of Pan.

It has been asked: Why did the very earliest American Indian skulls "look like present-day Caucasians with an Indian cast"?[22] It is because the Ihins settled the New World 24 kya, and proceeded to blend with the native stock. Thus did the early AmerInds in Washington state (near the Columbia River) end up with somewhat Caucasian features. The morphology of 9,000-year-old Kennewick Man's skull is also non-Native, resembling the Ainu and Polynesian type—not Siberian. Walter Neves (University of Sao Paolo) found links between the skulls of Polynesians and North American Paleo-Indians, just as anthropologist Marta Lahr matched Patagonian skulls to South Pacific islanders. Barry Fell, the Harvard paleontologist, was able to match skulls of Tennessee's ancient little people with lookalikes among the Negritos of the Philippines: all *H. sapiens pygmaeus*.

"The dwarfishness [of the world's little people] . . . who have retained so many traits in common . . . suggests that all of these far-flung groups may be linked in a common ancestry."[23] Workers have connected the prehistoric Tennessee, Kentucky, and Ohio mound builders to populations in Malaysia, Fiji, and the Sandwich Islands, their common artifacts "clearly of a Polynesian character. . . . The North Americans, Polynesians and Malays were formerly the same people, or had one common origin."[24]

Forensic evidence may bear this out: blood types indicate a strain of DNA in American Indians that shows relationship to Pacific Islanders,[25] not Siberians or Europeans. Why is there scarcely any B blood type among AmerInds, which is so common in Eastern Asia, whence their supposed ancestors made the trek across a Bering land bridge? Neither does Inca blood type have Old World affinities, even though the royal Inca looked like white men. Rather, a genetic clan in South and Central America has been found to be closely associated with Polynesians and curiously* absent in Siberia and Alaska. Tools, too: On the California coast, adzes and ax heads of Pacific origin were found.

Engravings of animals in Mexico, dated 22 kyr, are curiously "very much like the Cro-Magnon [art] in France."[26] (And we won't need the Bering Strait to make this connection.) Indeed, Europe's Solutrean spearheads are almost identical to the Clovis points in the American Southwest, and they are nothing like Siberian ones. These European cousins (and cousins they are) produced the cave art of France between 20,000 and 30,000 years ago. Cro-Magnon people appear to have had knowledge of a decimal system that we mistakenly believe was invented only a few thousand years ago,"[27] just as their Mayan cousins, according to the Popul Vuh, were "the true race of men . . . who possessed all knowledge, and studied the four quarters of heaven and the round surface of the earth." In Peru "the shy, furtive and gentle" white people

---

*Curiously, because that's where the AmerInd ancestors allegedly passed from the Old World to the New (i.e., across Beringia). A new book by Dennis Stanford and Bruce Bradley, *Across Atlantic Ice,* finds that spearheads and knives in parts of prehistoric America have no correlation with those found in Siberia, or even Alaska. None of the Indian tribes, by the way, trace themselves back to Siberia.

with beards are considered "the oldest race now alive,"[28] like the white Tapuyos who were thought to be refugees from a *former* civilization.

Up against twentieth-century academic opinion, Brown proposed that a lost continent in the South Pacific inhabited by white men had lent its culture to Peruvian civilization. Notable are similarities in language; significantly, it is the *oldest* Peruvian languages that have the greatest affinity with Polynesian ones, just as the statues from the *earliest* period on Easter Island bear a striking resemblance to the huge statues at San Agustin, Colombia, whose Stone Age temples and tombs also have counterparts on Easter Island: South American ceremonial platforms are almost identical to those on Easter Island.

The migration, it has been surmised, must have taken place in very early times; "and there can be no question of these immigrants having transmitted a higher civilization."[29] Would the influx of these proto-Caucasians explain the presence of "white Indians" throughout the Americas? Pierre Honore, like Brown, thought that people from Polynesia (not Africa or Asia) must have reached America, remains of which migration he found in both Brazil (Xingu River) and Ecuador. All along the upper tributaries of the Amazon one hears reports of pale-skinned tribes *still* inhabiting the ruins of cyclopean cities. Ten-thousand-year-old skeletons in Brazil show different skull types than modern Indians or people of Northern Asia. British explorer Percy Harrison Fawcett photographed white Indians all across Brazil's back villages, including the big, handsome, red-haired Tahuamanu people, their tallness answered by the white, bearded dwarfs (four-foot-tall pygmy men) along the Amazon, this tribe, said to be "Greek" in type, living near one of the dead cities of the interior.

The ancient songs of the Tupis of Brazil recount "destruction by a violent inundation . . . a long time ago . . . only a very few escaping."[30] The Tupi people are leptorrhine and brachycephalic (narrow nosed and round headed, like the Ihins), as are other "people of higher culture in North America."[31] This race, thought Fawcett, was of Pacific origin. Did they survive that overwhelming disaster known as the great deluge? The name Tupi, we might note, corresponds to Tapi in Aztec tradition, which recounts that the Creator taught this pious man, Tapi, to build a boat in order to escape a great flood of waters.

Daniel Brinton, the esteemed American folklorist and linguist, reported that the Tupis of Brazil "were named after the first man, Tupa, he who alone survived the flood . . . an old man of *fair complexion* [e.a.], *un vieillard blanc.*"[32] Paul Radin, who together with Brinton was among America's outstanding early ethnographers, described the Arawak tribes of the Caribbean and coastal areas as "men of light color with long hair and beards."[33] Harold Gladwin adds that the Arawaks "may have been Polynesian," just as comparative analysis suggests that "the Polynesians have, at some remote period, found their way to the [American] continent."[34]

Figure P.6. Arawak. Arawaks are sometimes thought to be Polynesian in origin.

Figure P.7. Filipino Negrito. Negritos, as we go on, will prove to be a stunning blend of the second and third races of men.

Figure P.8. As a graduate student I did my fieldwork among an Arawakan tribe in Venezuela: the Guajiros. These pictures were taken in 1969.

THE JAPANESE CONNECTION

Why is 42 percent of Japanese vocabulary Malayo-Polynesian? Why is the practice of tattooing so widespread: found in Sumatra, Polynesia, ancient America, and Africa (the Nigerian Yorubas practice this art, which along with their short bow, is non-African in provenance)? Prehistorian Leo Frobenius thought that these related customs came from a lost Pacific center, whose survivors spread to Asia, then to the West.

Why does Japanese mythology include so many elements reminiscent of Polynesia?[35] Here on the western side of the Pacific, local traditions recall that the islands of the Pacific arose after "the waters of a great deluge had receded."[36] The hazy outline of that early period comes into better focus with the archaeological discovery that the population of Japan swelled at least *20,000* years ago; (more like 25 kyr, according to Jeffrey Goodman[37]). It was at this time that all the arts of Japan improved, as seen in excellent microblade tools and the "elaborate and sophisticated" ware of the acclaimed Jomon potters.

These Jomon people were the dominant race in the Japanese islands during the "reign of the gods." In fact, they *were* the gods, or rather, the deified ancestor, appearing in sculptures with European features. Their fine carvings, done in relief, are considered the *oldest* in all the world: "and of all people, ye [the Japanese] shall be reckoned the oldest in the world." Intriguingly, Jomon ware has been compared not only to the work of the Alaskan Ipiutak (pre-Inuit) across the pond, but also to identical work along the upper Amazon and Ecuadorian coast.[38] Jomon pottery has also been found in the Pacific itself (at Melanesia).

The earliest people, say the Japanese, were the white-skinned Yamato (Yamato later became the name of a Japanese dynasty). "Many thousand years ago, the islands of Japan formed a distant colony of Lemuria. . . . The Yamato enjoyed a sophisticated culture."[39] In the prehistoric tombs of Japan are found images of the Yamato, called *haniwa,* curious clay figures of little people, with faces of a Caucasian nobility, said to have brought with them from the motherland a developed civilization. The Japanese flag, the Rising Sun, still embodies the sacred emblem of that drowned land.

These Jomon or Yamato folk, who also built megalithic stone circles, were diminutive (*tsuchigumo*): remains, four and a half feet tall, have been found in association with their wares.[40] Some of their genes survive today in the little Ainu people, a living remnant of those long-lost whites (whose original name Ihin/Ine permutes to Inuit and Ainu).

The "undersized" Ainu stock occupied much of Asia at one time; today only a few remain—occupying northern Japan (Yezo). Although there has been a lot of intermarriage, the Ainu are quite different than Mongoloids, sporting luxuriant beards and wavy hair, their faces of a Caucasian cast. Neither were the ancient Jomon of typical Asian descent; "they were proto-Caucasoids, fair-skinned with prominent noses and full, light-colored beards."[41]

> The fleet of two ships [from Pan] carried to the north was called . . . in the Wagga [Panic] tongue Zha'pan, which is the same country that is to this day called Japan, signifying, Relic of the continent of Pan, for it lay . . . where the land was cleaved in two. . . . Thus was settled Japan.
>
> OAHSPE, THE LORDS' FIRST BOOK 1:55

The Japanese have their own recollections of a golden age and its ancient deity who gave happiness to the people of Okinawa "from beyond the sea." The western shore of that homeland was recently rediscovered beneath the waters off Okinawa. Along the East China Sea, a startling discovery was made in 1994, spread over more than three hundred miles of ocean floor: the remains of an ancient city, "a civilization lost in the sea," containing stone circles (à la Jomon), shafts, and massive monuments. To prehistorian Frank Joseph, the ruins are a remnant of a large continent in the Pacific Ocean, with

*Figure P.9. Underwater ruins off Japan at Yonaguni, Ryukyu Islands. This formation is called The Turtle.**

counterparts on the Peruvian coast. He compares the sunken buildings off Okinawa to Pachacamac (near Lima) with its broad plazas and sweeping staircases.

Is it sheer coincidence (the skeptic's primary weapon) that the most striking facsimiles of Japan's lost horizon appear at the *other* end of the vast and empty ocean? Hugging the western shores of the Americas, sites in Peru, Ecuador, and California have provided some of the richest sources of early *H. sapiens* (see appendix F). The same time depth of 20,000 to 25,000 years BP (that archaeologists estimate for the flourishing of Japan) obtains at *Peru's* earliest toolmaking site. Are we satisfied calling this a coincidence? Parallel evolution?

Some stone circles like those of the Japanese Jomon (rediscovered in the Yonaguni ruins) are found in the British Isles as well as

*Vernon Wobschall (editor of Oahspe Standard Edition, in a personal communication) wrote about a "rock specimen off Yonaguni that has carvings on it that look remarkably like those found in the Panic language tablets in the Book of Saphah."

in North America at Lake Michigan, where calendrical/astronomical devices were built, according to the Ojibway, by the ancient ones. Such circles are also found in Peru. The entire gestalt of "Stone Age" masters is found in Peru, witnessed by highly skilled masonry, megalithic ruins, bridges, canals, wonderful temples, a Great Wall (not unlike China's), and the extraordinary "Nazca lines." Now with Jomon-style pottery found on the coast of Ecuador, and early Ecuadorian skulls resembling those of Melanesia (where Jomon pottery has been found), one can reasonably suppose that *both coasts of the Pacific* were settled by refugees from sunken Pacifica (Pan). "Words cannot express adequately the degree of similarity between early Valdivia [Ecuadorian] and [Japanese] Jomon pottery," marveled archaeologist Clifford Evans. "Many fragments of both are so similar . . . that they might almost have come from the same vessel."[42]

The Asian elements found in the Americas "are common to the Aztec, Incas and Mayas, and to many other people, not because they were borrowed from China, but because all these people inherited them from a great, vanished civilization."[43]

The subject of Pan (Panology) requires a book, really, not a chapter or short prologue. But absent that book, let us consider this foreword an appetizer. More to the point, it is the necessary backdrop to the chapters that follow, for without an understanding of the children of Noah, our own lineage remains truly a "mystery."

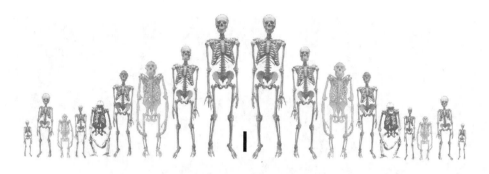

Our Knowledge Is Skeletal

The Truth behind the Bones

O ye of little wisdom; how ye are puffed up in judgment, not knowing the race whence ye sprang!

OAHSPE, BOOK OF APOLLO 5:11

MEET PROTOMAN: ASU

To begin this journey at the beginning may I introduce you to your long-lost ancestor, Asu man?

Naked and unashamed, Asu was the first manlike specimen to walk the Earth, though he might also scamper about on all fours and spend a lot of time in the trees as well. His long arms, curved digits, and upward-facing shoulder joint suited him to arboreal life, while the human angles of his knee joints supported his upright gait. Known as *Ardipithecus ramidus* to the paleontologist, he was about four feet tall with a biped's placement of the foramen magnum, meaning neck and

A

B   C

Figure 1.1. Three artists' impressions of first man.
(A) Illustration by Joy Walker. (B) Illustration by Karen Barry.
(C) Illustration by Ernst Haeckel.

head were in line with spine, rather than jutting forward (like apes).

Yes, Ardi/Asu had a very small brain; he was not a knowing creature, not sapient, not a thinking man and did not use fire or tools. He did not speak and, as one study surmises, was "barely capable of babbling."[1] His diet was of fruit, nuts, seeds, berries, vegetables, roots, and probably bugs. Being the first race of man and in an aboriginal state, he was dun colored and, as the Vedic scriptures of India define the term *asu* (meaning "animalistic"), "lived and moved in the great phenomena of nature."* He appears again in the Old Testament as Esau, Jacob's twin, covered with red, shaggy body hair (animal-man). Sumerian Asag may also be a cognate; similar to the Hindu *asura,* Asag was a demon cursed in a manner not unlike Jacob's twin Esau.

Concerning this Asu or Adam, the Chukchee (Siberia) have a story of the beginning, when the Creator made the first man—an animal-like, hairy, and four-legged creature with very long and strong arms, great big teeth, and claws. Fearing this man would destroy all living things, the Creator contrived to slow him down a bit, make him less dangerous, so he had him walk upright and *shortened his arms:* "with Mine Own hands, molded I . . . the arms . . . no longer than to the thighs."[2] Indeed, the first fossil hominid was long armed† and powerful—helping him scuttle along on all fours, as the morphology of remains in Africa and Asia attest. These creatures, manlike but rough-hewn, were the very first race of human beings.

Asu man was devoid of the spirit part, just as in Buriat anthropogenesis, where first man was without a soul. "A humanlike being that lacked a soul,"[3] he was without the spark of the divine, "like a tree, dwelling in darkness."‡ Hence beings of *wood* as the Mayan Quiche

---

*The Vedic name Asu was later changed in Persia to Adam. Just as Sanskrit *asura* means lower self, the name Asu appears later in the biblical word *asui* (Genesis 16) also meaning "animalistic."

†Note, however, that long arms are not necessarily apelike: Miocene apes, discovered by Louis Leakey, had short arms, as did the protogibbons of the Tertiary.

‡The tree metaphor extends to Norse mythology in which the first man, created by the Norse god Odin, was called Ask and was formed from an ash tree.

say, were the race preceding their own, for they had no soul, no reason, and did not remember the Creator. In the same vein the first man in Chaldean memory had "lived without rule, after the manner of beasts."

Yes, *like* a beast but *not* an animal, for all the animals (unlike man) have instinct fully supplied. But Asu man was a blank, "the nearest blank of all the living, devoid of sense."[4]

> *The low condition of the first race of man [Asu] is known; but still he was a man and not a monkey, nor any other animal.*
>
> JOHN B. NEWBROUGH,
> "COMMENTARY ON OAHSPE"

It is my understanding and the premise of this book that Asu man together with Ihin man (the little people, see chapter 2) are the mother lode, the two Ur races of mankind, our true common ancestors. These chapters are about their offspring—the races of man. To some extent, we all have a combination of Asu-Ihin blood. And *Australopithecus*, the very earliest *hybrid*, represents the first infusion of Ihin blood into the Asuan race, accounting for "modern" features appearing even in the most archaic of hominids. Table 1.1 summarizes the mixed features of australopith (Au); for Asu was quickly upgraded by those early gene exchanges with the Ihins.

WAS *AUSTRALOPITHECUS* OUR ANCESTOR?: ARGUMENTS PRO AND CON

Gridlock: Maybe (hopefully) Neaderthal-as-ancestor has been put to rest, but the jury is still out on *Australopithecus*.

Those Who Argue against Au as Ancestor

Australopithecine specimens from Tanganyika and South Africa were long regarded as off the main evolutionary line. Paleoanthropologist

TABLE 1.1. SUITE OF AUSTRALOPITHECINE TRAITS, DISPLAYING MIXED GENES

Primitive (Asu blood)	Manlike (Ihin blood)
Heavily muscled body	Slender frame
Shoulder joint faces more upward: climber	Vertical foramen magnum (opening in occipital bone of cranium): walker
Pelvis not modern, funnel-shaped torso	Broad sacrum and lumbar curve, shortened hip bone
Toes curved, long midtoe, heelbone without tubercule	Lower leg and ankle well shaped; "leg and foot bones essentially manlike"*
Hands with power grip, fingers curved	Hands with precision grip; Olduvai hand bones modern
Large jutting face and jaw, receding forehead, large brow	Flat, smooth face
Large molars and canines; palate archaic	Manlike dentition and enamel; some lack projecting canines
Skull vault low, thorax broad at base	Small occipital lobes, thin-walled skull
Small brain, less than 600 cc	Prominent frontal lobes

*Coon, *Living Races*, 248.

Richard Leakey (son of famed Louis Leakey who shifted the focus of anthropogenesis to the continent of Africa) judged Au to be only a relative of our forebears, one who reached an evolutionary "dead end." Louis Leakey, like his son, rejected Au as our ancestor, voting instead for *Homo habilis*, the next higher type with his modern lightness of skull. The elder Leakey, some say, tended to deny that anything (including Peking Man and Neanderthal Man) other than his own (African) finds was ancestral.

Anatomist and anthropologist Sir Arthur Keith dismissed Australian anatomist and anthropologist Raymond Dart's finding of *Au. africanus* in Taung, South Africa in 1924 as "just another fossil ape." In his last book Keith (wisely) changed his mind—twenty years after Dart's finding.

Several authorities rejected Au on the basis of brain size. England's curmudgeonly zoologist Lord Solly Zuckerman is remembered for his brisk dismissal of australopiths "as having anything at all to do with human evolution." "They are just bloody apes," howled Zuckerman, underscoring his lifelong rejection of the australopiths as human ancestors. He didn't even think they were proper bipeds. And Earnest Hooton, distinguished professor of physical anthropology at Harvard in the first part of the twentieth century, also said "no" to Au as ancestor because of brain size.

Others who did not think Au was on a direct line to humans were archaeologist and anthropologist Brian Fagan and evolutionary zoologist Charles E. Oxnard. Oxnard's computer analysis in 1975 presumably settled the matter, making Au an extinct ape unconnected with human ancestry: "they must have been upon some side-path."[5] (We'll look at these "side paths" in a moment.)

More recently, in 2008, paleoanthropologists Rob DeSalle and Ian Tattersall noted that numerous australopiths came and went over a period of over two million years without significant change in body structure, and therefore, no known Au anticipated human beings.[6] Besides, there is no one to fill the one-million-year gap between the australopith Lucy (*Au. afarensis*) and the first *Homo* (*H. erectus*).

Those Who Argue in Support of Au as Ancestor

Although Au were not yet known in Charles Darwin's time, the father of evolution nonetheless predicted that such a type, once found, would prove to be "our semi-human progenitors" (*The Descent of Man*). He envisioned "the divergence of the races of man from [such] a common stock." Raymond Dart, the first to discover any australopithecine (the Taung Child in South Africa, 1924), thought Au was indeed an ancestor, especially considering the position of Taung's foramen magnum, indicating upright posture.

Among other supporters of Australopithecus as ancestor of man are John Robinson, also of South Africa, who placed all the more gracile Au material within the genus *Homo*; British-American anthropologist and humanist Ashley Montagu; and Donald Johanson. Johanson, the discoverer of Lucy in 1974, saw Au as the root stock of all future hominids, even though they were "not men"—yet. Still, "you and I have in our teeth and jaws some characteristics that we get almost directly from Lucy," said Johanson. "We see her, in a sense, as the mother of all humankind. . . . Not everyone has agreed."[7]

There have been some fence sitters. Sir Wilfrid E. Le Gros Clark, the great English anatomist and paleontologist of the past century, thought the genus Au as a whole might include our ancestral stock, following Robert Broom, the Scotch–South African physician and internationally renowned vertebrate paleontologist, who thought that Au—based largely on dentition—was a direct ancestor of mankind. In the 1930s, Broom (see figure 8.7 on page 327) discovered Au in Sterkfontein, South Africa, nicknamed Mrs. Ples (short for *Plesianthropus transvaalensis*, or near-man from the Transvaal), who was generally accepted as human (or almost human) by the 1950s.

Le Gros Clark (1955) had to agree with Broom, but only provisionally: though Mrs. Ples's pelvis was modern, the small brain bothered him. Carleton Coon, chairman of the Anthropology Department at the University of Pennsylvania till 1963, also thought Mrs. Ples was a direct ancestor, but rejected the *East* African australopiths. Paleontologist and geologist G. H. R. von Koenigswald, like Le Gros Clark, was inclined to accept australopiths based on dental form—but felt the geology was too recent and the teeth were actually too big.

THE FIX IS IN

Pursue thy studies, O man, and thou shalt find that supposed exact science is . . . only falsehood compounded and acquiesced. . . . Is the man who finds the vertebra of an insect, not said to be scientific? But he who finds the backbone of a horse, is a vulgar fellow. Another man finds a route over a mountain or through the forest, and he is scientific! Why, a dog can do this.

OAHSPE, BOOK OF KNOWLEDGE 1:36–7

With the origin of man our subject is history, really. We only *make* it science, clothe it in science, in order to study it. Evolution is a descriptive body of knowledge decked out in hypothetical models, formulas, theories, and best guesses. Though its *methods* are no doubt empirical, it remains an interpretive field of study. Which is fine. "The ambiguous nature of fossil evidence," as one critic sees it, "obliges paleoanthropologists to pursue the truth mainly by hypothesis and speculation . . . a science powered by individual ambitions and so susceptible to preconceived beliefs."[8]

Mark Isaak, one of the Darwinian fraternity, declares that "[t]he theory of evolution still has essentially unanimous agreement from the people who work it."[9] Does a majority vote among paleontologists add up to the truth? These experts of course are trained professionals, trained to look at the evidence through the Darwinian lens.

This is the simple formula: supposition plus consensus equals fact. "There is something fascinating," Mark Twain once commented, "about science. One gets such wholesale returns of conjecture out of such a trifling investment of fact."

The various family trees for man are admittedly provisional, indeed "highly speculative"—in Charles Darwin's very own words. His expositions are dotted with such terms as *feasible, plausible, in principle.* As one author went to the trouble of counting, Darwin uses the phrase *we may suppose* and similar qualifiers more than eight hundred times in his two major books.[10] Some have complained of his "constant hedging" and "guile."[11] Darwin's approach, though certainly rational and data-based, is, in its most important aspects, profoundly conjectural. It is still, and will remain, only an opinion that man has a phyletic relationship to the apes—that our species, *Homo sapiens sapiens,* has evolved from a different, earlier species.

Paleoanthropologists seem to make up for a lack of fossils with an excess of fury, and this must now be the only science in which it is still possible to become famous just by having an opinion.

J. S. JONES, *NATURE*

There is only one opinion that *is* shared by all: "The alternative to thinking in evolutionary terms is not to think at all"[12] (biologist Sir Peter Medawar). So now evolution is the sacred cow (or is it sacred ape?), and to oppose it is something close to blasphemy, sacrilege. "People made a religion of them [my ideas]," Darwin himself once fretted.[13] Evolution has become (oxymoronically) a secular religion—or call it scientific naturalism. A new orthodoxy for a material age, fundamentalist Darwinism's central dogma is faith in mutation and transmutation, the process that allows one species to turn into another species (also called speciation). The official creation story of modern humanism, it is our shared origin myth, maintained and glorified by its own scientific priesthood and bully pulpit. Richard Leakey and many others besides have trumpeted evolution as the greatest scientific discovery of all time. And so it has come to pass that of all scientific gospel, evolution is the most sacrosanct. And like all myths, it contains a few truths but is riddled with a mass of errors.

BARE BONES

Today East Africa's Turkana Boy (*Homo erectus* or *Homo ergaster*), discovered in Kenya in 1984, and Lucy (Johanson's famous Au find) are among the most partially complete skeletons of any pre-Neanderthal hominids—with more than 50 percent of Turkana's bones recovered and 40 percent of Lucy's. It is unusual to collect more than a few fragments. Peking Man was named and classified (1927) on the basis of a single tooth! The first human ancestor discovered in the New

World was also identified on the basis of a single tooth, which turned out to be a pig's molar—this "Nebraska man" was actually an extinct wild peccary. As William Howells, in *Mankind So Far,* cautioned, there is "a great legend . . . men can take a tooth or a small and broken piece of bone . . . and draw a picture of what the whole animal looked like . . . If this were true, the anthropologists would make the FBI look like a troop of Boy Scouts."

A bone, a femur, a tooth, a mandible, a pinky fragment. All of evolution is a grand and daring reading between the lines of the past. In graphic reconstructions of our ancestors, everything depends on the *angle* used to put these fragments together. Paleoanthropologists Franz Weidenreich's and Ian Tattersall's reconstructions of the same Peking Man look quite different. Teasing out a picture of our ancestor is, to one critic "deceptive evolutionist artistry,"[14] while Harvard's Prof. Hooton in *Up from the Ape* warned, "To attempt

Figure 1.2. Mr. and Mrs. Hesperopithecus (1922), body structure based solely on what turned out to be the molar of an extinct pig!

to restore the soft parts is . . . a hazardous undertaking. You can with equal facility model on a Neanderthaloid skull the features of a chimpanzee or the lineaments of a philosopher. . . . Put not your trust in reconstructions. . . . The faces usually being missing . . . leaves room for a good deal of doubt as to details. . . . These alleged restorations of ancient types of man have very little, if any, scientific value." Without having found any nasal bones of *Au. afarensis*, the artist's reconstruction gives a superplatyrrine (flat) nose, very much like our simian "cousins."

Neanderthal is often represented with his big toe diverging in apelike fashion, which he never had. Artist Peter Schouten gives *Homo floresiensis* a gorilla face; after all, his brain, like the ape, was only one-third the size of ours. Though wild-looking, *H. floresiensis* did not really resemble an ape.

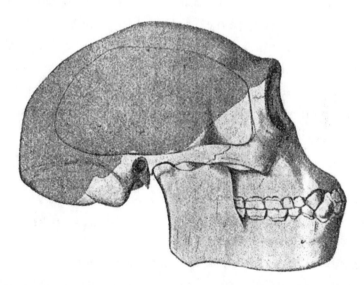

Figure 1.3. Trinil Homo erectus reconstructed. Note that the shaded area indicates the only portion known. As stated in Fossil Man, *French paleontologist Marcellin Boule thought such reconstructions "far too hypothetical. . . . It is astonishing to find a great paleontologist like Osborn publishing attempts of this kind."*

DEAD ENDS AND HOMELESS FACTS

Paleoanthropologist Tim White and colleagues thought *Ardipithecus ramidus* (who corresponds to Asu man) was, if not our direct ancestor, at least a close relative ("sister") of that ancestor. All such early types (human though they are) that do not quite meet our idea of ancestral are labeled sisters, cousins, collaterals, dead ends, or divergent offshoots from the main line of descent.

So what becomes of human evolution if Neanderthal and Java Man and Au and Ardi are not man's ancestor but merely an extinct side branch or "failed attempts at becoming human"?[15] No, they are not failures, just—as these chapters lay bare—race mixtures. Louis Leakey believed that most lineages have their dead branches, which quietly moved on to extinction. He thought that Au *left* the *Homo* line about 6 or 7 mya, and that most of our collected fossil types are simply "aberrant offshoots" from the main stem. American anthropologist Loren Eiseley concurs, deferring to the school of thought that "the true origin of our species is lost in some older pre–Ice Age level and that all the other human fossils represent side lines and blind alleys."[16] Now that we have washed our hands of all these unqualified ancestors, what is left of our family tree? The answer is—a bush (in which the ancestor, nonetheless, still hides).

> *The complications of interbreeding make it impossible to draw neat branching lines of descent on the family tree.*
> JOSEPH THORNDIKE JR., *MYSTERIES OF THE PAST*

And so, with ever more hybrids (rather than ancestors) turning up, paleoanthropologists take care of the problem by calling it diversity or variability and changing the family tree to a bush—but *still* in an evolutionary context, thoroughly blindsiding the factor of crossbreeding, and all those hybrids. As we have seen, numerous evolutionists concur that Au is off the main evolutionary line to *H. sapiens;* but as these irrelevant branches continue to multiply, it leaves us with a family tree that is all branches and no trunk! So many also-rans but no winners. No ancestors. Indeed, almost all the nodes or branching points remain unidentified, with big question marks printed at the point when species first appeared.

A tree with no trunk might get knocked down with the first storm; or as the delightful British author Norman Macbeth so aptly put it, "these forms, being ineligible as ancestors, must be moved from the trunk of the tree to the branches. The result is that the tips are well populated while the trunk is shrouded in mystery . . . we see forms that purport to be our cousins, but we have no idea who our common grandparents were."[17]

Ian Tattersall of the American Museum of Natural History breezily mentions that the tree is really "quite luxuriantly branching." Yet in transforming our family tree into a bush, what has really been laid bare is the multiplicity of crossbreedings that make up the human lineage. The molecular revolution (see chapter 6), we might also note, forced us to scotch the tree and go for the bush, since "genes move quite freely between the branches."[18] Can we interpret that to mean intermixtures? If so, we do not need evolution at all to account for the ascent of man! All we need is the *blending* of types.

First it was an evolutionary "ladder" (linear model), and when that didn't work, it was changed to a "tree." But there were problems with tree so they changed it to "bush"; now, as problems arise with bush, they're changing it to "network"—where will it end?*

Louis Leakey considered Neanderthal, Java Man, Peking Man, Au, and others as mere evolutionary experiments that ended in extinction. Others say the evolution of a successful animal species involves trial and error and failed experiments. What is this experiment that everyone is bandying about but an empty conceit probably personifying the

---

*Although "network" is the latest concept to be canonized, the idea is actually more than fifty years old, presented by Ashley Montagu as a network of crisscrossing lines.

materialist's god of nature? "It gives one a feeling of confidence to see nature still busy with experiments," writes Loren Eiseley,[19] while William Jungers, a paleoanthropologist of the State University of Stony Brook, New York, says, "We're far from the only human experiment." After all, the apes, except the bipedal ones, were "failed evolutionary experiments."[20] But we are not an experiment and the apes are not a failure. Science engages in experimentation—not nature or the universe. Pray, who is the experimenter?

GOING IN CIRCLES

Evolutionary arguments tend to be jargon happy, model obsessed, and insufferably pedantic: Is it really necessary to say lithic technology instead of stoneworking? If not empty rhetoric, the arguments are couched in forbiddingly technical language and nomenclature ("pompous polysyllabification," as one critic saw it). There are often surprise changes in the terminology, which forever keeps you off balance in the paleo world. For example, East Africa's *Paranthropus boisei* was originally called *Zinjanthropus boisei*, then changed again to *Australopithecus boisei*. But it was also known as Olduway Man, Dear Boy (a pet name), and Zinj, as well as *Titanohomo mirabilis* and FLKI. (One of the Leakeys' stage names for this creature, Nutcracker Man, had to go: as it turned out, he did not eat nuts after all; he ate grasses and sedges.) Today Olduvai is a safari destination, sporting a stone plinth marking the site of Zinj's discovery.

Ndangong man of Java, depending upon which book you read, may be called *Homo Javanthropus, Homo soloensis, Homo sapiens soloensis, Homo primigenius asiaticus,* Wadjak Man, *Homo sapiens,* or *Homo neanderthalensis soloensis.* Or try researching Swanscombe Man. He is one and the same as *Homo sapiens protosapiens, Homo marstoni, Homo swanscombensis,* and *Homo sapiens steinheimensis.* In my own research, it took months before realizing *Homo rhodesiensis* was the same fellow as Broken Hill Man, Kabwe Man, and Rhodesian

Man (and it doesn't help that Rhodesia was changed to Zimbabwe).

Embarrassing gaps and jumps in the record are called "systemic macro-mutations" (swiftly turning a liability into an asset). I am certainly not the first to point out the circularity of Darwinism's basic premises. Example: "The process of human evolution [is] a series of adaptive radiations . . . the basis for each radiation was a mixture of adaptations to local conditions and geographic isolation."[21] Translation: species evolved by evolving. The buzz phrase *adaptive radiation* is itself circular, the concept founded on nothing more than the assumption of evolution itself. "Adaptive radiation strictly speaking refers to more or less simultaneous divergence of numerous lines from much the same adaptive type into different, also diverging adaptive zones . . . a rise in the rate of appearance of new species and a concurrent increase in ecological and phenotypic diversity."[22] Say what? Do these words mean anything at all? Bipedal locomotion (which I discuss in chapter 9) is given as another (meaningless) instance of adaptive radiation. The Cambrian explosion is another. Every big set of changes is an adapative radiation. Why? Because the experts say so.

Circular logic underlies the extremely important business of fossil dating: The number of mutations, they say, gives us age; yet age tells us the number of mutations! This molecular clock (discussed in chapter 6) is said to measure the time since humans and chimps last shared a common ancestor, assuming of course they did indeed share a common ancestor. But the word *homology* does just that: homologs are *defined* as protein sequences that diverged from a common ancestor. Correspondence of structure (homology) then indicates a common ancestor. And this is circular: evolutionary descent supposedly explains similar organs in different animals, and these similar organs are then cited as proof of evolution. "Homologous organs provide evidence of [descent] from a common ancestor."[23] In other words, comparable organs, say hand and paw, can be traced to a common ancestor. Sheer supposition! (Things could be alike for a different reason—just as the tuna, porpoise, and mako shark look much alike though of vastly different ancestry—but more on homology in chapter 8.)

Dating follows another loop of circularity: if a hominid is dated before the Mindel interglacial period, call it *H. erectus*; if it's *H. erectus*, date it before the Mindel.[24] And how's this for tautology?: "Small changes operating by degrees were the main instrument of change."[25] In other words: change was caused by changes. And how does evolution explain change? By natural selection (chapter 9). What is natural selection? That which produces change!

Fitness, as genetics defines it, is effectiveness in breeding. But there is a problem here because today, the least "fit" (in Darwin's thinking) reproduce the *most,* that is, the poorest, most disadvantaged people in the world have the highest birth rate. American sociologist Elmer Pendell (political correctness be damned) has recently opined that the decline of our institutions and way of life is caused by the higher reproductive rates of those who should reproduce least. It has also been pointed out that in time of war, it is the *most* fit who perish.

One critic, taking a closer look, finds circularity in the competitive exclusion argument, which assumes only one group can inhabit or dominate a niche: "Count the number of species in a given habitat to determine the number of [ecological] niches the habitat contains, and lo and behold, there is one species for every niche."[26] In *Origin,* Darwin informed us that "variations will cause the slight alterations"; in other words, the cause of variations are variations.

One writer assures us that *H. erectus* "clearly migrated out of Africa because that was where people learned how to hunt large animals."[27] Logic? One team found (from DNA) that the Neanderthal population must have at one point shrunk to as low as just a few thousand individuals. Sure, they eventually went extinct! One critic has recommended that evolutionists "take a course in logic."

Strange and presumptive logic is seen in assertions such as Neanderthal *had to be* a big game hunter "for their bodies demanded" many calories in the cold.[28] But he was not a big game hunter and probably not cold, either (see chapter 9). Along the same lines, *H. erectus had* to have fire, otherwise he could not have come out of Africa (as theory requires). Neither of which is the case.

Here we note that evolutionists keep telling us that things arose because they were *needed*. I wonder if it's that simple. This is Darwin country's particular metaphysic. But do things really happen because they're needed? Did pale skin and freckles come about "under pressure in northern latitudes to evolve fair skin to let in more sunlight for the manufacture of vitamin D"?[29] (a problem we will take up later on). Most insidiously, evolution itself "needs" deep time (long dating) for things to evolve (to which question chapter 6 is devoted).

The would-be syllogism is: Species lacks (whatever); species needs (whatever); therefore, species acquires (whatever). Ashley Montagu, for instance, accounts for the increase in human brain size in this manner: "The brain would have been enlarged because of the *necessity* [e.a.] of a large enough warehouse in which to store the required information."[30]

That human knowledge or language allowed us to advance as a species is a bit like saying that we have legs allowed us to walk. In this connection science writer James Shreeve finds Jared Diamond's logic "dizzyingly circular" for he, Diamond, defines the Upper Paleolithic by cultural invention, which depends on language as the spark or prime mover behind the creative explosion at that time. How do we know, asks Shreeve; the answer—"because the Upper Paleolithic is defined by invention." Diamond's argument betrays a "total lack of evidence for the crossing of the language Rubicon at the beginning of the Upper Paleolithic."[31]

All told, modern-day evolution has perfected the art of assuming what you wish to prove, circularity is deeply embedded in the heart of this "science." We search in vain for the causation, the forces exerted to make man evolve from a lesser to a more sapient state, for *H. sapiens* did not evolve from anything. *H. erectus* did not evolve from anything. Neanderthal did not evolve from anything. But each came about through the process of crossbreeding.

HOW THE MIGHTY ARE FALLEN

Evolution makes better murderers.

WALTER MOSLEY, *WHEN THE THRILL IS GONE*

One fly in the evolutionary ointment is stasis, meaning no change over time. We see this stasis, for example, in Mayan art and architecture, which show "no development with the passage of time, ending up exactly as it began."[32] In the Old World, tools remain crude between the time of Zinj (Au) and Peking man (*H. erectus*)—a long stretch with no indication that man was evolving or progressing.

> *The history of the human race . . . gives no countenance to*
> *any doctrine of universal and general progress . . . but sustains*
> *rather a doctrine of predominant natural tendencies to*
> *degeneration. . . . There is no invariable law of progress. . . .*
> *As a matter of fact, degeneration and disintegration seem as*
> *likely to take place as real progress and advancement.*
>
> GEORGE FREDERICK WRIGHT,
> *ORIGIN AND ANTIQUITY OF MAN*

Indeed a certain trend toward *de*volution is noted in many different traditions, which mutually speak of a lost golden age, a lost utopia, having fallen to barbarism; and this is where *retrobreeding* shows its face. "In times past the same countries were inhabited by a higher race."[33] In Malaysia "a white race relapsed into barbarism" in the jungle interior.[34] Protohistorians of every stripe can attest to such relapses, as in the Pacific, in places such as Ponape and Easter Island, where a dramatic regression of culture has taken place.

> *In some nameless, distant past mankind must have*
> *ascended a long way up the ladder of civilization, only to*
> *relapse into chaos and barbarism.*
>
> PETER KOLOSIMO, *TIMELESS EARTH*

Future studies of the protohistorical world might well reveal that retrobreeding, followed by war, destroyed both antidiluvian and post-diluvian civilizations, "and war spread around about the whole earth" (ca 70 kya).[35] Much later, just before the flood, men were "descending in breed and blood . . . and they dwelt after the manner of four-footed beasts."[36] Then, after the flood, these histories follow with Ihuan inter-necine wars (ca 20 kya).[37] Later still, around 10 kya, "the Parsieans [high culture] were tempted by the Druks [barbarians], and fell from their high estate, and they became cannibals."[38]

Hopi history tells of the third age* with a mighty civilization, full of big cities; Hopi artist and storyteller White Bear spoke of big cities, nations, and civilizations that once were, but war and other evils destroyed them. Likewise does archaeology reveal extensive trade networks once centered on the Mississippi Valley mound culture. Canadian Indians knew of "shining cities" before "the demons returned"; there, where once splendid cities stood, "there is nothing but ruins now."[39]

All was lost. The complexity and beauty of the *earlier* layers at Guatemala's Tikal was puzzling to archaeologists; things did not get simpler, less elaborate, the deeper they dug, quite the reverse. By the time of the Spanish conquest, Mayan civilization was already run down. And the same befell the great societies of the Old World:

> *The palaces and houses with their goodly apartments*
> *fall into ruin. . . . Serpents hiss and glide amid broken*
> *columns.*
>
> KALIDASA, SANSKRIT POET, 350–420 CE

Speaking of India, the erudite British archaeologist Godfrey Higgins observed that the science and learning of the Hindus "instead of being improved, has greatly declined from what it appears to have been in the remote ages of their history."[40] Here in India and

---

*The Hopi third age corresponds to the Oahspe's third era (ca 48 kya): "the people of earth dwelt together in cities and nations. This was the third era" (page 1, verse 5).

Pakistan, the *lowest* strata at Mohenjo Daro, an ancient city of the Indus Valley civilization, produced implements of *higher* quality and jewelry of greater refinement than in the upper layers. This mysterious civilization has been compared to the equally mysterious Easter Island, where the oldest cultural level is again the most advanced: the statues of the second period have a degenerate style compared to the earlier ones; just as Egypt's first pyramids were indeed their finest. The masterpieces of Mesolithic art (Cro-Magnon), possibly the oldest art in the world, were never matched in the later Neolithic. Rather than straight-line evolution, we are faced with cycles, the curve of civilization as a waxing and waning thing.* In chapter 3, I come back to these relapses of the Ihuan race.

> In case after case, the oldest stone remains are the grandest and most perfectly executed; what followed later are crude imitations.
>
> RICHARD HEINBERG,
> *MEMORIES AND VISIONS OF PARADISE*

"The tribes of men . . . prospered for a long season. Then darkness came upon them, millions returning to a state of savagery."[41] In *Mysteries from Forgotten Worlds* Charles Berlitz noted "the retrograde tendency of American and numerous other ancient cultures . . . as one goes farther back in time, one finds more advanced cultural patterns in preceding eras."[42]

Evolutionists seem to be entirely indifferent to the now substantial body of literature on high culture prior to the Magdalenian. Nevertheless, independent studies of the deep past reveal a golden age superseded by decline and savagery. If nothing else, the repeated fall from high civilization puts the lie to steady evolutionary progress from primitive to advanced.

---

*See my book *Time of the Quickening* for an in-depth study of the cycles of time.

Her people build up cities and nations for a season . . . but soon they are overflooded by [darkness], and the mortals devour one another as beasts of prey. . . . Their knowledge is dissipated by the dread hand of war . . . and lo, her people are cannibals again. . . . As oft as they are raised up in light, so are they again cast down in darkness.

OAHSPE, SYNOPSIS OF SIXTEEN CYCLES 3:6–10

I wonder why evolutionists, so big on competition (see below), deny the obvious cannibalism of prior ages? Why did Cro-Magnon disappear ca 12 kya? The questions are related, for the Ihuans (Cro-Magnon), inhabiting the wilderness, "ate the flesh of both man and beast."[43] As we go on, we will encounter other instances of retrogression in the Ihuan–Cro-Magnon race, which turns up in the record with a considerable sampling of archaic features (low skull, long arms, large orbits), indicating that this perfectly AMH race *retrobred* with Neanderthaloids, thereby descending lower and lower.

Just as the Paleolithic was coming to a close, ca 12 kya, "tribes of Druks and cannibals covered the earth over"—which corroborates ufologist/prehistorian Brinsley Le Poer Trench's observation of cannibalism in Egypt around the time of Osiris. It was probably on that same horizon that the Greeks had their Polyphemus, a notorious cannibal. The cave of Fontbregoua in southeast France contains bones bearing cut marks, dated to around the same time, attesting to the presence of "savages and cannibals all over these great lands; enemies slain in battle [were] cut up for the cooking vessel."[44]

But man eating was nothing new. Raymond Dart found evidence of cannibalism among Au, almost our earliest progenitors. In South Africa charred bones of human victims were found at Klasies, as well as at Bodo (Ethiopia). "In sights ranging from South Africa to Croatia to the U.S. Southeast, similar evidence indicates that cannibalism may long have been part of the human heritage."[45]

Cannibalistic giants in the Americas may have been large Druks—the red-headed goliaths who terrorized the Indians of Nevada until fairly recent time. Sarah Winnemuca, a Paiute, wrote about a tribe of barbarians that once waylaid her people, and killed and ate them. The 1,000-year-old Anasazi site in the Southwest did indeed produce bones with cut marks. We'll come back to these huge and troublesome Druks.

Just as Judeo-Christian tradition has the sins of the giant Nephilim bringing the deluge upon the world, Greek, Arabian, Roman, and Peruvian traditions likewise suggest that troublesome giants were the cause of the flood: "[T]he land of Whaga (Pan) was beyond redemption. . . . They have peopled the earth with darkness . . . and cannibals."[46]

Archaeologists sifting through the remains of Druks and Neanderthals have come across unmistakable evidence of man eating: for example, in Spain at Atapuerca's Gran Dolina and Sidron caves (dated 43 kya), where cut marks on bones indicate that the edible marrow was sucked out from defleshed, charred, and splintered bones, and at Germany's Ehringsdorf site as well as at Steinheim, where skulls show heavy blows to the frontal bones. In Europe, Krapina (Yugoslavia) and Monte Circeo (Italy) are two important Neanderthal sites with the telltale remains of cannibal feasts: bodies and bones smashed open to get at marrow and brains.

But for some unknown reason, archaeologists are squeamish about announcing these incontrovertible finds, despite the work of Earnest Hooton, Franz Weidenreich, Robert Braidwood, and other top dog anthropologists, who found cannibalism at Zhoukoutien, China—judging from heavy blows on the *Sinanthropus* pate. Though this Peking Man was a headhunter who ate brains, Ashley Montagu stated: "It has been asserted that Peking man was a cannibal. . . . This is possible but unproven. . . . Except in aberrant cases, it is highly unlikely that man has ever resorted to eating his own kind except *in extremis* or for ritual purposes."[47] British archaeologist Paul Bahn interprets the Krapina and Monte Circeo finds as the result of a tragic roof fall, or perhaps a landslide, or the use of dynamite during excavation. And if that doesn't fly, the Fontbregoua (France) and Krapina bones are understood in a ritual context: the defleshing of bones taken "as a stage in a mortuary practice." Bahn also discredits the Monte Circeo evidence, despite the presence of a chisel-like object used to scoop out the brains, every skull of which shows fractures at the *same spot* around the right temple. This, he concludes nonetheless, must have been a hyena den, the tooth marks consistent with those of hyena. Cannibalism, he says, was "rare or non-existent," and his interpretation has "demolished the myth."[48]

But headhunters and cannibals are well known even from the recent past. The Aztecs, not too long ago, practiced human sacrifice and cannibalism, and when the Portuguese first got to South America, some of the local natives were cannibals, including the Yaghan and "Indios bravos" of Brazil's Matto Grosso. In Borneo, the Dyak's longhouse still displays the skulls of people whose brains they ate, the practice also known in Indonesia and once widespread in Sundaland: Ngandong skulls of Java were smashed at the base like the Dyak ones. Until the eighteenth century the Ebu Gogo on Flores ate human meat. Until recently in New Guinea, too, tribesman ate brains of their enemies, the practice prohibited eventually by the colonial powers (in the 1880s). Other examples of cannibalism in modern times have been documented among Fijians and other Melanesian tribes. In the seventeenth and eighteenth century Easter Islanders were found to have resorted to cannibalism and the Tongans in the South Pacific were also man-eaters. Even in today's Africa are the fierce cannibal people of the Congo, the Fang. Uganda's atrocities and war crimes of very recent days include rape, mutilations, and cannibalism. During India's drought and Skull Famine (1792) there was widespread cannibalism. Man of both the past and present was not above feasting on his fellow man.

A PAEAN TO MOTHER NATURE

When men first cottoned to the idea of evolution (*On the Origin of Species,* 1859), the world was good and ready for it. The universities were still strongly beholden to the Church of England; natural science, in other words, was still in thrall to the tenets of religious gospel. Science *needed* to be secularized. Its time had come. It was a world that was reaching beyond the straitjacket of theology and past-thought, that wanted to know the unadorned truth, the facts of man's beginnings, free of dogma and doctrines passed down by scriptural fiat. It was time for old fallacies to bend with the winds of change.

The Darwinian revolution, toppling old and fixed ideas, was inevitable, refreshing, and necessary. It served its purpose. But having freed itself from the shackles of theology, the scientific mind proceeded to embrace the opposite fallacy, the materialist/reductionist dictum. No sooner was anthropogenesis wrested from its biblical confines and made an article of science, than Darwinism itself became the new authority—the new gospel, the new dogma.

Victorian society, in the midst of the grinding Industrial Revolution, was having a love affair with the phenomena of nature, and Darwinism was the ultimate paean to nature per se, Mother Nature's power. The progressive Victorian mind was also ready to jettison the "fixity" of species (immutability), as one more token of ancient dogma. Evolution appealed also to the entrepreneurial class, the educated, and the rising bourgeoisie, somehow affirming their sense of rank and higher worth, their qualification as the "fittest." Darwin's transmutation of species, moreover, only reinforced the optimistic nineteenth-century belief in progress, but more than just progress, it vindicated the upper-class white Europeans "who were taking over the world."[49]

It is a curious thing that modern thinking has struck down almost every Victorian theory—except evolution. It's staying power depends on this: Without it, science would have to admit it doesn't know how human beings got here, a naked confession of ignorance. With atheism (or humanism) the predominant "religion" of today's intelligentsia, the Weltanschauung of this secular age, the only real alternative to evolution—the divine hand, the work of the great unknown—is not acceptable. Many scientists are agnostic or possessed of split mind (forever undecided), a kind of cognitive dissonance.

BLUEPRINT FOR A COMPETITIVE AGE

Evolution, in our own age of profit, has granted us a subliminal charter for exploitation, privilege, and discrimination. The standard word for a species' use of the environment (the ecological niche) is *exploit*. Animals and men are routinely described as "colonizing" their habitats (even if they've always been there, see chapter 11).

A foundation myth, to give a quick definition, is a story or legend that legitimizes a social custom or value system. The Darwinian struggle for existence, conceived as the war of nature, helps us to understand and justify our own warlike age. Richard Dawkins, Darwin's principal sharpshooter today, uses the arms race as an analogy for survival of the fittest and, by inference, makes it quite normal for one race to dominate another. In a personal letter, Darwin once wrote that the more civilized races have "beaten [the others] in the struggle for existence. Looking to the world at no very distant date, what an endless number of the lower races will have been eliminated by the higher civilized races throughout the world."[50] Regarding this outlook, I am reminded of an Oahspen verse: "Now, behold, I have left savages at your door, and ye raise them not up, but destroy them. Showing you, that even your wisest and most learned have no power in resurrection [upliftment]."[51]

Capitalists have long used social Darwinism to justify unfettered market competition, on the plea that competition is crucial for progress and for weeding out those who don't keep up. Progress, in such a scheme, is only possible by eliminating imperfections from humanity—best accomplished through competition. The complete title of Darwin's famous book was *The Origin of Species by Means of Natural Selection, or The Preservation of Favored Races in the Struggle for Life.* Buttressed by such ideas of the favored or fittest *nations*, expansionist policies have only thrived on survival of the fittest and the belief in Manifest Destiny, all cloaked in the garment of science and enlightenment. "In prim Victorian England," noted Jeffrey Goodman in *The Genesis Mystery*, "Darwin's thoughts about dark-skinned natives prevailed, providing a footing for racism and . . . imperialism."[52] This idea of survival of the fittest, of the favored, which nicely lends itself to a master race theory, finds Darwin, in chapter 2 of *The Descent of Man*, accounting for animal and plant migrations according to the "dominating power" of those favored species in a "higher state of perfection." Well, no wonder: Darwin's father always told him that "the race is for the strong."

Not a few have taken exception to the competition model, which seems to rationalize one group (the fittest) monopolizing the niche. American politician William Jennings Bryan, in the 1920s, believed that the Darwinian law of competition was just the sort of ideology that allowed such evils as robber baron capitalism and German militarism. Hitler, a convinced Darwinist, leaned on the evolutionary biology of Ernst Haeckel. Haeckel, Germany's best-known Darwinian propagandist, judged the different human races to be as distinct from one another as different animal species, with the Teutonic/Nordic people, of course, as the pinnacle of evolution. Consider the devastating results of this racial-state idea and predatory nationalism.

A recent writer, chronicling the words of a Holocaust survivor, quotes him as stating: "If we present a man with a concept of man which is not true, we may well corrupt him. When we present man as . . . a mind-machine . . . as a pawn of drives and reactions . . . we feed the nihilism to which modern man is prone. . . . The gas chambers of

Auschwitz were the ultimate consequence of the theory that man is nothing but the product of heredity and environment." The chronicler then makes mention of "the famous German Darwinist Ernst Haeckel . . . [who] blasted Christianity for advancing an anthropocentric . . . view of humanity. . . . Today the atheistic Darwinian biologist Richard Dawkins argues that, based on the Darwinian understanding of human origins, we need to desanctify human life, divesting ourselves of any notion that humans are created in the image of God and thus uniquely valuable."[53]

The central mechanism of evolution, natural selection, always seems to entail one group holding the advantage over another. In evolutionary thinking, someone always has the advantage. It is obvious, at least to Ian Tattersall, why *H. sapiens,* "intolerant of competition . . . and able to do something about it," came to be the only human species on Earth. When they arrived in Europe, supposedly from Africa, and found Neanderthals, within ten thousand short years, the Neanderthals were gone, presumably as a result of *failing* in competition with the newcomers. Known as the replacement model, this idea of conquest or subjugation or even genocide is just one spinoff of the misguided and self-condoning theory of competition. It is likewise assumed that *Homo habilis* went extinct because "he could no longer compete."[54] In some circles, australopiths are regarded as "failed" humans. We often find this word *failure* (as well as success) in the literature: "The successful development of one group spells doom for another," said C. Loring Brace. But even if Darwin and his followers were right about competition, so what? It does not follow that a different species *resulted* from it, and it cannot be used successfully to explain species disappearance or extinction. The idea that poorer species are replaced by better ones remains unproven. Through the evolutionary lens, even extinction is a failure (failure to adapt), which is not only "impossible to prove"[55] but actually circular—extinction is the failure to survive!

With evolutionary patois set in the competitive mode, technology, it is argued, gave early man "an astonishing advantage" over other hominids, Tim White remarking that man now had "the ability to *exploit* a broader range of habitats, eventually enabling our ancestors to leave Africa and *colonize* most of the globe . . . [becoming] the ultimate *victor* [e.a.]."[56]

Using survival alone (self-preservation or even group preservation) as the measure of success sends the wrong message to students of life. With no morality or ethics or quality in the equation, life is sterile, stripped of purpose and meaning, trapped in the darkness of matter. Just survive, as if the end (survival) justifies the means, which are selfishness and competition. The implication is that aggression, outdoing the rival, is not only inherent but also the key to success.

NOT A DOG-EAT-DOG WORLD

Overlooked in all this is nature's give and take, her symbiotic side—two species engaging in intimate, mutually beneficial behavior. Plants, for example, get phosphorus and other nutrients from fungi, which settle on their roots; while the fungi get carbohydrates from the plant. Plovers pick leeches from crocodile's teeth, offering dental hygiene in return for food; just as the whale shark waits calmly as pilot fish swim in and out of its mouth—cleaning its teeth. Bees, while collecting nectar, pollinate dozens of species of flowers. "Symbiosis . . . pops up so frequently that it is safe to say it's the rule, not the exception. . . . More than 90% of plant species are thought to engage in symbiotic couplings."[57] Neither could any of us humans exist without the bacteria that live in our gut, digesting food and producing vitamins. Evolutionists call all this coadaptation, but are at a loss to explain how this evolved *in stages*. How could two different species evolve separately, yet depend on each other in such intricate ways for survival?

As Richard Milton has pointed out, in *Shattering the Myths of*

Darwinism, "the overwhelming majority of creatures do not fight, do not kill for food and do not compete aggressively for space." Food sharing, it is widely believed, is what made us human. We hear that the last common ancestor of hominids and African apes was characterized by relatively little aggression. We have not found much evidence of primate-versus-primate strife. Chimps, our "nearest" cousins, are quite peaceful; most apes are vegetarians. Altruism, it is further revealed, even between different species, is not uncommon. Dominance *nonetheless* remains a central theme of Darwinian evolution. Has it endured simply as a charter for Western opportunism and unrelenting hegemony over the Third World? For ours is a philosophy of winners and losers—what a dark view of man, making self-preservation at the cost of others the supreme law. Symbiosis/cooperation is called by Darwinists the *problem* of altruism! But it is not a problem.

> *Within the same species, powerful . . . safeguards prevent serious fighting. . . . These inhibitory mechanisms . . . are as powerful in most animals as the drives of hunger, sex or fear. . . . The defense of territory is assured nearly always without bloodshed.*
>
> ARTHUR KOESTLER,
> *THE GHOST IN THE MACHINE*

> *Our ancestors were strongly cooperative creatures. . . . Social cooperation [w]as the key factor in the successful evolution of Homo sapiens. . . . We are a cooperative rather than an aggressive animal.*
>
> RICHARD LEAKEY AND ROGER LEWIN, *ORIGINS*

We will not be a "successful" species until we achieve harmony, trust, understanding, fellowship, self-discipline, and reciprocity, which are the keys to enlightenment. Science without a moral rudder, without the guidance of humane principles or the consideration of goodness itself, will run riot on the future of mankind. If man is to be a successful species, he will learn to *overcome* his competitive bent, seeking unity and common cause all the days of his life.

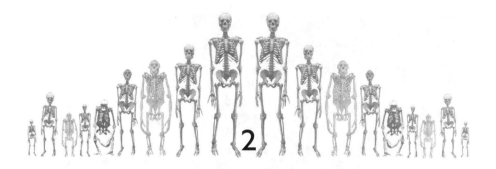

RETURN OF THE HOBBIT

Our Ancestors the Little People

"Our ancestor was much smaller than contemporary humans," noted Richard Leakey and Roger Lewin in *Origins*. Who was that little ancestor? This chapter is about the real missing link in the fossil record: *Homo sapiens pygmaeus,* Ihin man, the little people, known in some places as the Old Ones, elsewhere as the Immortals.

HOBBIT HULLABALOO

Thanks to good old media hype, 2004 was dubbed "The Year of the Hobbit": When a fossil woman in her tiniest dimensions was discovered on the Indonesian island of Flores, anthropologists struck a pose of humility, averring "how much we still have to learn about human evolution."[1] But Flores woman (dubbed Flo or Lady Flo) goes more to helping us *un*learn human evolution. As one website pegged the problems surrounding this diminutive race: "Little people raise big questions."[2] Hmm, they said about the same thing when Africa's extremely short Lucy (*Au. afarensis*) was found in 1973: "This little midget . . . will mess up everything"[3] on the family tree. Others said "these little

Figure 2.1. Australian little people as seen in two Wanjina figures fully clothed with the distinct anatomical features of foreign-looking beings, with pallid, triangular faces, big eyes, long narrow noses, and short arms. Courtesy of Vesna Tenodi.

Figure 2.2. Ihin man. Drawing by Ruth (Skorjenko) Wobschall.

remains could rewrite the story of modern human evolution."[4] *Rewrite?* Actually, it could kill it.

The problem was that Flo, at three feet seven inches, seemed too little to be *Homo*/human. As for Africa's Lucy, her primitive jaw was at least 500 kyr *younger* than a previously collected older, but more *modern*-shaped jaw: something more modern came *before* her, upsetting of course the evolutionary scheme of things.* But that's where the little people come in, *Homo sapiens pygmaeus*, whose ancestry is Ihin, a race almost as old as Ardi/Asu and actually older than Lucy and her Au (australopith) family. At Hadar in East Africa, where Lucy was found, there was another larger hominid (*Homo*, thought Donald Johanson, if only because the molars are reduced and the front teeth rather modern). But again, these mod features were turning up *too early* in the record. Someone along the lines of modern man (AMH) must have existed in earliest times. But who?

Figure 2.3. Donald Johanson (left) and Tim White (right) in the halcyon days (1978), announcing Lucy, their wonderful discovery, to the world; fossils on the table are mostly Hadar specimens. Courtesy of Donald Johanson.

---

*Other anachronisms are collected in appendix E.

The origin of Homo sapiens *must be sought for in a much
more remote past than we could ever have supposed.*

MARCELLIN BOULE, *FOSSIL MEN*

When England's prestigious *Nature* magazine broke the hobbit
story late in '04, the media loved this prehistoric shrimp, following up
with headlines like "Tiny-human find becomes huge news." But then,
as everything about the Flores hobbit stood to upset the evolutionary
apple cart, news stories began to reflect the controversy: "The Hobbit
Wars Heat Up," "Hobbit Hullabaloo," "Hobbit Bone Wars."[5]

The first challenge was how to pigeonhole this pre-Neolithic creature who weighed little more than sixty pounds and was an astonishing
composite of archaic and modern features. Overlooking the obvious race
mixing, the binomial scientific name given her—*Homo floresiensis*—
placed her in the human genus (*Homo*) but kept her at arm's length
from *sapiens,* the true human group, by assigning her to a separate species (*floresiensis*).

Also called LB1 after the caves (Liang Bua) in which these bones
were found, this hobbit did not belong to any separate species, others
argued, but was merely a pathologic version of our own species, the creature beset not only with "dwarfism," but also, supposedly, with some
brand of microcephalic disorder or cretinism, the brain being only one-
third the size of us moderns. But let's remember that pathology had
also been pinned on the first-discovered Neanderthals, the first *Homo
erectus,* and so on. *All hominids,* * as I understand the human family, are
in the same species, the only differences being racial (subspecies).*

Well, the pathologic argument was shot down soon enough: hobbit
brains look nothing like those of microcephalics. Besides, the argument
could hardly hold when thirteen additional specimens were found, all
with the same chimpsize brain as LB1: 400 cc. Flores's hobbit was cer-

---

*The word *hominid* technically includes all manlike creatures, both members of and
candidates for genus *Homo.* I will not be using the newer term *hominin,* as its very definition is based on evolutionary biases.

tainly different from any other AMH (anatomically modern human); these people, in fact, rather resembled *H. erectus:* similar shape of crania, long arms, thick bones, hunched shoulders, short legs, large flat feet (see figure 2.4) not especially suited for running, bulging eyes, prominent brow, sloping forehead, and, the proverbial clincher, no real chin. As a matter of fact, *H. erectus* did coexist with *H. sapiens* in this region 60 kya.[6] Indeed, the first-known abode of *H. erectus* was Java (Flores, east of Java, is situated between Bali and Timor). Now isn't it pretty obvious Flo carried some of that Java Man's (*H. erectus*) genes?

But even more primitive than *Homo erectus* in some ways, Flores's

Figure 2.4. Artist's rendition of hobbit. Drawing by Karen Barry.

hobbit has been compared to one of our earliest ancestors, Au: "Maybe hobbits had descended from *Au. afarensis*—Lucy's kin," even though hobbit's "face and teeth are all wrong for australopiths," grappled William Jungers, who nonetheless noted that Flo's "suite of primitive skeletal characteristics, such as her apelike wrist bones and her flaring pelvis, bears an uncanny resemblance to . . . Lucy."[7]

Others agreed that perhaps Flores's hobbit descended from Au (both were so short). Or perhaps she was a dwarf form of *H. erectus*. Skeptics, though, pointed out that hobbit's pea brain was way too small to be a dwarfed version of any *Homo erectus* (whose brains are twice the size). As Carleton Coon once said, "the racial dwarfs are a subrace . . . without known *erectus* ancestors."[8] And the reason is because the little people actually came *before* the Druks (*H. erectus*). They were, as we will soon see, essentially ancestors, not descendants, of the *H. erectus* race (and of the Au race as well).

Absolute size, it turns out, has little to do with mental powers. For example, despite Neanderthal's impressive cc, his gray matter shows a want of convolutions, and he is back-brained and has skimpy frontal lobes. Although hobbit's brain-body ratio was more like a chimp, her tiny brain was intricately folded and unusually complex; the temporal lobes were really wide, and convolutions were just in the regions associated with executive functions.

In the hobbit were combined Lucy-like (Au) genes with pithecanthropine (*H. erectus*) genes—and modern ones as well. Flo, a happy hybrid, was definitely "modern" in teeth, temporal lobes, and aspects of the skeleton. These little people were skilled hunters of the dwarf elephants (called stegodons). They made blades, perforators, points, and other sophisticated tools for the hunt, and they made fire for cooking. Though the brain was small, the *organization* of their frontal lobes and cerebellum—the way the neurons were connected—was humanlike, not apelike. Their brain, say experts, was actually highly evolved.

What is the point of arguing whether hobbit's ancestor was

TABLE 2.I. ROUNDUP OF BRAIN CC'S (CUBIC CENTIMETERS)

Hominid/Primate	Brain Size*
Ar. ramidus (Asu)	350
Au	400–600
Au. robustus	500–550
chimpanzee	300–400
gorilla	460–550[†]
H. erectus	700–1,200
H. ergaster	800–900
H. erectus pekinensis (Sinanthropus)	915–1,075
H. floresiensis (hobbit)	400
H. habilis	510–750
H. heidelbergensis	1,200–1,300
H. neanderthalensis (Amud)	1,200–1,740
H. neanderthalensis (Ehringsdorf)	1,480
H. neanderthalensis (La Chapelle)	1,620
H. neanderthalensis (Shanidar)	1,200–1,600
H. rudolfensis (Skull 1470)	800
H. sapiens (Cro-Magnon)	1,350–1660
H. sapiens (earliest)	1,250
H. sapiens (modern human; smallest, nonpathological)	790[‡]
H. sapiens (modern human; world average)	1,350–1,450[§]
H. sapiens (modern Tiwi and Andaman Islanders)	1,000
H. sapiens (average Australian Aboriginal)	1,287

*Overall, intellectual capacity has not shown any correlation with brain size.

[†]750 cc in male gorillas is not unknown.

[‡]Lower limit, according to Sir William Turner, though Daniel Lyon, a five-foot-tall, normal, nineteenth-century man, had only 624 cc.

[§]Examples of individuals with larger than average brains are: Ludwig van Beethoven, 1,750 cc; Otto von Bismarck, 1,965 cc; Daniel Webster, 2,000 cc; Aleut skull, 2,005 cc; and G. W. von Leibnitz, 1,422 cc. 5.

H. erectus or Au or *H. sapiens* when we find contributions from *all* these phenotypes combined in little Flo? The simple fact is that she is a marvelous hybrid, not a "freak born with a tiny brain."[9] These freaks, according to Australian paleoanthropologist Peter Brown, of the University of New England, who analyzed the hobbit remains, were "surprisingly human." In fact, they were a lot more human than the preposterous chimp-faced reconstruction supplied by *National Geographic*.[10] Who told the artist to draw in a gorilla face? Evolutionists, of course. Even Java Man, older than Flo, though rugged, did not look like an ape.

Figure 2.5. Peter Brown photographing hobbit skull.
Courtesy of Peter Brown.

Flores natives have not forgotten their own Ebu Gogo, true wild men or cryptids, covered with body hair and only four and a half feet tall, who "ate anything" (Ebu Gogo roughly translates into "grandmother who eats anything"), raided their crops, and may also have made off with their children. This is not Flores mythology (as a Google search will tell you). It is Flores *history:* In the eighteenth century, the Nage villagers of Flores disposed of the devious little Ebu Gogo by tricking them into a cave and firing at them. The same tactic, as we will see, was used in Sri Lanka to exterminate the equally wild Nittevos.

Long armed and pot bellied, Ebu Gogo were probably the last of the hobbits' mixed descendants, carrying the genes not only of *Homo flore-siensis*, but also those of the ancestors of the light-skinned, very short, Rampasas folk, still living on Flores. Some researchers noted that Flo had the same rotated teeth and receding chin as the Rampasas pygmies living there today, quite near the LB cave.

HOMO SAPIENS PYGMAEUS

Yes, quite a bit of *H. erectus* blood flowed in the veins of our hob-bits, even though they lived as recently as 12,000 years ago. Flo's race is said to date back 95,000 years (others say only 38 kyr) and to have survived up to 12,000 years ago. But these dates can be problematic: Even *H. erectus*'s successor, Neanderthal man, was supposed to be gone from Earth by 28,000 BP. It seems, though, that both *H. erectus* and Neanderthal garnered a bit of immortality by planting their genes with more advanced humans who "succeeded" them.

Ah, but until scientists twig to the persistent crossbreeding among all these ancient stocks (not different species, just stocks), they will keep discovering "previously unknown species" like Flo, and keep ask-ing misfired questions like: How did hobbit get so small? The favorite (and flawed) answer is that they are a scaled-down version of *H. erectus*, according to the principle of island dwarfing; or they got small due to "global warming caused by massive volcanic activity . . . [which] led to dwarfing." Or Flo is a "stunning" example of isolation at work "minia-turizing" a sequestered race.[11]

Their littleness is not an anomaly at all and does not need to be explained away, though every analysis I have come across has the little people *shrinking* to their current size, *acquiring* reduced stature. Unbiased research points to an original gene pool for littleness, belong-ing to the race of *Homo sapiens pygmaeus*, which began with the Ihins, the first AMHs whom we will come to see as a people of learning and peaceful ways, and who, though now extinct, once dominated the world population. "Now, for the most part, all the people had become Ihins."[12] (at around 45 kya).

The shrinkage theory *completely ignores* the primeval existence of *Homo sapiens pygmaeus* (the name coined by zoologist Ivan T. Sanderson), the original little people. Instead, the experts account for the short stature of these living and extinct races as *downsizing* or simply an irrelevant oddity that needs to be explained away as:

- anomalous mummies, or perhaps the skeleton of a small bear
- hoax specimens, or simply those of children
- tendency toward pedomorphism (retention of juvenile traits in adults)
- congenital abnormality, growth deficiency
- an adaptation to jungle life: easier to move around dense forest, climb trees for fruit, etc.
- squatting in canoes
- inbreeding (due to isolation)
- lack of sunlight (in forest)
- crowding (on islands)
- shifting the reproductive age forward: having babies younger, hence stunting growth, under "evolutionary pressure" to reach adulthood faster
- island-dwarfing due to limited food supply; nutritional stress leads to smaller bodies

> *The stature of early man is often near the upper limits of pygmy stature.*
>
> HARRY SHAPIRO, *PICK FROM THE PAST*

No, the little people were not "derived" or shrunken from anything; they were always around, even before *H. erectus*, even before Au. This is why Earnest Hooton proposed "evolution of the Negro from the pygmy,"[13] rather than the other way around (which posits pygmies as "shrunken Negroes"). We perceive Ihin genes in Africa's Negrillos (pygmies) in their browridges—smoother than the forest Negroes. The pygmies are racially different from the Bantu and Sudanic people. Colin Turnbull, in his charming ethnography, *The Forest People,* lists a good number of pygmy traits "uncharacteristic of the Negro tribes." Most notably, the pygmies are more brachycephalic (round headed) than the dolichocephalic (long headed) Africans, and they are lighter skinned, not so woolly haired, straighter nosed, and not prognathous.* As Roland Dixon put it, "the admixture of this element [brachycephaly], which throughout the centre of the continent seems to be derived from the Pigmy peoples, leads to a diminution in stature."[14] The combination of very short stature *and* brachycephaly is again seen in New Guinea's little people as well as other Negrito groups. The Ihins were short, brachycephalic, gracile, and big brained.

Though the genes of *H. sapiens pygmaeus* are represented everywhere in the record, they remain unrecognized by the bone people. Only their gracile (delicate) build is worth a mention; but gracility did not develop from some previous condition, as evolutionists suppose. It was not some random gene change "selected" (by the environment). Gracility was part of the *H. sapiens pygmaeus* package from the start, and *H. sapiens pygmaeus* was on the scene from the beginning. And this is the reason why little fossil men (like Au) are *older* than their robust counterparts. In South Africa, for example, "the smaller form [*Au. africanus*] seemed to be from deposits that are *earlier* [e.a.] than those containing the larger form,"[15] (the smaller ones from Taung, Sterkfontein, and Makapansgat, and the larger robust ones from Kromdraai and Swartkrans). *Au. africanus* also had better shaped frontal and temporal lobes than *Au. robustus.*

The brain of East Africa's Skull 1470 was, surprisingly, much bigger than *H. habilis's*, even though 1470 was quite a bit older. It was

---

*Prognathus indicates structural prominence of the mouth region.

another out-of-sequence disjunction, whereby Richard Leakey had to conclude "that there were several different kinds of early man, some of whom developed larger brains earlier than had been supposed."[16] But both Hooton and Weidenreich wisely understood gracility and advanced brain as part of the *same original package*. No "development" here: the little people, *H. sapiens pygmaeus*, were anatomically modern in every respect from the start, save height.

Kenya's (Lake Rudolf) enigmatically modern 1470 man (*H. rudolfensis*) then is a disturbing case of AMHs *older* than they should be (Ihin older than the upgraded Asu who received their genes). Richard Leakey announced his 1470 find as a "surprisingly advanced" specimen, springing up 2.9 mya among ultraprimitive hominids.[17] Opponents howled that 1470 was actually younger; the volcanic tuff in the region, retested, made him more recent, 1.9 mya, therefore a *H. habilis* type, thus avoiding an embarrassing reversal to the scheme of ever-improving evolution. Paleoanthropologist Ronald J. Clarke, for example, matched one Olduvai *H. habilis* palate (OH 64) to the jaw of 1470 and voilá, "*Homo rudolfensis* [conveniently] disappears."[18]

But 1470 had good chopping tools, which no other contemporary creature had. The expected heavy bones or visor brow (of the primitive type) just were not there; 1470's skull was also too modern for *H. erectus* (let alone *H. habilis*), and his cranial capacity was well beyond *H. habilis*'s. How could his moderate brow and flat face be ancestral to *H. erectus*, with his massive barlike browridge and prognathous face? Not terribly likely that 1470 evolved into a more archaic form! Reversal? Devolution?

Compared to *H. erectus* or *H. ergaster*'s (3733) heavy visor, "it is a very strained theory that posits that we moved from 1470 to 3733 and then back again to lighter brows, fuller foreheads, higher domes."[19] Broca's area, a region of the brain linked to speech found in modern humans, was also present in 1470. Its leg bones, moreover, were almost indistinguishable from *H. sapiens. Yet this man was a contem-*

Figure 2.6. The devolution of evolution.
Cartoon by Marvin E. Herring.

porary of Au; he lived *before H. habilis, H. erectus,* Neanderthal, and Cro-Magnon. But he was too large to be a *H. habilis* and had a bigger brain.* The features of 1470 were, in short, a remarkable mixture of both primitive and advanced. The solution to the problem (if there is a problem) is extremely simple: crossbreeding. Only mixing gets us past this tangled web (see chapter 5, where we again see Ihin retrobred with Au, producing such types as 1470). Richard Leakey himself declared: "Either we toss out this skull [1470] or we toss out our theories of early man. . . . [for] it leaves in ruins the notion that all early fossils can be arranged in an orderly sequence of evolutionary change."

The stirring question of lost races and lost civilizations has engendered a new breed of researchers: protohistorians, most of whom

---

*Why is small-brained Skull 1813, a Kenyan *H. habilis,* the same age or even *younger* than 1470 who has the *bigger* brain and larger body? Why are *H. habilis* and Au thinner skulled than *H. erectus*? Reversal? Answer: Not to worry; they are only hybrids with different ratios of Asu and Ihin blood. And the time difference between them is actually rather insignificant.

are considered fringe voices. In this area of knowledge, the flow of information has been embargoed, blocked, ridiculed, and dismissed with contempt, largely because it is a threat to the ruling paradigm of onward-and-upward evolution. *H. sapiens pygmaeus* are the bodies buried in the foundation of the house of evolution. Only indirectly have we been able to find these ancestral little people. The problem took me on a book-length search (*The Lost History of the Little People*). They were, as this winding journey revealed, the first truly upright even civilized race of man, the first honest-to-goodness humans. They did not evolve. (Chapter 7 goes into their remarkable genesis.)

> *Man came on earth fully and perfectly developed . . . [He]*
> *was a special creation and not of nature's making . . . but*
> *required education and mental development.*
>
> JAMES CHURCHWARD,
> *THE LOST CONTINENT OF MU*

We find the little people in folklore, we find them in language, we find them in scripture, we find them in elfology, and we find them in the fossil record. A tiny yet well-shaped (not dwarfish) and genteel people, the Ihins were civilized almost from the start. Darwin's sometime friend the Duke of Argyll, along with Archbishop Whately, thought "man came into the world as a civilized being." A similar belief was later held by Arthur Keith and his famous student Louis Leakey, both of whom believed (despite the Darwinian paradigm) that AMHs (anatomically modern humans) somehow appeared very early in the record. Steinheim Man (Germany), discovered in 1933, seemed to corroborate this idea, as well as Galley Hill Man (United Kingdom), Olmo Man (Italy), Atapuerca Man (Spain), and other AMHish specimens *earlier* than Neanderthal. Yet another example of an early AMH is Hungary's Vertesszollos Man, dated by some as old as 700 kya, though possessing a very large brain (1,500 cc) and rounded occiput (like *H. sapiens*). He lived at the *same time* as *H.*

erectus. The claim was that he *was H. erectus*, even though his skull was entirely out of *H. erectus* range.

Also consider this: With the appearance of AMHs, many scholars thought physical evolution may be considered as finished. Therefore, since AMH is so very early in the record, how can evolution be seriously applied to man at all?

When the Ihins first appeared, the only other creatures on Earth were Asuans (Ardi). Au (australopith) was not yet on the scene. The Ihins came into being as sapient men; they were the first on Earth to know religion and consciousness, the first to speak words and to congregate in cities, and the first to wear clothes and to labor, practicing the arts of agriculture long before the Neolithic Revolution. They were AMH *from the start*, with upright posture, opposable thumbs, large brains, and small teeth; they never did swing from trees.

But they are a lost and forgotten race and their great antiquity has been trifled with. Today, when AMHs are discovered in unexpectedly early strata, the verdict is likely to be "intrusive" burial: meaning, we must take them as much more recent, accidentally reburied in older deposits. This is a misstep, which was exposed in the 1980s when it became clear that some *Homo* finds "indeed predate many australopithecine fossils." One example comes from Ethiopia, paradoxically revealing that "*afarensis* [Au] becomes more modern the older it gets."[20] Here, the 3.6-myr fossil (KSD-VP-1.1), although 400 kyr *older* than tiny Lucy, is more *Homo*-like, especially in humerus, ulna, and scapula. All the same, they lumped it together with Lucy's bunch.

Earlier, then, does not necessarily mean more primitive at all. A great array of fossil men who are AMH in some respects, but considered too remote in time to be so advanced, have been pushed back into the shadows—saving Darwinism from the wrecking ball.

Close to a century ago, Prof. Dixon made it plain that "as far back of the Neanderthaloids as these are back of us, there existed men . . . comparable in most respects to the peoples of European type today."[21] For a hundred years that fact has been swept under the rug. We keep finding a more advanced type (AMH) on the *lower* branches of the family tree. Why? Because, back in the beginning, the little Ihins, our ancestors, were fully human. Indeed, another question lurks: Why do the first AMH specimens resemble the European type?[22] Because the little people were Caucasian in their features.

The AMH remains of little people, *H. sapiens pygmaeus,* have been recovered all around the world: in Pennsylvania, Ohio, Kentucky, Tennessee, New Mexico, and Vancouver, British Columbia, as well as in Germany (Bonn), France (Montespan), Scotland (Hebrides), Switzerland (Dachsenbiel), Belgium (Spy), Egypt (Baderian), Africa, the Philippines, Japan, and Sundaland. All and any AMH traits recovered from "surprisingly" archaic horizons can be traced to Ihin genes, which are equally the source of all the gracile (and large-brained) types from Au on forward.

The little people were finely made and shapely of limb, their signature being distinctively dainty hands and feet, which show up in many parts of the world (see table 2.2).

Sterkfontein fossil feet in South Africa, like today's Bushmen's,

Figure 2.7. Laurence van der Post with the Bushmen.

proved to be quite tiny (under four inches long); the anklebone of these specimens (named Little Foot) was "extremely humanlike," even though they are the *earliest known hominid* in South Africa: Little Foot was 1 myr *older* than most of the other Sterkfontein Au (*Au. africanus*).[23] The anachronistic difficulty of this specimen's too-early modernity goes away if we allow it to represent one of the earliest Ihin mixes with Au.

Donald Johanson wondered how his famous Lucy (*Au. afarensis*) got such modern feet. In some *H. habilis,* too, as well as in the acclaimed 3.7-myr Laetoli Au footprints of Tanzania, a small and "improved" foot is seen on these little people (only four feet seven inches), featuring a rounded heel, uplifted arch, and forward-pointing big toe, all typical of the perfectly modern foot. All it takes are a few of the right *H. sapiens pygmaeus* genes!

TABLE 2.2. WORLDWIDE INSTANCES OF SMALL HANDS AND FEET

Where	Who	Description
Brazil	Tapuyas	Delicate, small hands and feet
Chile	Yaghans	Tiny hands and feet
Far East	Mongolians	Small hands and feet
Italy, Switzerland, UK	Neolithic skeletons	Remarkably small feet
Kalahari Desert, southern Africa	Bushmen	Tiny hands and feet
Mexico	Maya	Small hands and feet
New Guinea	Pygmies	Small, dainty, graceful feet
United States, Arizona	Hopi women	Exquisitely molded little feet
Worldwide	Protopygmies	Small feet, barely five inches long*
Yemen	Zeranik people	Small hands and feet

*According to Coleman and Huyghe, *The Field Guide to Bigfoot, Yeti, and Other Mystery Primates Worldwide.*

In America, when the Ongwee hybrids appeared, possessing 50 percent Ihin blood, they too inherited the modern foot: "His instep is high; he can spring like a deer. . . . He flees to the plain and the forest on his swift feet."[24] Even primitive Dmanisi Man (whom we will take a better look at in chapter 11) is a real mongrel, blending Au, *H. habilis, H. erectus,* and Ihin traits, the latter including modern feet—in contrast to the typically large and clumsy feet of *H. erectus.* Resulting from gene mixing among the races, possessors of a primitive foot may evince little people genes in their short stature, but also may retain *H. erectus* genes for large feet. These include:

- Most Au: flat arch
- Flores hobbit: large, flat feet
- Andamanese, Andaman Island: very large feet on very short people
- Bogenahs, Panama: pint-size folk with large feet
- Veddas, Sri Lanka: unusually flat feet

The Veddas of Sri Lanka show a baseline of Caucasian blood (Ihin genes), and so they are classed. Prof. Coon identified a major substratum in Southeast Asia and Sundaland as "Veddoid," representing an ancient mixture of Caucasoid and Australoid people. Very short and pale, the Veddas are handsome and well built, their hair wavy to straight; some of the men sport elaborate beards. Who, then, were their Caucasoid ancestors?

Moslems throughout the world regard Sri Lanka as the Garden of Eden, the birthplace of human civilization. Ancient works on the teardrop island may be a clue to the Vedda's civilized forebears: the water tanks in Sri Lanka are of great antiquity, a vast well-made system for irrigation. "The race which constructed these tanks has passed away, and the country where . . . there once existed a highly civilized and skillful engineering people, is now the abode of wild Veddahs."[25]

Today, DNA analysts would have us believe that the origin of

Figure 2.8. The Veddas are a striking mix of racial traits.

the Caucasians is recent, dated after the demise of the Neanderthals (a subject I return to in chapters 10 and 11). But Marcellin Boule, Arthur Keith, Henri Vallois, and other leading European paleontologists of the previous century recognized a Caucasian antiquity much deeper than Neanderthal. More recent scholars, though, think they have overturned these venerable theorists, some of whom believed in the *primacy of AMH*—which today's experts smugly call "a moribund approach to human ancestry."[26]

Nevertheless, the well-sculpted, gracile form of the original (now-extinct) little people offers confirmation to every paleoanthropologist who has ever thought that somehow, despite the evolutionary paradigm, earliest man was of the oxymoronically "modern" type (AMH). And he was. After all, the earliest fossil skulls of Au were gracile. And with AMHs standing at the earliest point in human history, the need for evolution simply evaporates, not only because a modern form in fact predates so many primitive ones, but also because evolution is thought to stop happening once man has culture to ensure his survival (we'll return to this idea that physical evolution is over). And the Ihins had culture.

Unlike his primitive fireplug contemporaries, the arms of early AMH were not long, his legs not short, his figure not burly. His symmetry was with him from the beginning. As an Au, Lucy had some Ihin genes; her hip structure was so refined as to make it hard for her to climb trees, and her family, the gracile Au, had bigger brains than the robust Au—who, indeed, are not as old. For gracile is *older* than robust in both South and East Africa. The puzzle of Au appearing as two distinct kinds (gracile versus robust) now dissolves. No, selection pressure did not change gracile to robust; it is a fruitless argument. East African robusts like Zinj (*Au. boisei*) were less humanlike, though younger, than *Au. africanus*. All these impossible evolutionary reversals are proof enough that Darwinian phylogenesis doesn't pan out. Such anachronisms (appendix E) as we constantly find are only the result of Au *retrobreeding* with even more archaic mates.

Figure 2.9. Jehai Negritos, showing the symmetry of the Ihin.

Not only in Africa but also America do we encounter Asu and Ihin genes thoroughly mixed (as we will see in chapter 11). We might expect such mixing wherever "giants and dwarfs" are found together in proximity. Early men of Argentina, for instance, were divided into a dwarf race with a strong chin and a short, broad, and smooth skull, living side by side with a larger people. According to James Shreeve, these are "two very distinct races of man."[27]

A LOST RACE

To recap: The little people—*H. sapiens pygmaeus* (the second race)—came *before* Au. The only race older than *H. sapiens pygmaeus* is the unripened, insapient Asu man (*Ar. ramidus* or Ardi, the very first race). I believe Kenya's Kanapoi specimen, *Au. anamensis,* demonstrates the earliest blend of the two: morphologically somewhere between Asu (Ardi) and Lucy (Au), Kanapoi shows quite a modern knee joint and humerus/elbow, even though he is one of the very oldest known hominids (4.3 myr). How could he have had more modern features than his successors? It is true, he was chinless, with curved fingers, short legs, and a primitive mandible. But the tibia was almost indistinguishable from *H. sapiens.* An anomaly. But they simply lumped it with Au.

Done deal, but a pseudosolution: Falsifying or refuting evolution, commented Marvin L. Lubenow in connection with these Kanapoi discrepancies, "is like trying to nail jelly to the wall."[28] For evolutionists, it was simply unacceptable that such an old specimen could be associated in any way with *H. sapiens*, even though Kanapoi archaeology revealed such sapient items as potsherds and hut circles. Viewed as a hybrid, Kanapoi would simply be an Asu upgraded by Ihin genes.

But let's move on, away from jellylike anthropogenesis. In reconstructing the past, Indonesia and beyond is rightfully seen as a key area. Indeed, here are most of the world's extant little people.

Their great antiquity can be read by the number of indigenous languages still spoken in Indonesia, the Philippines, Malaysia, Papua-New Guinea, and Australia: estimated at over 1,400, this is more than a quarter of all the world's languages.[29] Such linguistic diversity betrays a formerly far-flung, extremely old and indigenous population, meaning the little people of insular Southeast Asia and Sahul were most probably here in situ from the beginning. They did not come from somewhere else.

And they were always little, despite the fatuous shrinkage theory that pegs small stature as an adaptation to life in the tropics or closed habitats or amid crowding. The "shrunken Negrillo" (the pygmy of the Congo) actually has few traits genetically linking him to black Africans. No, they are not shrunken Africans. The argument that they are shriveled versions of a taller race actually stands fact on its head, for the little people everywhere arose *before* the larger races, the "giants"; just as the gracile Au are older than the robusts. Chronologically, the giants were an offshoot of the third (Druk) and fourth (Ihuan) races, much more recent than the second race (the little people). Welsh tradition holds that the Manx (a fairy race of little people) were the original inhabitants of the land and lived there long before the giants. In the Bible, too, the giants are introduced thousands of years after Adam, the first man.

We find the same sort of reckoning in America, where, after the (Algonquian) hero Glooscap created the world, he formed "the smaller human beings." The Wyandot Indians, for their part, said the little people were old enough to remember the flood; while the Choctaw, so rich in little people lore, held that a race of diminutive folk, teachers, lived on Earth before them. And who were those little people? They were the mysterious American mound builders, who left their own tiny bones and diminutive sarcophagi in Ohio, Tennessee, and Kentucky. Such skeletons, deposited in coffins not more than four feet long, were found near Cochocton, Ohio, in an ancient cemetery situated on *elevated* ground (i.e., artificial mounds). "They are very numerous," said the earliest report,[30] "and must have been tenants of a considerable city. . . . All are of this pigmy race." The city of Lexington, Kentucky, thought historian George W. Ranck, was built on the metropolis of a *lost race* that flourished centuries before the Indians who are themselves "a tall people; the [early mound builders] were short . . . rarely over five feet high."

In the South, the Creek Indians reported that the *oldest* mounds were built many ages prior to their arrival. Most Indians said the mound builders were a different race. In *The Secret: America in World History before Columbus,* J. B. Mahan observes that the priest caste inhabiting the Georgia mounds were of European mold: "the people of the mound cultures are clearly Caucasian in type." In some AmerInd traditions, the little people on the mounds are remembered as the teachers of their tribal elders. They had been mentors to the Crow Indians: "Our forefathers claim the little people lived there once, it was a sacred place many years ago."[31]

Figure 2.10. Chiseled Crow features suggest a measure of Ihin genes, for the Native Americans had an Ihin foremother.

Diminishing the chances of finding their remains was the Ihin-related practice of cremation, which brachycephalic people introduced into Europe in the early Neolithic; brachycephals and mixed-head forms fringe the areas in which cremation was most prevalent. By the time of the Bronze Age, it became the principal method of disposing of the dead. In the New World, cremation is evident in regions dominated by the Ihins, such as Tennessee, Ohio, and Wisconsin (with cremation pits near the ancient copper works of the mound builders). We also find cremation at Lake Mungo in Australia, with its early gracile people.

In South American traditions, it is said that men of fair coloring came to the Andes long before the Inca: "In the very ancient times the Sun God, ancestor of the Incas, sent them one of his sons and one of his daughters to give them knowledge. The Incas recognized them as divine by their words and their light complexion."[32]

Priests and wisdomkeepers, the Ihins in North America introduced the worship of Gitchee Manito to the Ihuans. Some western tribes declared these priests and mound builders were people who "came from another world [chapter 7] and dwelt on earth for a long season, to teach them of the Great Spirit and of the Summer Land in the sky."[33] Genetically different, being born with the veil,* the Ihins were constitutionally open to spirit. "The race which provides the meaning of world history . . . spread over the whole earth . . . a blue-eyed, fair-haired race which . . . formed the spiritual features of the world."[34] They are one and the same as the round heads of antiquity, the "broad-headed [brachycephalic] people apparently with a higher civilization."[35]

---

*According to the Rosicrucians, the psychic ability of this lost race was based on a kind of third eye in the forehead. But the veil of the Ihins was not an extra organ; it was a caul, discussed in chapter 3 of my book on the little people, *The Lost History of the Little People.*

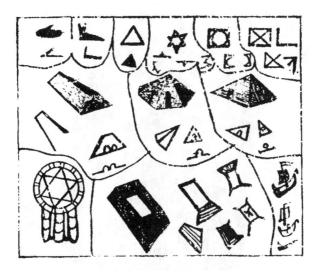

Figure 2.11. Tablet of the mound builders, from Oahspe.
As noted in Peter Kolosimo's Not of This World, *bone fragments, on analysis,*
show the mound builders "did not belong to the red-skinned races
but to . . . a race similar to the [Europeans]."

Called the Alpine race by Prof. Dixon, the clearest remnant today of these brachycephalic people is in the high Alps. "The brachycephalization of Europe . . . may be summed up as a contest between the older longheads and the later roundheads."[36] Thus, even in Africa, the Ihin-blooded little people, the pygmies, are morphologically of the Paleo-Alpine type, as are the remnant people of Hiva (lost Pan).

In America, stretching from Panama to the southernmost tip of the continent, are ensconced remnants of this lost race along with their mixed descendants, from the San Blas White Indians of Panama to the small-footed Yaghan at Tierra del Fuego. Many of the latter (in the western part of Cape Horn) are only four feet seven inches tall. Likewise are Chile's Alakaluf people short statured and thought to be among the earliest South Americans, pushed ever southward by later people. Writer Bill Mack followed this indigenous little race from Mexico to Tierra del Fuego: "The tropical rain forests of Central and South America are inhabited by a race of dwarflike people"; known as

Figure 2.12. The Yaghan of Tierra del Fuego are mentioned from time to time in the writings of Charles Darwin, who, on board the HMS Beagle, *encountered these Stone Age people.*

Alux in the Yucatan, "in other parts of Central and South America they were called Sisemite or Toyo . . . . The villages gave them special local names."[37]

A similar pattern exists in Oceania, where the small people of Melanesia represent an older population than their tall neighbors.[38] On Malekula (New Hebrides), the pygmies in the mountainous interior are "the last remnants of an earlier racial stock, similar to that found in the interior of the larger land-masses in the western Pacific. . . . Practically nothing is known about them."[39]

This distribution, this geographical pattern, is typical of the world's little people. In New Guinea, stature goes down with altitude: "[S]ome of the highland tribesmen are small enough to be called Pygmies."[40] Here in New Guinea (as well as Southeast Asia), the people tend to get *shorter and lighter* as you penetrate the uplands. As one leaves the coastal plain of Sri Lanka, for example, and climbs the central mountains, the people grow "shorter, stockier and somewhat lighter in skin color."[41] Hawaii's legendary Menehune, also secluded in the uplands, were a singularly short race, who brought culture and lasting engineering works to those islands.

What does all this tell us about our ancestors? Armand de Quatrefage, the nineteenth century's eminent professor of anthropology at the Musee d'Histoire Naturelle of Paris, concluded that the Negrito race (little brown people) once inhabited a vast domain of Indo-Oceanic Asia, extending from New Guinea all the way west to the Persian Gulf, and from the Malay archipelago north to Japan. Significantly, the Onge Negritos of the Andamans (in the Bay of Bengal) are considered genetically *ancestral* to Asians. Along the same lines, Jared Diamond, in *Guns, Germs, and Steel,* suggests that the Negritos of New Guinea may be the ancestors of the Papuans, that is, the little ones came before the tall. Likewise were Australia's Negritos anterior to the tall Aborigines, providing the initial population for the whole of Greater Australia.

Great minds think alike: Professor J. Kollman of Basle considered the Negrito and Negrillo populations the oldest form of humanity: "from them the taller races have been evolved," rather than the other way around (the dwarfing or shrinkage theory). The dean of American anthropology in the mid-twentieth century, William Howells, identified the Negritos of the Philippines, Taiwan, Malaysia, Sri Lanka, and India as the *underlying population strata,* the submerged remnant of South Asia's first people. Just as India's Negritos have long been regarded as the oldest people on the subcontinent, the little Semang, Malaysia's last remaining Paleolithic people (living in remote forest areas), are known to be the earliest inhabitants of the Peninsula. Their name, Orang Asli, means "original people."

We were born just after the earth was made.

WELSH ELF PROVERB

TABLE 2.3. WORLDWIDE LEGENDS
OF LITTLE PEOPLE AS THE *OLDEST* RACE

Where/Who	Little People	Description
Africa/Wolof	Yumboes	White pygmies/fairies
Europe, eastern/Serbs, Poles	Ludki	Little people, lived before humans
Europe, western/French	Fees	The Old Ones, oldest beings on the planet
Ghana/Ghanaians	Small man	First hero of the race, made by Anansi
Mexico, Central America/Maya	Saiyam Uinicob	Race of dwarfs inhabiting the First World
Mozambique/Yao	Little man and woman	First people ever seen, a tale of the beginning
Northeast America, rocky heights/Seneca Indians	"Great little people"	Predecessors of the Seneca Indians
United States, Hawaii/Native Hawaiians	Menehune	Little people, first race in the region
United States, Nantahalas, North Carolina/Cherokee	Nunnehi	Little, white, and bearded people
United States, Southeast/Choctaw	Kowi Anukasha	Forest dwellers or little people, lived on Earth before Natives

It is among such dwindling and isolated groups that we might catch a last glimpse of the earlier editions of *H. sapiens*, in some ways more like the short and pale Ihin ancestor. One stunning example is the Filipino Abenlens, with lighter skin than the Aetas, a different language, and the signature graceful limbs of early *H. sapiens pygmaeus*. (The Aeta, themselves tiny folk, are considered the earliest inhabitants of the Philippines.) The remote Abenlens, living deep in the isolated Zambales Mountains of Luzon, are exceptional; unlike the regional Negritos (Aetas), they are shorter still and light, almost blond in complexion. Some are olive-skinned with light brown eyes; most possess delicate features and soft wavy hair. One reason the Abenlens' fair coloring and short stature are taken as a sign of Ihin ancestry is that they are highly reclusive and never mixed their genes with neighbors or conquerors.

> *In remote corners of the world, far removed from the great currents of migration, moderately pure remnants of older, more original races still survive.*
>
> ROLAND DIXON, *THE RACIAL HISTORY OF MAN*

TABLE 2.4. LITTLE PEOPLE IN REMOTE AREAS

Little People	Where	Comment
Ainu	Hokkaido, Japan	Earliest widespread moderns in Far East
Bushmen	Southern Africa	Driven to the Kalahari wastelands
Lacandon	Mayan jungles	Short and pale
Tapiros	Mt. Tapiro, New Guinea	Remnants of a once widely distributed race*
Tarifuroro	New Guinea, interior	Driven back by Papuans, at one time inhabited entire tableland
Maithoachiana gnomes	Kenya	Driven out by pre-Bantu people
Negritos	India and Polynesia, inaccessible mountains	Remnants of the earliest stratum†
Pygmies	Central African forest	Original people, overrun‡
"Race of dwarflike people"§	Central and South America	Driven back to interior forests

*According to explorer A. F. R. Wollaston.
†Flower, *The Pygmy Races of Men*.
‡Turnbull, *The Forest People*, 33.
§Mack, "Mexico's Little People," 40.

Once widespread but largely lost to history, driven back to refuge regions, persecuted and extirpated, "the pygmies fled to the ends of the earth."[42] Just look at the tips of the continents: people there are small and undersized, such as the near-Arctic Yaghans of Tierra del Fuego, where some of the women are only four foot three inches. Even in places like Italy, the shortest people are found in the most removed mountain enclaves.

NEGRITO PEOPLES THROUGHOUT THE WORLD

The localities in which the Negrito people are found in their greatest purity in inaccessible islands . . . or in the mountainous ranges of the interior . . . point to the fact that they were the earliest inhabitants.

WILLIAM H. FLOWER, "THE PYGMY RACES OF MEN,"
ESSAYS ON MUSEUMS

The worldwide distribution of Negritos tells us that, not only were little people a universal race, but so were black and brown people. According to George Frederick Wright, "[t]he original inhabitants of Europe were a long headed [dolichocephalic], dark-skinned race."[43] England's controversial Galley Hill fossil man, for instance, was very dolichocephalic, his build modern, his face "Negroid." Similar to Galley Hill is France's Combe Capelle specimen, Negroid in dentition, palate, skull, and length of tibia. Belgium, too, had its "Ethiopian" type in the Engis skull; some have found bones of the Negro type in Austria and Liege.

Dixon, who thought the most ancient types of mankind had been discovered in Europe, identified brown-skinned proto-Australoids as the substratum of Neolithic France and western Russia, his determination based on cranial and nasal measurements. Dixon also found proto-Negroid people in Silesia, Bohemia, Denmark, Italy, Greece,

Norway, and Yugoslavia—the latter including the Krapina folk with dentition (crowns) of the Negroid type.

There are brown and black races in almost every part of the world: the Makrani blacks in Pakistan, and in India the Dravidian Negritos, dark with frizzy hair, dolichocephalic skulls, and thick lips, as well as jungle tribes like the Kadar and Paniyan Negroids. There were Neolithic platyrrhine proto-Negroids in the Ganges Valley, extending up to the Persian Gulf. In Iran, near Susa (the capital of Susiana aka Elam), a black Ethiopic people once lived and there are black people in southern Arabia as well.

In the Americas, too, Dixon found "Negroidal" skulls in New Mexico, northern Arizona, Tennessee, New England, and along the Great Lakes (in Ohio).[44] Other sources allude to prehistoric blacks in the Andes and Peru (sculpted at Marcahuasi), Ecuador, Patagonia, and the Brazilian highlands,[45] the latter region being home of the living Botocudo people at Minas Geras; their ancestors (best known from the 11,500-year-old Africoid woman of Luzia) were the underlying stratum populating this whole region (before being driven back by the Tupi). The Botocudos are dolichocephalic and prognathous and are probably descended from the people of the Lagoa Santa caves. Spanish and Portuguese expeditions of the late Middle Ages encountered tribes of black people (non-Indians) in both Central and South America. The Africoid Olmec heads of Mexico are a famous example. Blacks are also depicted in surviving Codices, as well as in images at Teotihuacan, Vera Cruz, Tula, and in temple paintings at Chichen Itza, in the terracotta heads at Jalapa, and carvings at Villahermosa.

In the Orient, Dixon found "Negroid survivals" in the extreme southwest of China, among the hill folk of Burma and the Mon-Khmer tribes of Cambodia. There are also blacks in Nepal and a Negroid factor dominant in the interior of the Celebes. True blacks live in the uplands of the Sandwich Islands, just as Easter Island, Melanesia, and Fiji all have their indigenous browns and blacks with woolly hair and broad

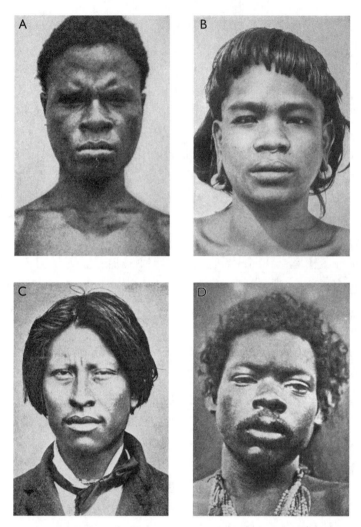

Figure 2.13. Black people are found in every division of Earth. (A) New Guinea Melanesian (B) Philippine Igorot and Batak (C) Guaycuru/Caduveo South American (D) Palliyan Indian

noses. Negritos, we have come to see, were once dispersed throughout Southeast Asia. Much of the Java, Thai, and Sumatra fossil remains are those of pint-sized persons.

Figure 2.15. Sumatra: Malaysian blends. These Negritos have always been here.

Figure 2.14. Australian Negritos.

The wide dispersal of the surviving Negritos indicates that they once peopled the entire forest of the Sunda Shelf.

JOSEPH THORNDIKE JR., *MYSTERIES OF THE PAST*

Negritos were in all parts of the world; black and brown people were indigenous to every land, and were all of mixed blood, for the Ihin had merged with the dark ground people. Dixon's Paleo-Alpine type, for example, is an Ihin–*Homo erectus* mix: short, brown-skinned, with straight black hair. Among the Filipino Negritos (almost purely brachycephalic), are "an overwhelming majority of the Paleo-Alpine type." The Ihins' Caucasian genes, in some cases, remain pronounced among the world's Negritos; in Malaysia, the Negrito iris, for example, is unusually light in color. Much of Sundaland is a living museum of these ancient blends.

TABLE 2.5. LOCATIONS OF NEGRITOS

Where	People	Comment
Argentina pampas	*Homo sinemento*	Fossil man with "likeness to the negrito type of Asia and Africa."*
Australia, rain forest	Aborigines	Standing at four feet six inches
Bay of Bengal	Andamanese	Flat noses, black skin; represent the original "primitive inhabitants of a large portion of the earth's surface"†
Borneo, northern	Cave skeletons	Bulbous foreheads (like Negrillos)
China/Tibet	Kunlun people	Many were enslaved
Flanders	Negrito Fenlanders	Predate the Celts
India, Gujerat	Bhils	"Aboriginal stratum"‡
India, Cochin region	Pulaiyan and Kadar	Standing at four feet six inches
India, Farabad Hills	Chenchus	Forest people
India, Malwar	Kurumba	Dravidian people
Indochina	Minh-Cam	Diminutive cave skeletons
Mexico, Monte Alban	Danzantes	Dancers carved on monuments
Myanmar and Vietnam	Moken	Oldest known human population in Southeast Asia
Thailand	Krabi	Plus pint-sized fossil men in region
Philippines	Aeta	Once commanded all of Luzon
Sundaland	Negrito groups	Considered some of the oldest tribes on Earth

*Hrdlicka et al., *Early Man in South America*, 243.
†Flower, *The Pygmy Races of Men*, 6.
‡Coon, *Living Races*, 206.

Figure 2.16. Danzantes, thought to be chieftains of hostile tribes; sculptures represent the conquest and humiliation of slain captives.

Figure 2.17. Figure from American mounds. Illustration by Jose Bouvier.

THE SECOND RACE OF MAN

Asu was the first race, Ihin the second. No other races are older than these. One Batek (Negrito) tradition recounts that after the first Batek was created from brown soil (Asu), there next came a pair of beings from white soil (Ihin), who were *tua*—a European-type.* The Batek like to stress that they were created first and were thus "the original humans."

Although the western mind holds Adam as an iconic Caucasian forbear, the name *Adam* in the oldest traditions denotes the first race, Asu, a being made of soil or clay, brown like the earth. It was not until the second race (the little people) appeared, that white and yellow skin entered the gene pool. There was little pigment in Ihin skin—at best, a paucity of melanin. This second race of man arose everywhere, covering the Earth over[46]—pale, tiny, and brachycephalic.

Indeed, Hooton found the earliest men of the modern type to be brachycephalic Caucasians. No, brachycephalic is not, as supposed, the result of mutations, or derived from anything. It was the original head shape of the little people.

The Ihins came into the world with white or yellow hair: one last vestige of the early Ihins survives in the platinum blonds of the east Baltic region. Nevertheless, modern anthropology (the out-of-Africa theory in particular) asks us to believe that white people are a fairly recent strain, having come from dark Africans through some mysterious process of "de-pigmentation" or "mutations toward blondness . . . in cool regions."[47] Theory runs something like this: the Neanderthals were originally brown, having adapted to tropical lands; but with the use of clothing (in their supposed move to northern places during the early Wurm glaciation), the epidermal pigment melanin was no longer "needed" to filter ultraviolet light. Just so, mutations arose that were selected for depigmentation.

---

*The little whites recalled in Maori oral history were also called Atua; the name is actually found all over the world, most often with the connotation of "civilized" man.

Adam the Mud Man

In Arabic tradition, it was a host of angels, at a place near Tayef, who found suitable earth and kneaded it into the human form. In Hebrew, Adam means "red earth" or "the one formed from ground," similar to *edama* in proto-Phoenician. We find this primitive Adam embedded in Meshe'adam, the name of the Azerbaijani Yeti/Bigfoot, signaling the most brutish form of man. *Adami* in Akkadian also meant "red clay," the material from which first man was legendarily fashioned. Thus in the Mesopotamian account of creation was the first race created from dust of the earth—and called Admi or Adami or Adamu.

This reckoning is almost universal. If the Inca said man was made from dirt, the Mexican Popul Vuh has first man made of mud. In North America too, the Creek and California Indians said first man was fashioned from clay. First man (Tiki), agree the Maori of New Zealand, was formed of red clay. Tahitians also hold that humankind was first created out of red earth, *araea*—as do the Siberians as well as Malaysia's Kenta Negritos who relate that Creator drew the first ancestors from the earth. This in turn matches the Batek account, wherein the soil was molded into little manikins.

In a Slavonic version, human beings made of clay are said to have inhabited Europe long before their own ancestors. We find this earth/clay anthropogenesis in Australia, New Guinea, New Britain, Palau, Tahiti, Vanuatu, Egypt, Africa, Russia, Borneo, Burma, and India.

There are, however, pale people indigenous to tropical regions (for example, tribes along the Amazon, the Southeast Asians, and the Abenlens), which no amount of "natural selection" can explain. Furthermore, we now have DNA-tested 43 kyr Neanderthals, at El Sidron Cave in Spain, who happened to be pale and red haired—

TABLE 2.6. LOCATIONS OF WHITES
ON EVERY CONTINENT*

Where	White People	Comment
Africa	Kalahari Bushmen	Tiny and yellow
Australia	Tribes in central desert and west central	"Blondism" common among Aborigines
Brazil	Tahuamanu and Assurini	"Milk-white" Indians with red hair and blue eyes
Brazil	Tapuyos	White, bearded refugees from an older civilization
Chile	Mapuche Indians	Pale and bearded
China	Miao-Tse	Pale mountain people
Ecuador	"Laron" dwarfs	Little, white people
Egypt	Ancient Badarians	Small and fair
Himalayas and Mongolia	tribal groups	White-rosy skin, gray-blue eyes
Honduras and Guatemala	White Indians	In the interior
Indonesia	Rampasas	Near Liang Bua caves
Malaysia	Senoi, Semang	Sakai fair-skinned natives
Nepal	Newars	Fair, Caucasian features
New Guinea	Tarifuroro	Yellow-skinned, light-haired pygmies
Panama and Surinam	San Blas Indians and the Akurias	Small and very pale
Papau New Guinea, East New Britain	Tolai people	Blond, blue-eyed children sometimes born
Peru	Chachapoyas	White race
Peru	Royal Incas	Fair people with wheat-colored hair
Peru	Upland Quichuas	Light-skinned, small people
Sri Lanka	Veddas	Short, some very pale, heavily bearded

Where	White People	Comment
U.S., Alaska	Native Americans	White Eskimos and blond Inuit
U.S., Arkansas and Florida	Native Americans	Fair skin[†]
U.S., California, North Dakota	Yurok	White, bearded people (forebears)
U.S., North Dakota	Mandan	White, bearded people (forebears)

*In addition to these documented peoples there were also many legendary white people including the Wusuns of Chinese annals, the Sarasvati and Dorani of Indian legend, the Attua of Maori legend, and the mound builders and the Yunwi Tsunsdi of Native American legends.
†Keith, *Ancient Types of Man*, 148; Chouinard, *Forgotten Worlds*, 166.

some even freckled. While this could reasonably be attributed to crossbreeding with Cro-Magnons, the preferred explanation has their pale skin resulting from pressures in northern latitudes to evolve fair skin. Other explanations for blondism in the "wrong" places include bleaching in saltwater, or for some accidental reason, or "new alleles" for hair and eye color that supposedly arose through mutation under some kind of selection pressure favoring light color.

Yet the San Blas and other white Indians are not pale due to any mutation or de-pigmentation or environmental pressure. Light skin is neither a mutation, an accident, or any other "change"; rather, it is an original and very early element of the human gene pool. Nor did it first appear in greater Europe, ca 40 kya, as so often stated, for aboriginal whites have been found on every continent. Indeed, the blondest people of the North African Rif, the Nordic-type Berbers, are said to be the country's most *ancient* inhabitants. Blondism in Barbary, Morocco, Canary Islands, and at Mt. Aureps (Algiers) all derive from Ihin-Hamitic genes, owing to the westward migrations of Noachian peoples.

All the sons of Noah, refugees from Mu, were of the Europoid type, and they dispersed to every part of the world, carrying the name Mu into California, in the tribal names of Pi-mu and Li-muw. These people, as sixteenth-century explorers documented, were fair and of fine form and demeanor (at Santa Catalina, Santa Cruz, and Channel Islands). Interestingly there are many other examples of Mu-named peoples and places including:

- Mu people of Kauai, Hawaii (pale skinned)
- Muza (Arabia), Mu (Crete), Mukawa (Japan): all port cities and landing places
- Murutic, spoken by lighter-skinned tribal groups in Northeast Borneo
- Murrian people of Australia (lighter skinned)
- Muysca culture of Colombia, with their legends of a great flood and white "gods"
- Mu'allakah, the Syrian village famed for its Noah's ark
- Murias, a city from which the founding Irish migrated
- Nah-mu, the ancient Egyptian word for the yellow race
- Mulatto, meaning an admixture of white and black

In the ancient world, the name *Tua* (as in Batek and Maori usage) designated people of the European type who, though they lived long ago, were quite civilized. The Sumerian tablets, for example, tell of man coming directly from the gods, being at once intelligent and civilized. This man, it seems, was "sophisticated from the start."[48]

> *Was there ever a time when . . . man did not know how to think, to produce, to create? . . . Man always spoke, fashioned, invented, from the very beginning.*
> ROBERT CHARROUX, *THE GODS UNKNOWN*

In Ireland and Scotland, the Daoine Sidhe (translated as the people of the mounds) were Ihins, hailed as the oldest families— "perfected men" or the "gentry." During the Neolithic, a remnant of this ancient community of wee bodies lived in the Orkneys. A gen-

Figure 2.18. Spain's Lady of Elche, sometimes called the Spanish Mona Lisa, has a classic Ihin face; the female sculpted on this twenty-one-inch limestone bust has also been compared to the Aztec goddess of rain. Photo by Luis García.

teel and knowing race, they were remembered as inhabitants of the "Islands of Wisdom," for they were a living sermon before the wandering tribes. To the Maya as well, their race founders were men who "could discern everything in the world . . . Their wisdom was great. . . . Verily, they were wondrous men."[49]

> *Anatomically modern humans go so far back in time that*
> *it becomes impossible to explain their presence on this*
> *planet by current Darwinian theories of evolution.*
> MICHAEL CREMO, *HUMAN DEVOLUTION*

Convinced that "modern" forms existed extremely early in the record, Louis Leakey followed in the footsteps of paleontologist Johannes Ranke who, in 1899, argued that the human races were unchanged throughout prehistory, and that there have *always* been AMHs. Why, for example, did the Negrito-like Grimaldis (on the Riviera) have large brains for such small people, like the Inuit at 1645 cc? In this connection I'd like to quote Hooton's keen observation that "small people usually have relatively larger brains than tall people."[50]

> *A day will come when there will be discovered a human being of small stature . . . having a brain-box relatively large in comparison to the total body-bulk. . . . This will be the true Eoanthropus [Dawn Man]. Perhaps this day is near at hand.*
>
> MARCELLIN BOULE, *FOSSIL MEN*

Near at hand—but it is hard to get a grasp of these long-ago mods, as they (queering the evolutionary mold) have been essentially removed from the Darwinian canon. All these anachronisms in the record (appendix E) must somehow be kept under the radar or otherwise held in abeyance (with vague terms like anomalous, a mystery, intrusive). Early mods don't stand a chance of attaining full citizenship in fossil land. Galley Hill Man, as an example, is an old mod on English turf, and, as such, judged to be an "intrusive" burial, new chemical tests allegedly proving that he was not very old after all—his date changed from 1 myr to 200 kyr to 100 kyr and finally to the Holocene.

Such unwelcome specimens are sometimes labeled aberrant or pathological—any term of dismissal will do. A modern chin on a too-old specimen must be a "pathological bone growth"! Any fossils "with a well-marked chin cannot possibly be very ancient."[51]—but see the very nice chin on Miramar (see figure 11.12, page 451).

Also appearing earlier than expected, France's Fontéchevade Man was mod in crown and brow; to invalidate him, evolutionists said it was an "immature individual," because, *if not*, mods were in Europe *before* Neanderthals (a no-no). Italy's Castenedolo Man, another problem, was changed, incredibly, from a Pliocene to a Holocene date after it was decided that it was an intrusive burial (in a fifty-foot deposit). But there are too many specimens of this early AMH (stubbornly *predating* "archaic sapiens") to be intrusive. Too, there was in America the mod-

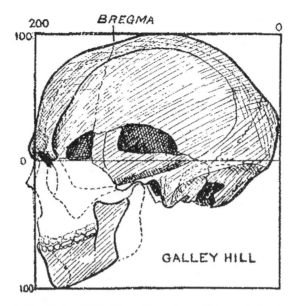

Figure 2.19. Galley Hill Man. Note the well-formed forehead.

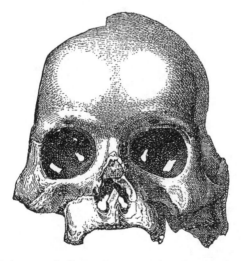

Figure 2.20. Calaveras skull. Hotly contested and rejected as an early AMH, this California specimen found in 1866 at a decent depth of thirty feet (in the middle of a mountain) rested in Pliocene gravels; the skull had the appearance of the modern type. In keeping with this find, West Coast Indians like the Yurok said that white-skinned folks occupied the land before them; today some of their remains are still kept in California museum collections.

looking Calaveras skull of California, as well as the Del Mar and Black Box specimens (50 kyr).

By evolutionary standards any mod feature on deep-buried specimens immediately casts doubts on their antiquity—this disrupts their linear view of the succession of human "species." But, as we will find in the following chapter, this kind of self-affirming logic won't hold up to scrutiny. The progression of human *races* is more entwined, meandering, than it is linear; a coexistence rather than a succession.

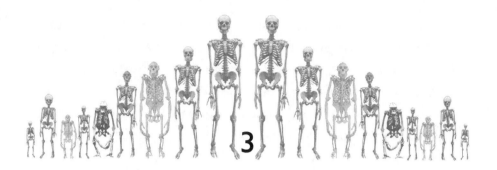

CHEEK BY JOWL

Coexistence of the Early Races

The succession of human races is disorderly.

ARTHUR KEITH, *ANCIENT TYPES OF MAN*

NEIGHBORS NOT ANCESTORS

We would like succession to be orderly, just as evolution claims that, over time, species A turns into species B. Theory also says species A *disappears* rapidly after being *replaced* by its successor (species B), which follows the (actually defunct) idea that at any one time, no more than a single species of man existed.[1] This single species hypothesis argued that at any given moment, only one hominid species existed—due to (the also fading notion of) competition for the niche.

According to these principles, the australopiths and pithecanthropines would have completely disappeared by the time modern man took over. Yet there was, as we have already seen, a definite overlap of hobbits and mods for at least 25,000 years. "What was the nature of their interaction?" one anthropologist asked, answering his own question: "We

111

have absolutely no idea." Equally baffling is William Howells's assertion in *Getting Here* that "at no time were . . . *Homo erectus* and *Homo sapiens* both in existence"; yet he adds, "Naturally, there are qualifications."

It strikes me that the halting transition from old, faulty ideas to new correct ones is usually marked by a small army of euphemisms. Euphemistically called "collateral hominids," highly different stocks of contemporary races, as we explore in this chapter, have had the same address almost from the beginning. These fossil men (who are supposed to be from different ages) actually lived at the same time—sometimes together. This is not good news for evolution.

> *Throughout the Pleistocene several of the key "sequential"*
> *types, not just one, were about, and none showed any signs*
> *of evolutionary change.*
>
> JEFFREY GOODMAN, *THE GENESIS MYSTERY*

This means of course that *H. sapiens* and his (supposed apelike) antecedents were flat-out *contemporaries*—which certainly muddies the water of phylogenesis, whereby species A basically gives up the ghost in the process of "turning into" species B. As Darwin saw it, species are *supplanted* by their better modified descendants; the daughter species will "take the place of . . . [and] exterminate its parent form."[2] Parricide!*

But that's not what happened—at least in the *human* kingdom, where archaic and advanced types lived side by side, and "the nature of their interaction" was quite intimate at times.

It is only recently (and reluctantly) that the coexistence of the races has been allowed to come out of the closet, *Discover* magazine announcing that "the fossil record tells us that . . . multiple species of hominids typically shared Earth's environment." Au lived cheek by jowl with *Homo*—which doesn't exactly inspire confidence in the orderly succession

---

*One critic sees "a particular fitness for Darwin" in all this, considering his "deep identification with his dead mother and submission to his overwhelming father . . . a classic Oedipal situation"—the offspring hoping to kill off the parent. (Hyman, *The Tangled Bank*, 40.)

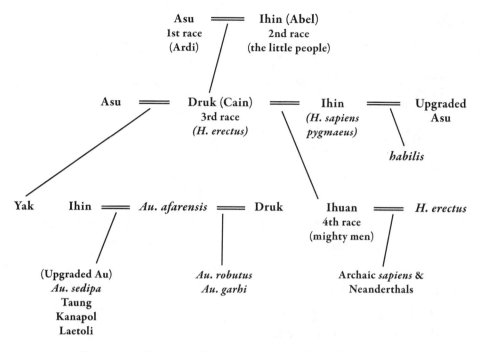

*Figure 3.1. Top of the family tree: history of interbreeding—
a whole lotta "interaction." Horizontal lines represent interbreeding, vertical
lines represent offspring. According to Oahspe the relative proportion
of the different races of men around 65 kya was: Ihins 100, Ihuans 3,000,
Druks 5,000, Yaks 5,000.*

of hominids, as old-style (or even new-style) evolutionism requires. *Homo
habilis* was neighbors with *Homo erectus*, Neanderthals were neighbors
with *Homo sapiens,* and so on down the line. Paleoanthropologist and
wife of Louis Leakey, Mary Leakey was honest enough to say that this
coexistence makes it unlikely one evolved from the other. Indeed, the
coexistence of our various forebears badly shakes the Darwinian family
tree. At the same time, it opens the way for a more correct reckoning:
contemporaneity and amalgamation of the races.

"When species overlap in time, it is difficult to imagine one as an
ancestor of the other," comments one team of evolutionists,[3] while one
critic observes more bluntly that "coexistence makes evolutionists very
nervous."[4] It should. Although contemporaneity of hominids makes it

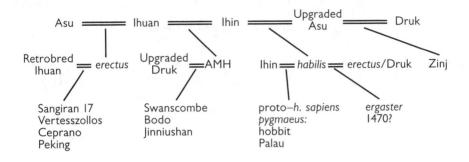

Figure 3.2. Suggested lineages of some well-known fossil men. Horizontal lines represent interbreeding, vertical lines represent offspring.

far less likely that one species was ancestral to another (its neighbor), the hardliners simply retort that they see no reason why ancestral species cannot coexist with daughter species. Some of the old type may remain, even when others of the same stock have evolved. Overlap of ancestral and daughter species does not bother us.

Arthur Keith, too, thought that certain archaic types *could* persist even after their cousins (or collaterals) branch out into new, more advanced forms. Nothing, he thought, is stopping a population, say, of *Homo erectus*, from surviving long after *Homo sapiens* appeared—as in Australia where *H. erectus* lived long after mods came on the scene.

But not everyone buys into this rationale. This reasoning, for one thing, contradicts the competitive exclusion model, which entails the survival only of the fittest, the ones with the advantages. For some, contemporaneity is grounds for disqualifying a phylogenetic relationship altogether. For instance, Charles E. Oxnard, in *The Order of Man*, would not allow Au on the human line. Why? Because some of them were *contemporaneous* with *Homo*. Multi-disciplinary scientist and author Carl Sagan said about the same: "Since *H. habilis* and *A. robustus* emerged at the same time, it is very unlikely that one was the ancestor of the other."[5] *H. habilis*, reasoned Goodman, was "a contemporary, and not a descendant, of *Australopithecus*."[6]

Every shred of evidence that has been adduced to credit phylogeny from a more primitive ancestor comes, frankly, from the interbreeding of the world's peoples and the splendid scrambling of their genes. Early and unstoppable mixings of the races of men preclude any need for evolution. Steady hybridization is the simple key to history's racial mysteries. The peopling of our world is about mingling, not evolution.

ROCK BANGERS AND THEIR NEIGHBORS

In the mid-twentieth century, Columbia University's T. Dobzhansky and University of Michigan's C. Loring Brace led a generation of researchers in the single species hypothesis, arguing that "there has been only one hominid species at one time, and . . . the hominids of different time levels are lineally [phylogenetically] related. . . . Since the australopithecines evolved straight into the Pithecanthropines, the two could not have been present at the same time." Competition within an ecological niche, it was argued, is fierce enough to preclude the coexistence of different types; only one will make it.

According to this competitive exclusion principle, two species, almost by evolutionary definition, cannot coexist, for the stronger or better, as Darwin maintained, "drives out its brutish ancestor. . . . Extinction at the hands of a successor is inevitable."[7] One species, then, preempts the niche; the other just fades away. But as the evidence against Darwin's scheme was mounting, Brace et al. had to recant.

Anthropologists have only grudgingly accepted the coexistence of different hominids in Africa. Let us visit Africa, turning the clock back 70,000 years: by the rivers "both Asu [pre-Au] and [Ihin] dwell. . . . Asu burrowing in the ground to avoid the heat by day and the cold by night"; but the Ihins were "inspired to make villages and to hide their nakedness." This passage indicates the coexistence of these two basic types: the first and second race of man.[8] In early Africa, Zinj (almost a monster, probably a Yak) and *H. habilis* (almost a man) lived in the same age and so did *H. habilis* and *H. erectus*. How could *H. habilis* have "evolved" to *H. erectus* if they were contemporaries? Louis Leakey saw no evolutionary connection between the two, for they lived side by side in East Africa; and when his son Richard found Skull 1470 (a more advanced type) coexisting with Au and *H. habilis*, it was another blow to both the single species hypothesis and to evolution itself.

"The clincher," says Jeffrey Goodman, "for *Homo erectus* types and fully modern man running about Africa together comes from Border Cave," as well as from Saldanha, Lake Eyassi, and Bodo specimens, which "make it clear that *Homo erectus* and *Homo sapiens sapiens* coexisted in southern and eastern Africa for a long period of time."[9]

All kinds of different "neighbors" have been uncovered by the spade on African soil: At least four species of hominid, *Discover* magazine said in November, 2000, "coexisted on the shores of . . . Kenya's Lake

TABLE 3.1. CONTEMPORANEOUS HOMINID SPECIES IN AFRICA

Hominid	Contemporary	Where
Ar. ramidus (Asu)	H. sapiens pygmaeus	Along African rivers*
Au. africanus	H. habilis/Skull 1470	East Africa
Au. africanus (ER 1808)	H. erectus	Lake Turkana, Kenya, and Olduvai
Au. boisei (Zinj)	H. habilis	East Africa
Au. boisei/Au. robustus (KNM-ER 406)	H. erectus (ER 3733)	South and East Africa†
Au. robustus (OH5)	H. habilis (OH7)	Tanzania
H. habilis	H. erectus	Several sites‡
H. erectus (Omo II)	H. sapiens (Omo I)	Omo River, Ethiopia

*Wolpoff and Caspari, *Race and Human*, 17.
†The Swaartkrans site, famous for its robust Au, gave up a different type that coexisted with *H. robustus*: SK 847, with a short, narrow face, delicate curved cheekbones, mandible not massive, more *Homo* than any Au.
‡See Lubenow, *Bones of Contention*, chart on pp. 352–53, which lists eighteen *H. habilis* sites contemporary with *H. erectus* ones.

Turkana," which explains why toolmaking industries were different from place to place in this region. "Early toolmakers cast off their rock-banger image"[10]: Some of these implements were made by "fairly unintelligent hominid[s]" while a *nearby* site indicates a more advanced people "capable of mass production."

> *At all periods some groups were doubtless more advanced than others.*
>
> ALES HRDLICKA, ET AL., *EARLY MAN IN SOUTH AMERICA*

Who exactly made those sophisticated tools? "A more advanced hominid genus *co-existed* [e.a.] with Australopithecus in Africa."[11] So here we are in Africa, looking for our evolutionary ancestor among Au or *H. habilis* or *H. erectus* but finding instead that they actually were all contemporaries. "Some form of *Homo* lived at the same time as some form of *Australopithecus*."[12]

This revelation was only strengthened with the discovery of Tanzania's very old (3.6 myr) Laetoli Man, a four-foot seven-inch primitive but *mixed* creature. Quite a bit more archaic than his modern-looking feet are his skull and teeth, the canines projecting. Yet the now-famous Laetoli footprints, preserved in lava, seem quite modern. Discovered by Mary Leakey, the prints show feet with the big toe neatly in line, not sticking out, not splayed (see chapter 12). Oddly, Lucy, who is younger, had feet more primitive than Laetoli's. Now, given the early date assigned to Laetoli, Mary Leakey thought this is too old, it *has to be* strictly Au—never once considering that Au might have coexisted and indeed blended with some more "modern" type.

Things fall apart all over again on the next rung of evolution: *H. erectus* in Europe. Germany's *Bilzingsleben erectus* was not so much a *predecessor* of AMHs, as their *contemporary,* coexisting with so-called archaic sapiens such as Germany's Steinheim Man. Primitive Heidelberg Man, too, lived at the same time as those modernish Galley Hill Men (although certain "adjustments" have since been made for their date, see chapter 6).

Figure 3.3. From Homo erectus *to mod in Asia, some continuity is discernible. Drawing by Karen Barry.*

Even in the New World, we find mod and Druk who lived side by side: Brazilian skulls, for example, fell into "two different races, some being smaller and relatively well-formed [Ihin types], and others larger, of a most unfavorable form (*d'une forme des plus desavantageuses*), with a forehead more sloping than that of many apes."[13]

In the Far East, as well, Louis Leakey found *H. erectus* and *H. sapiens* to be roughly contemporary in China. Here, continuity of types from *Sinanthropus erectus* to mod seems to indicate extensive Ihuan-Druk mergers. These Chinese fossils, in other words, show no evolving, just a blending of stocks that bridges earlier hominids with modern Chinese; the pattern (which I get back to in chapter 11) is significant and known as continuity.

INTIMATE RELATIONS

All these contemporaries take the wind out of any evolutionary scheme. No missing links here or anywhere, just bedfellows. "We know indeed," wrote Marcellin Boule almost a hundred years ago, "that men of relatively higher organization . . . existed from very early times in Europe *simultaneously* [e.a.] with the Neanderthal type."[14] Evidence abounds that *H. sapiens* and *H. neanderthalensis* shared the same general turf in the Middle East as well for thousands of years. And some of them "shared" more than turf (as we will see in chapter 5).

For the lower races sharing, of course, spelled an improvement. But for the higher races, it was a loss. The Ihuans (AMH) were slumming again; their "affairs" in Palestine are key, for at Mt. Carmel we find them sharing the Neanderthal's turf, tools, hunting habits, and shelter types. This was 39 kya, which date is significant for it marks the second wave of Ihuans (Cro-Magnons), who would soon spread through Europe (around 36 kya) and intermingle (back-breed) with the "barbarians" (Druks and Neanderthals). Result? A degraded Cro-Magnon (Ihuan), but an upgraded Neanderthal.

A sure sign of such Ihuan retrogression, in other words, is the combination of modern morphology and a bigger brain with primitive behavior and tools. Neanderthal tools in such situations, were no less sophisticated than the implements made by their AMH contemporaries—both in Africa and the Middle East. Even earlier, mid-Paleolithic industries (ca 100 kya), fairly crude, were produced by virtually AMH people at Skhul (Israel), Klasies (South Africa), Krapina (Eastern Europe), and the Crimea. Although Israel's Qafzeh people are "strikingly modern" in appearance, their tools are Neanderthal. The nearby Skhul (with its AMHs) and Tabun caves (with its Neanderthals) were at the same *cultural* level. Their tools were "almost indistinguishable in workmanship. . . . The two sites were contemporaneous."[15]

Thus do Mt. Carmel's old caves show *Homo sapiens* and *Homo neanderthalensis coexisting—and cohabiting.* (Neanderthals at Tabun are the same age as mods at Skhul and Qafzeh.) This presents a problem to evolutionary thinking: Why did two different groups of humans occupy the same region, using the same tools, for 40,000 years? Did they get along? Did they fight? (see competitive exclusion, above). Did they do better than get along? We might well suggest this, since their skeletal remains betray "every variety of intergradations"[16]—a marvelous intermixture, fusing Neanderthal traits (such as the prodigious browridge) with modern ones (high forehead, well-sculpted chin). The fact that they shared the same burial practices* and were even buried together indicates they lived as one, or in proximity. Only reluctantly did physical anthropologists classify these specimens (Skhul and Qafzeh) as early moderns, belatedly conceding that they coexisted for a very long time in the Near East with Neanderthal people.

Mt. Carmel, as far as I can tell, gives us a classic case of retrobreeding, for here, the AMHs used only Mousterian (Neanderthal) tools. They were devolving rather than evolving. No *sapiens* explosion here in the Near East, where the (AMH) Ihuan Cro-Magnons violated the rule of endogamy, which enjoined: Marry only with your own kind.

Israel, it seems, continued to be a great mixing ground; some of today's "Laron" (midget) cases are found here, suggesting a further type of racial blending at the Mt. Carmel archaeological site: Let's fast-forward to the year 13,000 BP. Here, a small pre-Neolithic race, some less than five feet tall, was found and named Natufian. As I see it, the Ihins at this time broke their vows and lived clandestinely with the world's people, begetting offspring in great numbers[17]—among them, the Natufians. This mixing of tiny Ihin and large Ihuan (and Ghans) took place in the Age of Osiris (just prior to the Neolithic); many of the little people "married with the Ihuans . . . and [they] were lost as Ihins."[18]

---

*Neanderthal use of hearths, burials, and pigments suggests the lower was improved culturally by the higher breed.

Yet not all was lost. Possessed of "vastly superior" tools, the Natufians may have been the bringers of the agricultural revolution to the Levant. Steven Mithen dates these Natufians to 14 kya. A sedentary people, some tall, some short, (the short ones at Kebara), the Natufians were farmers. They lived in villages all year-round, and made regular use of mortar and pestles, as well as sickles.[19] They were also fishermen, their bone hooks the oldest known in the world. In these people, who possess the distinctive gracile build of Earth's first civilizers, we might recognize the principal culture bearers of Mesolithic Palestine.

These late Mesolithic mixings in western Asia began with the advent of the Ghans, around 18 kya, when Ihuan women laid the honey trap, seducing Ihin men, their offspring resulting in the Ghan race, also known as *Homo sapiens sapiens* (see appendix B). Later still, perhaps 15 kya, when the Ihins in turn mixed with these handsome Ghans, the Natufian race was born.

The mighty Ghans, whose glory days filled the Mesolithic, would be familiar to us as the builders of enormous megaliths and as founders of the Sun Kingdoms. These are the powerful and ambitious people who brought an explosion of culture, technology, and monumental works to the lands of Europe, Eurasia, North Africa, Oceania, Asia, and Peru. Tall and stately, the Ghans were a proud race who made themselves the elite, the aristocratic class of protohistory. In this connection, Hooton once observed: "[I]n every part of Europe the socially superior and economically elevated classes show a disproportionately large number of tall persons."[20] In Upper Paleolithic times, this tall man was the Ghan (late Cro-Magnon) end product of the "beautifying" reign of Apollo (see chapters 4 and 9).

Culminating in the Magdalenian peak time of the great representational cave art of Europe, the Ghan is best known to us from the great paintings at Altamira, Spain—which petered out in time due to retrobreeding. These were the clever people who introduced small *objets d'art,* oil paint and brushes, engraving, frescoes, crayons, oil lamps, bodkins, and awls.

What does the fossil record tells us about the different racial groups coexisting in the Upper Paleolithic? Backpedaling to around thirty thousand years ago, we find mods like Cro-Magnon Ihuans shared the planet with at least three other human types: *H. floresiensis, H. neanderthalensis,* and the recently discovered (2010) Denisova, or X Woman, found buried in a cave of the Altai mountains in Siberia. The Denisova evidence suggests that *H. sapiens* (Ihuans) had interbred with them.[21] Publication of the Neanderthal genome in 2010 supported—some say proved—the idea that an AMH man "enjoyed intimate relations" with other races, including Siberia's Denisovans. "It is possible," reported an article titled "Pinkie Pokes Holes in Human Evolution," that Denisova "was descended from the hybrid spawn of an ancient tryst between her ancestors and Neanderthals."[22] *The Barnes Review* said that "to discover that the Denisova hominin was a hybrid . . . would change the view of man's prehistory."[23]

Was X Woman a Yak (i.e., an Asu/Druk blend)? Known to us through genetic analysis of a single finger, X Woman's mtDNA (mitochondrial DNA) belonged to neither a Neanderthal or early mod, but *someone* whose forebears had cohabited with *Homo.* It is "a totally different humanlike creature. . . . All three probably came in contact." Modern technology (beautiful bracelets discovered nearby) could not possibly belong to Denisova Woman, but only to her more cultivated neighbors. It's a big step to argue that the Denisova hominid created them, argued one worker: "If you find a Coca-Cola bottle near a mummy's tomb, you don't assume that the mummy invented Coke."

HYBRIDS, HYBRIDS, NOTHING BUT HYBRIDS

The tall AMHs of the Solutrean (ca 18–25 kya), known in Europe as Cro-Magnon, were Ihuans—impressive souls, "red and brown and tall and majestic."[24] A cross of Druk and Ihin genes, these Ihuan men and women were remarkably robust. "The great Ihuan race [were] half-breeds betwixt the Druks [*H. erectus*] and the Ihins,"[25] even though the chosen had been commanded to marry among themselves. But the Druks "burnt with desire . . . and my chosen came unto them, and they bare children to them" (producing the second wave of Ihuans). Later (third wave), the sacred little people again mixed with the ground people (Druk), for the latter came to them in the winter as beggars; and the chosen were tempted, and it came to pass that the Ihuans were again born into the world.

"Now these Cro-Magnon," reasoned Boule, "who seem suddenly to replace the Neanderthal people in France, must previously have existed somewhere, unless we can imagine a mutation so great and so sudden as to be altogether out of question."[26] Thus is Cro-Magnon's sudden appearance called an evolutionary saltation (meaning a jump). But wasn't it simply a new wave of hybrids? A hybrid race needs no implausible mutations or saltations to explain it. The second wave of Ihuans appeared suddenly in Europe 39 kya—not because they came from elsewhere or "mutated"; the Ihuans were produced, quite simply, by crossing, and no less than three times: 70 kya, 39 kya, and 18 kya.

The Ihuans have degenerated by marrying with the Druks.

OAHSPE, BOOK OF FRAGAPATTI 39:1

But each round was lost through retrobreeding: Cro-Magnon's final "disappearance" ca 12 kya came about through the back breeding of the last (third) wave of Ihuans, explaining then the mysterious disappearance of the great Lascaux cave artists of Magdalenian France.

"I raised up Ihuan and I gave them certain commandments, amongst which was not to cohabit with the Druks [*H. erectus*], lest they go down in darkness. But they obeyed not my words, and lo and behold, they are lost from the face of the earth."[27]

The Ihuan's sudden appearance (as well as disappearance) is found to correspond with the two pulses of European cave art, around 35 kya and 17 kya, being the second and third rounds of this powerful, gifted race. Thus in the French and Spanish caves do we find sterile layers between the Aurignacian and Magdalenian, reflecting the lapse between the last two waves of Ihuans, during which time the art reverts to unprepossessing work: simple outline art, stick figures.

Retrobreeding will also explain why tool use in the Southeast Asian Paleolithic shows a regression, reverting to older-style stonework. Blade tool technology was invented and then forgotten at least twice before the Upper Paleolithic—probably representing the first two waves of Ihuans who later back-bred.

The advent of Ihuans around 39 kya (second wave) could also account for the beginning of the end for Neanderthal—displaced now by Ihuan/Cro-Magnons, who hunted them down without mercy. Starting 39 kya, when this worldly mod culture appeared suddenly in southwest France, the region was thickly populated with Neanderthals. This time is known to the archaeologist as a period of transition to new tool industries in both Europe and Africa, the *sapiens* explosion, the acclaimed Great Leap Forward, when Cro-Magnon—as if out of nowhere—appears on the scene changing everything.

Shortly after the second round, ca 38 kya, mods appeared in Kurdistan and north of the Caucasus in Central Asia. According to genetic studies, the Altai region (home of Denisova) was the hub of "population expansions" radiating to Eurasia as well as to the Americas, and presumably accounting for Cro-Magnon's abrupt debut in Europe ca 35 kya. But even if this new race of modern men did expand out of the Altai (southern Siberia), we are still in the dark where they came from in the first place! This was not an invasion, nor was this Ihuan wave a migration or colonization. Rather, this was the second round of Ihuans, who came into the world through racial amalgamation.

But no sooner did the Ihuans/Cro-Magnon come into the world than they themselves backslid; thus do we encounter AMHs *without* the expected tools or behavior, signaling that in these mixings the lower type tended to dominate, culturally. We can see this at Borneo's Niah

Cave where Deep Skull is AMH all right, but without the expected modern behavior: only crude tools.

The same scenario—AMHs with simple pebble tools—is also evident in Malaysia (Penang site). And it was the same in Europe (Krapina, Crimea) and in the Near East (Mt. Carmel) and Africa (Klasies River and Border Cave), where the use of passé Paleolithic tools by mods is supposedly difficult to explain. Why did behavioral modernity lag behind anatomical modernity?, paleoanthropologists ask, bewildered. The problem, though, disappears with an understanding of the dynamics of retrobreeding.

I think the archaic sapiens tell the same story, i.e., the predominance of the lower type (*H. erectus*) over the higher (Ihuan) type with whom they mixed—at Heidelberg, Solo River, Sima de los Huesos, as well as at Vertesszollos, Steinheim, Bilzingsleben, Fontéchevade, and Swanscombe. (Note that the careful term *archaic sapiens* has put out a lot of fires; it became a euphemism for Neanderthal—when no genetic continuity could be found between Neanderthals and mods.[28] Archaic sapiens are a catchall for all the Ihuan retrobreeds and other mixes; they have been dated anywhere from 800 kya to 5 kya! Chapter 6 goes into these ludicrous dates.)

Ihuans slumming with *H. erectus* probably explains why we find Ihuans with long arms and short legs and Cro-Magnon skulls that are long and pentagonal, with proportions that are "disharmonious." For *H. erectus,* this crossbreeding spelled an upgrade, for when the Ihuans mingled with Druks, it resulted in the tall, improved *H. erectus* types we so often find.[29] This was a very old Ihuan habit—back breeding. They couldn't help themselves. It finds Cro-Magnon in many places with dolichocephalic skulls (yet high forehead), a hybrid with Neanderthal,[30] their stage of culture unimpressive.

Thus is the record full of seeming throwbacks, sterile layers, and even anachronisms—but nothing is out of sequence, really, for there *is* no phylogenetic lineage for man, only persistent mixing of the races. The idea of Neanderthal as a withered side branch of man is completely off, when we realize he was merely an *H. erectus*—Ihuan cross, a downgrading of the AMH type. Even Thomas Henry Huxley, English biologist and "Darwin's Bulldog," saw Neanderthal as a reversion toward a more primitive type. Keith also thought the human skull sometimes went through a process of retrogression, the Neanderthal race "a great step backwards." Boule, for his part, called it "a degenerate species." Finally, Hooton labeled it "degradation."

The word *retrogressive* appears also in the problem of *H. erectus* "evolving" from the owners of Skull 1470, which is to say, *more archaic* men evolving out of surprisingly advanced ones. Do all these retrogressive trends fit evolution? Not at all.

> There is no evidence to suppose that the Cro-Magnon
> people grew out of the Neanderthal . . . for, at all ages
> of human history, undeveloped and advanced tribes and
> nations have existed contemporaneously.
>
> CHARLES BERLITZ,
> MYSTERIES FROM FORGOTTEN WORLDS

Cro-Magnon (Ihuan) and Neanderthal (a Druk-Ihuan blend) were contemporaries. In Portugal, their affairs resulted in a hybrid race. Recent digs in Romania and Czechoslovakia have come up with more of these blends—part *H. sapiens,* yet sporting distinctive Neanderthal features such as great noses and molars and occipital buns (protuberance at the back of the skull), suggesting, all told, an "intermixing between modern humans and Neanderthals."[31]

Cro-Magnon himself was brawny and brainy (1,660 cc). He had a high forehead, strong jaw, and broad barrel chest. Notwithstanding his modern form, he was mixed, judging from his heavy brow, large teeth, beak nose, large orbits, wide cheekbones, broad face, and long arms. All these traits are part of his Druk heritage, although they could equally reflect some mixing with Neanderthals. Especially in Europe, Cro-Magnon shows a mixture of races in his proportion of limbs and in certain features of the face. Not too civilized, the back-bred Cro-Magnon was a hunter of wild cows and horses.

A MIXED BAG

Ethiopia's Omo 2 and Omo 1 are fossil men found at the same level, apparently contemporaries. Yet the two are very distinct people, one archaic, the other modern. Nevertheless, evolutionists call this a single population with an unusual degree of variability. *Variability*? What a great word; putting out another fire, this buzzword (or smokescreen) serves only to blur two distinct races living cheek by jowl!

Has evolution stood the test of time? Does it explain man's ascent? Or is mixing the answer, the Rosetta Stone of the fossil record? Says geneticist Marcus Feldman, "a lot of mixing has gone on," but in the next breath he assures us of "a consistent picture of modern human *evolution*"[32] [e.a.]. Oxymoron? Or split mind? Don't you think we have seen enough mosaics to appreciate that gene exchange (the nice word for nookie)—not evolution—gives us the intermediate types of the fossil record, the so-called transitionals of step-by-step evolution?

All these bogus "transitionals," propped up to prove the meta-morphosis of lower into higher hominids, are nothing but the hapless offspring of exogamous unions, which is *the* factor that produced the tremendous variation so typical of interbreeding populations. This per-plexing variability that appears at every turn resolves at last into the incessant crossbreeding of different stocks. And nothing more.

> *There is no doubt that new races can arise from such miscegenation.*
>
> EARNEST HOOTON, *UP FROM THE APES*

Darwin (in *The Descent of Man*), denied the fact of interracial unions (Hooton's unpleasant term being miscegenation), speaking instead of tremendous *variation within* racial groups. He was mistaken in saying the races "graduate into each other, independently . . . of their having intercrossed." How could they manage that—without intercrossing? This he did not answer.

As we saw above, in the case of Israel's Qafzeh and Skhul specimens, neither transitionals nor variability fit the bill; the bone people were obliged to confess that "there were two groups of humans in the Near East."[33]

Things happened in America, too. Kennewick Man is a manifest composite of Caucasian (chin and skull), Mongoloid (shovel-shaped incisors), and Polynesian (shape of eye sockets). Today, if we met such a man, we would recognize his mixed heritage at once. Why should it be any different for the fossil record? Invoking bogus variability serves only to mask the factor of pervasive race mixing, which after 150 years of research is still being swept under the rug.

But this is just what the evidence is telling us—if allowed to speak for itself. Hybrid man is a fire getting ready to burn down the house of evolution. How to put out the fire? Variability is one ploy. (The constructs of isolation, specialization, saltation, and immigration, as we will see, also come in handy.)

In the Upper Paleolithic, some kind of Caucasoid (not Mongoloid) people, a group like the Ainu of Japan or the Uighurs of Western China, inhabited North China (at Choukoutien). The tribal name Ainu simply means "human being," but specifically of the modern (versus primitive) type, for the Ainu were infused with Ihin blood conjoined with something more archaic: the northern Ainu people are hairy, prognathous, dolichocephalic, and heavy browed, with eyes wide apart. Different from the indigenous Mongoloid races, the Ainu are white (not yellow) and wavy haired (not straight).

The classic Ainu skull has the Cro-Magnon large face; there are no Oriental epicanthic folds, and some Ainu eyes are of a greenish color. Their fingerprints have more loops than whorls (a Caucasoid characteristic); their sticky earwax is also Caucasoid. But they are much smaller than most Caucasians (early Ihin mixes).

If Earth is a melting pot now, it was then, too. Even the allegedly pure Nordic race "is primarily a blend of two radically different types."[34] In the valley of the Rhine, a hybrid man called Chancelade has been unearthed, his type found everywhere from France to Denmark, Lapland, Greenland, and Alaska. The Rhine skulls are so *variable* in appearance "that one would not hesitate to assign them to two races if they came from separate localities." But that is just the point: Groups exhibit wide variability, not because they are any kind of evolutionary transitional, and not because of gene mutations, but because they do, in fact, represent different races. The Upper Paleolithic is full of such Cro-Magnon blends, most notably the Chancelade type who is nonetheless lumped together with Cro-Magnon—with little or no notice of his size change or other differences such as face height and shape of orbits. Azilian skulls from Bavaria also "present an extraordinary mixture of types," just as the bones found in French, Italian, and Belgian caves represent "products of the crossing of races."[35]

It was not just a Caucasoid Cro-Magnon enjoying Europe in the Upper Paleolithic. This Chancelade Man, discovered only a few miles from

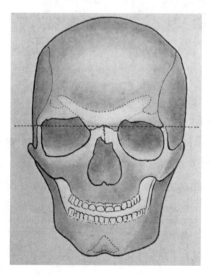

*Figure 3.4.
Cro-Magnon
skull.*

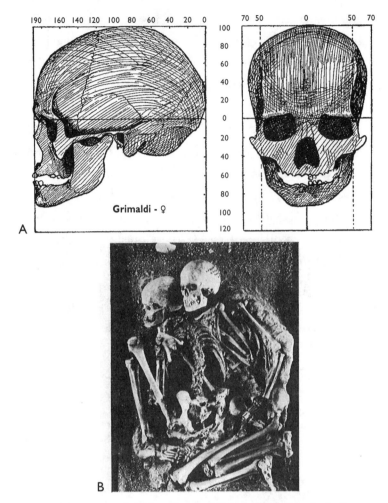

Figure 3.5. (A) Two views of Grimaldi Woman's skull.
(B) Grimaldi Woman and Boy skeletons found in cave.

a Cro-Magnon site, is described as Eskimoid in type. The two Grimaldi skeletons (male and female) found in the French Riviera, like the couple at Obercassel (a hybrid of Cro-Magnon and Chancelade), were described as Negroid. These Grimaldi people were there in France before and during Cro-Magnon's stay. Different from the Cro-Magnons, they were also unlike the robust Neanderthals (whose tools are at the lowest, Mousterian, horizon). But the Grimaldi people *were* similar to the Negrito Fenlanders, the Arthurian wild men, and the legendary black dwarfs of Europe.

Grimaldi Man was slender and gracile, none taller than 160 cm (Cro-Magnon was up to 190 cm); they were certainly less robust than either Cro-Magnon or Neanderthal. Grimaldi showed mixed traits—long arms, prognathous face with a wide nasal opening (Negroid) combined with high nasal bridge, smooth brow (Europoid), and tall braincase. The Grimaldi woman was small, her skull long and narrow—hyperdolichocephalic (unlike the flat Cro-Magnon vault), her face short, her teeth and palate quite large. On the other hand, she did have certain Cro-Magnon features, like her large brain (up to 1,580 cc). Since Arthur Keith's day, however, these intriguingly mixed Grimaldi people have been lumped together with Cro-Magnon, on the plea of variability, despite the fact that Grimaldi was an obvious blend of African, Negrito, and Europoid races.

HOPELESSLY MIXED

In the Malaysian isles, the Sakai people (see figure 3.6) are a perfect blend of two races: though light-skinned, the nostril is very broad and

Figure 3.6. Sakai people showing mixed traits.

winged. Sakai hair ranges from curly to woolly, yet most have straight or wavy hair. There is no prognathism. They simply are a composite race, like the Semang Negritos with their glossy hair, thick lips, and flat nose. "The Semangs have often been altered by crossing."[36] In this region, black people once crossed with a different population, resulting in considerable variation—half-breeds of all kinds. Many Negritos of Sundaland (unlike their namesake) have "long narrow noses of button type." Altogether, the Malaysian Negritos show "a great deal of fusion."[37]

The Malay people themselves are a marvelous blend of racial and ethnic types, the indigenous stock generously layered with Hindu, Persian, and Chinese genes. In a single Singapore photo, I noticed, cheek by jowl, a wide range of types. One individual was of the Malay type, though her wavy hair and other features betrayed some crossing. Standing next to her was an out-and-out Negrito. One companion had straight, lank hair; another, absolutely woolly. Here in Sundaland, given the mingling of Negritos with Malaysian, Papuan, and Asian populations, there is every shade of combination—*clinal* populations—one regional group blending smoothly into the next.

This easy mixing, if we go back far enough, will prove to be the origin of every creature in the hominid family.* Even among Malaysia's woolly haired Negrito tribes, 20 percent of the population has straight hair, many with a reddish tinge, a true admixture of bloodlines. The Jakun of southern Malaysia have so much Australo-Melanesian blood that they cannot be easily distinguished from Papuans, who are big, black, and bearded.

Papuan Negritos (four feet nine inches) of the New Guinea interior are black, well-formed, and elegantly proportioned, yet the body is covered with woolly hair; there is no prognathism. Here in New Guinea, one also encounters light-skinned natives, for example, the Tarifuroro, who are master horticulturalists tending terraced fields and pretty hedges. Southeast of New Guinea, New Caledonia also boasts signs of very ancient culture with its mounds, tumuli, cast pillars, canals, and

---

*The only pure races on Earth—Ihin and Asu—are long extinct.

terraces. Here is another mixed bag: Some of its people are hairy, with heavy brow and jaw, others are brachycephalic with light brown or even blond hair. At Easter Island, too, the natives say their ancestors had light skin and reddish hair. And these whites were as indigenous as they are.

Turning to India, the ancient pygmaioi were fully Ihin blacks, with their great beards and long hair. Still seen today in the proud Dravidians are the small stature and other features of the old Ihin race. Of exceptionally mixed ancestry, the Veddas of Sri Lanka are classed as Caucasian; some clans are very pale skinned and tiny, the men barely five feet tall, the women usually four feet ten inches. Well built, the men sport chinbeards. The dainty Veddas are a striking blend of modern and archaic types, with their flat feet, straight or wavy hair, small skulls, and long arms. Are their Caucasian ancestors the great engineers who built the water tanks and vast irrigation works of prehistoric Sri Lanka?

Coon noted a profound fusion of types among India's Kadar people of Kerala; an extremely primitive group, they are short statured and chocolate colored with frizzy hair; some are taller with wavy hair, more the Australoid type, although their hair is the texture of the Mongoloid. Other Kadars are largely Caucasoid in appearance. "Obviously the Kadar are a composite people."[38] No wonder de Quatrefages (of the nineteenth-century Paris Musee d'Histoire Naturelle in Paris) described India as a land of very ancient and extensive crossing between numerous and diverse elements.

Figure 3.7. An Andaman Islander, photographed in the 1890s.

Bloodlines are so hopelessly mixed among so-called Negritos (the world's swarthy little people, some with an Oriental cast), that it seems a fool's errand trying to "classify" them. South of India, the Andaman Islanders, for example, have the large feet and teeth of archaic stamp; and though their skin is dark, their "features possess little of the negro type,"[39] for they are brachycephalic and their jaws are not prognathous. These Andamanese Negritos compound the woolly hair of the Negro, the glabrous (hairless) body of the Mongoloid, and cranio-metric features in keeping with Egyptians and Europeans! In yet other traits, they resemble the Australian-Melanesian type. Geneticists waste their time, really, trying to determine whom the Andamanese most closely resemble. In a word, we are all hybrids—part of the great plan (see chapter 7).

Melu, a culture hero of the Bagobo, an Aeta people in the Philippines, is featured as a white man; the ancestors of these folk, according to a thirteenth-century Chinese account, had yellow eyes. There is still a trace of blondism among these Negritos. Small framed and small nosed, the tiny Aeta—the women very pretty—are thought to be the earliest inhabitants of the Philippines.

Especially pale are the Abenlens, living deep in the isolated Zambales Mountains of Luzon. They are a remote people, unlike the regional Negritos—and shorter still (no more than four feet six inches high). Some are light, almost blond in complexion; others are olive skinned with light brown eyes. Most possess delicate features and soft wavy hair. Ancient mixings of the Ihins with indigenous groups gave us all the world's gracile stocks and Negritos, parts of the Philippines a living museum of those early blends.

Australia: the nether continent's Aborigines are regarded as a (morphologically) primitive Caucasian stock. Yet with their dusky complexion, beards, and wavy hair (not frizzy), they are a fusion of Melanesians and Caucasians. As Hooton saw it, once upon a time, an archaic white stock fused with Tasmanian natives, producing the modern Australian Aborigine, "a composite race."

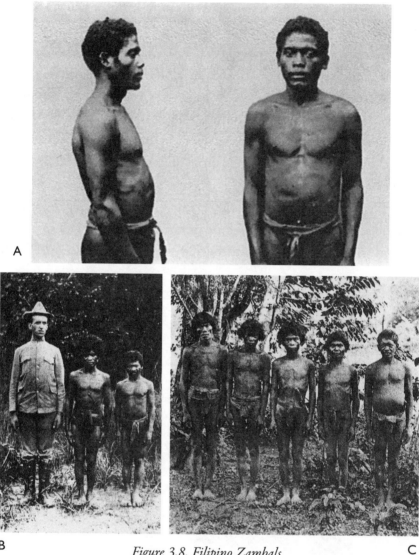

Figure 3.8. Filipino Zambals.
(A) A pure-blood Zambal.
(B) Taken in 1904, the photo shows the relative heights
of an American, a mixed blood, and a pure Zambal Negrito.
(C) Five mixed-blood Negritos of Zambal.
There are also white natives on these islands:
on Suuna Rii are people with reddish hair.
There are also red-headed women in the Malay Peninsula,
among the Semang and Pangan.

Bearing a remarkable resemblance to today's Australian Aborigine skull, Wadjak Man of Java, quite modern in type, had a brain volume around 1,600 cc—although the Australian has only 1,300 cc on average. Despite this, some still see an evolutionary line leading from Wadjak man to the Australian Aboriginal. Concerning this decrease in cranial capacity from the earlier Wadjak and Talgai type to contemporary Australian Aborigines, Hooton could only conclude that it was "by *hybridization* [e.a.] of proto-Australoids with the pygmoid Tasmanians."[40] (See figure 3.9.)

In South America along the Orinoco River, crania show that two quite different types of humans once used the same caves.[41] Judging from Peru (Mochica) burials, Caucasoids coexisted with

Figure 3.9. Artist's rendition of a Tasmanian. Drawing by Karen Barry.

Figure 3.10. This Guarani (Bolivia) man is of mixed type.

non-Caucasoid indigenous groups.[42] Here in Peru, each face depicted on ancient pots is strikingly different, and skins are of different shades—white, yellow, brown, copper, black, and with "such dissimilar physiognomies."[43] In Bolivia, too, the famous Tiahuanaco carvings show a conglomeration of types—people with high, low, or broad foreheads and with eyes that are narrow, slit, slanted, popped, and deep set. Noses vary from flat to hooked, snubbed, and fleshy. Every shape of face and profile is seen. (See figure 3.10.)

The hybrid theme was noted again by the point man for the Heye Museum, Alpheus H. Verrill, who, in the 1920s, observed Quichua and Aymara blends in the Andes. The Quichua, a docile and peaceful people, are short and light with tall foreheads (traces of Ihin genes). More yellow than brown, the Quichua have full lips, straggling beards, and high cheekbones—almost like Mongolians. In other localities, though, such as along the Pacific coast, Quichua groups have dark skins, round heads and broad noses. Yet others in this tribal family have thin lips and aquiline noses. Then again the Huancas are often pale olive with finely chiseled features, though their noses may be enormously hooked (Ihuan and Druk genes). The Aymara people on the other hand are more Ihuan than Ihin—bronze, tall, and slender, with sharp noses and more receding foreheads than the Quichua; but they vary almost as much as the Quichua—many possessing full beards, Polynesian faces, and hazel eyes! There is no end to the mixture of bloodlines here—and just about everywhere.

THE LONG AND THE SHORT OF IT

Some anthropologists (analyzing Flores's hobbit) suppose that evolutionary forces (involving limited resources) may have pushed some inhabitants toward dwarfism, while others were selected for gigantism. I don't think that is what happened. The early races varied greatly in size and stature depending on the amount of Ihin (short) or Druk and Ihuan (tall) genes, great size appearing as a feature of "hybrid vigor."

There were giants [Druks] in those days and in time after that; and my chosen [Ihin] came unto them and they bare children to them.

OAHSPE, THE LORDS' FIRST BOOK 1:29

If women, via mtDNA, pass on their shortness to daughters only, and if the Y chromosome is passed down from fathers to sons, it would only be the male children of a (female) Ihin and (male) Druk match that would be tall and robust. Hence the sexual dimorphism in so many of the early races, the men being much larger than the women: *Au. afarensis, Au. Laetoli*, Zinj, Dmanisi, Chancelade, Grimaldi, Tabun, Choukoutien, Palau, and so on.

No, "giants and dwarfs" is not just a myth; large Ihuans (and Druks) coexisted with little people all over the world. Even in today's world, giants and dwarfs coexist in certain remote spots. In Africa, the tiny Bushmen and the tallest race of the country, the Kaffirs, are close neighbors, just as the Mutua pygmies live side by side with the tall Watusi in Uganda. Likewise in Mongolia, the world's tallest man, Bao Xishun, seven feet nine inches (see photo), shakes hands with He Pingping, the world's shortest at two feet four inches. In Yunnan, Hubei, and Shaanxi provinces in China, there exist four-foot-high hairy people and seven-foot-high hairy people. In Europe, the tallest people are the Norwegians; yet the Lapps of Norway are the shortest Europeans. The tallest Indians in the New World are found in Patagonia and Tierra del Fuego, which regions are also inhabited by the shortest peoples in the Americas. (See figure 3.11 on page 141.)

In writing about the different-size races, Hooton, evidently without meaning to deal a blow to evolution's cherished natural selection (chapter 9), did just that: "The geographical distribution of stature groups is not at all in accordance with the supposition that climate and altitude have anything to do with bodily height. Among the very shortest peoples in the New World are the Yaghan and Alakaluf of Tierra del Fuego. . . . Adjacent to the Yaghan live the Ona . . . [with] prodigious average height. . . . Both these Indian tribes subsist largely

TABLE 3.2. GIANTS AND DWARFS
IN THE SAME LOCATIONS

Where	Small Ones	Tall Ones
Australia	Extant pygmies	Giants: 23-inch footprint*
Brazil	bearded white dwarfs	Tahuamanu, 6 feet, 5 inches; 7-foot, 7-inch skeletons
China	4-foot-tall wild men	Gigantopithecus; "the great giant race of Ihuans in Jaffeth"†
China	Han-Dropa	Sinanthropus, 10 feet tall‡
Denmark	Little Danes of Stone Age Mounds	Tall Danes of today
Ecuador	"Laron" little people of Loja	Tarija giants,§ 10 feet tall
France, Basque	Laminaks	Giant Tartaros; enormous skeleton at Chaumont castle (found in 1613)
France	Chancelade Men, four feet nine inches	8-foot, 6-inch bones, Field of Giants¶
India	Veddas, Bhils	Daityas Hiranyakashapa; 2-foot-long footprints; 11-foot-tall skeleton, Assam\\
Italy	Neolithic little men	10-foot, 3-inch skeleton, Valley of Mazara
Italy, Pompei	Dwarfs	Eleven-foot skeleton in mine shaft (1856)
Java region	Flores hobbit and Rampasas	Meganthropus at Trinil: "enormous," twice size of male gorilla**
Libya	Little people (Herodotus)	Giant Antaeus (who fought Hercules)
Mexico	Alux; small Lacandon (Mayan)	Cholula "deformed giants"; Aztec

*Von Daniken, In Search of Ancient, 178.

†Oahspe red-haired giants: see Book of Wars, chap. 28 and XLV:20; "land of Giants": this was China in the time of Abraham.

‡Weidenreich, Apes, Giants, and Man, 29.

§Kolosimo, Timeless Earth, 24–27; Norman, Gods and Devils, 135.

¶Kolosimo, Timeless Earth, 24–27; Near St. Romans, see Norman, Gods and Devils, 129.

\\Wilkins, South America, 194.

**Weidenreich, Apes, Giants, and Man, 61; von Koenigswald's giant Java man, Meganthropus at Trinil, 1939–1941, from a skull; lower jaw was massive; more recently Solo Man was also enormous.

Where	Small Ones	Tall Ones
New Zealand	Maori legends of little people	Giants, 10 feet tall
Norway	Mound folk; the *Edda*'s elves	Giants of Norwegian caves
Patagonia and Chile	Short Chonos and Yaghans	"Horse Indians" of the sixteenth century, 7 to 8 feet tall; Ona people
Persia	Kermanshah dwarfs	Zarathustra, 9 feet tall[*]
Philippines	Aetas, Zambal	Seventeen-foot skeleton at Gargayan[†]
Scandinavia (Norse epics)	Dvergar (dwarfs)	Loki, Jotunns; "race of giants side by side with dwarfs"[‡]
South Africa	Bushmen	Swartkrans Man,[§] nine feet tall
Turkey	Laron cases	Legendary giants of Troy
UK	Little Picts enslaved by	Giant Fians (Figda)/*Book of Lecan*
UK	Cornwall brownies	Giants of Cornwall, up to the time of King Arthur
U.S., Cascades, Alaska, Arctic	"Chancelade" type	Giants in the region[¶]
U.S., Dakotas	Mountain of the little people	The Tall Ones[\\]
U.S., Hawaii	Menehune	Tall Polynesians
U.S., Oklahoma	Prehistoric small men	Enormous footprints[**]
U.S., Pennsylvania	Basket Makers	7-foot, 2-inch skeleton at Gasterville[††] and Bradford County
U.S., Tennessee	Extensive pygmy graveyards	7-foot skeleton; 16-inch tracks[‡‡]

[*]Oahspe, Book of Wars 21:1 and 21:4.

[†]Kolosimo, *Timeless Earth*, 24–27.

[‡]Norman, *Gods and Devils*, 134.

[§]Kolosimo, *Timeless Earth*, 24–27.

[¶]Childress, *Lost Cities of North and Central America*, 453–54.

[\\]The Dakota Indians had a sacred place called the Mountain of the Little People or Spirit Mound. Ella E. Clark, in *Indian Legends of the Northern Rockies*, recorded this tradition: "our forefathers claim the little people lived there . . . the Elders say there were a small people, like eight-year-olds," living in the rocks near the Yellowstone; Deloria, *Red Earth, White Lies*, 156: Deloria heard from the elders that the Indians were once much larger and taller. They call them the Tall Ones.

[**]Steiger, *Mysteries of Time and Space*, 31–32.

[††]Corliss, *The Unexplained*, 13.

[‡‡]Corliss, *Ancient Man*, 682.

Figure 3.11. These two men hail from the same region of Inner Mongolia.

Figure 3.12. Basque drawing of the long and short men of yore.

by hunting and fishing. . . . The gigantic Ona are [even] said to have a somewhat less abundant food supply than the stunted Yaghan. . . . In Scandinavia are the Lapps with [short] stature. . . . Just to the south of them are the Finns, Norwegians and Swedes . . . of [high] mean stature. . . . Instances of very tall and very short peoples living in the same regions and under substantially dietary similar conditions might be multiplied. I know of no example of such juxtaposition of very short and very tall peoples in which the differences in size can be explained by invoking any environmental factor."[44]

Germany's little people (such as the four-foot eight-inch skeletons unearthed near Bonn) are answered by her giants tombs, Hunengraben. There was a tradition among the peasantry of the German states that God had created the giants to slay the wild beasts and great dragons to ensure the safety of the (sacred?) dwarfs. This curious legend is echoed in the Oahspe, which records that the large Ihuans (mighty men) were tasked with slaying all dangerous beasts, thus acting as defenders ("shield"[45]) of the Ihins and their followers, the faithists; for "My chosen on earth . . . are a harmless and defenseless people. Therefore. . . the barbarian [huge Ihuans] shall destroy all evil beasts and serpents; and the forests shall fall down before him"[46] (this last phrase referring to the planned extinction of the dangerous and outsized megafauna of the Paleolithic). Thus did the strapping Ihuans become the official guardians of the Ihins, the sacred people.[47]

But in a few years these giants, as various old legends recount, would themselves come to oppress the dwarfs, for they, the giants, had become "altogether wicked and faithless"—so wicked as to be the cause, in many traditions, of the flood. In my view, this "wickedness" reflects the retrogressive tendency of the Ihuans who repeatedly mixed with the lower races, losing all holiness.

In America, among the Arikara Indians, it is said that the Creator caused a flood to get rid of these unruly giants, but saved the little people by "storing them in a cave." Likewise does the Pima creation myth involve the salvation of the Elect from a great flood; just as the Skokomish (of Mt. Rainier) relate that the Great Spirit was displeased with the evil in the world, and secluded the good people before causing a universal deluge.[48]

AND CAIN SLEW ABEL

Thou killest my prophets.

OAHSPE, BOOK OF GOD'S WORD 1:9

Before the flood, in four great divisions of the earth, Vohu (Africa), Jud (Asia), Thouri (Americas) and Dis (Europe), they did not leave one alive of the Jhin race.

OAHSPE, THE LORDS' FIRST BOOK 1:25

The myths of Apollo slain by Python, Osiris slain by Typhon, and Bacchus slain by Titans may all personify the slaughter of the Old World's priests and sacred tribes (the little people), more than 24,000 years ago, at the hands of the degenerate hordes of Ihuans and Druks. Genesis 6 recalls such infidel giants, as does the Book of Enoch, the latter declaring that they, the giants, were at their worst in the (antediluvian) time of Jared.

As tradition has it, after Cain slew his brother Abel, "he went forth . . . with an impious race, forgetters and defiers of the true God. . . . All nations preserved the remembrance of that division of the human family into the righteous and impious."[49]

The depraved and godless are descended from Cain.

LOUIS GINZBURG,
LEGENDS OF THE JEWS

Abel, as prototype of the sacred people, was "able" to understand spiritual things and was capable of hearing the voice of God. But Cain slew Abel: meaning, the huge Druks wiped out the holy Ihins, which scenario portrays the extermination of the sacred people by the large, warlike *H. erectus* and barbaric Ihuans, those "ancient warriors that destroyed the chosen, before the flood."[50] If "the earth rose up against My chosen and sought to destroy them," this was later dramatized as the death of Abel: "the Druks fell upon the Lords' chosen and slew them, right and left."[51] Hence Cain's "mark of blood"—demonized in the Persian and Hindu word *druj*, the equivalent of Druk, referring to low spiritual rank.

"Those of the lesser light were called Cain, the Druk, because their trust was more in corporeal than in spiritual things."[52] The family of Cain, says Levantine tradition, resided in the field of Damascus; they were "towering giants." By biblical times, a few of those giants remained and are known to us as the enormous war chiefs of the Canaanites, fearsome and formidable, next to whom the invading Hebrews (of the Exodus) were as "grasshoppers."

But the story of warring brothers, Cain and Abel, is not limited to the Levant, for it was in all divisions of Earth that the Druks rose up against the prophets of God. In New Guinea mythology, for example, Abel, named Kulabob, is the peaceful and enlightened brother, an inventor of ocean-going navigation, planting, pottery, carving, and all the useful arts. He was also responsible for ritual and spiritual matters, while his brother Manup was a stocky man of the land, a hunter. As the legend unfolds, Manup goes after Kulabob whom he suspects of violating his wife. "We pause in this saga . . . to count up the shared motifs with . . . the Cain/Abel story-type," says Stephen Oppenheimer, who recorded these New Guinea myths in *Eden of the East*. "Kulabob and Manup were very different men. . . . The two brothers clearly belong to different cultures."*

---

*The descendants of Kulabob, fair-skinned, are regarded as "another race" by the Papuans. Oppenheimer, a medical doctor, may have confirmed this, in his discovery that an inherited form of anemia sets the Kulabob descendants apart from the other people of New Guinea. The learned Kulabob, we might add, retains the sacred Ku- in his name, see appendix A.

CHILDREN
OF ABEL

Without really meaning to, scientists gave the (hybrid) *Homo habilis* the same name, Abel—able man; for *H. habilis* was thought to be the first toolmaker—a handy man. *H. habilis* (possessing more Ihin genes than did Au) was clearly an improvement on Au in a number of ways: narrower cheek teeth, improved femur, more modern feet, increased brain capacity, higher forehead. *H. habilis,* standing as an improved Au, was upgraded by the infusion of little people genes. And this is why *H. habilis* was so short (shorter than Lucy) and had a more gracile skull than the later Neanderthal.

For the early Ihins, the little people, contrary to their own law did mix with the authochthonous Asuans (Ardi), their offspring showing various combinations, the most progressive of these offspring being *H. habilis*. Christened by Louis Leakey in the early 1960s, *H. habilis*'s actual stature was not discovered until 1986. Before then, in 1973, Louis's son Richard, lamented: "Unfortunately, we cannot be sure of his body size: A 650 cc brain might actually be large for an individual of extremely short, light stature"[53] (yet we noted in chapter 2 that little people are seen to possess relatively larger brains). But then the 1986 find, OH 62 (OH stands for Olduvai Hominid) proved just how small *H. habilis* was: some three feet tall, others taller at three feet five inches. OH 62, being the first find of postcranial *H. habilis,* turned out to be actually smaller than tiny *Au. afarensis,* his femur smaller than Lucy's, indicating more Ihin genes. Here in *H. habilis* was a creature upgraded from Au by AMH genes: bigger brain, thinner skull, more human in teeth, feet, and face. But his more humble genes gave him an apelike torso, curved hands, long arms, and short legs: his limb bones were actually more primitive than *Au. afarensis!* But so slender. Hence the fruitless but eternal debate: Was *H. habilis* the first *Homo?* Or just a big-brained Au? The evolutionary mindset never once considered that this mishmash of fossils could be understood as the product of crossbreeding—with its fantastic kaleidoscope of traits.

But let us move on now to chapter 4 where we will test the idea that environmental pressures can cause creatures to evolve into different species—the subject being Darwinism's premier concept: speciation.

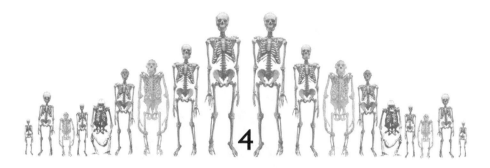

"I Do Not Believe I Ever Was a Fish"*

Debunking Evolution

Not one thing of all the living mergeth into another, but every one bringeth forth after its own kind. . . . Each and every living thing [is] created new upon the earth and not one living thing created out of another.

OAHSPE, BOOK OF COSMOGONY 4:19

SPECIATION

Charles Darwin's great brainchild was this: that with "favorable variations preserved . . . the result would be the formation of new species," and this is based on "the tendency in organic beings descended from the

---

*Quoting Benjamin Disraeli, one-time British prime minister, in *Tancred*.

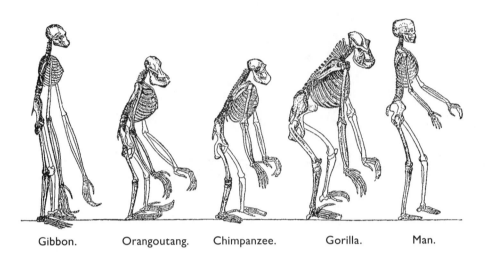

Gibbon.   Orangoutang.   Chimpanzee.   Gorilla.   Man.

Figure 4.1. Frontispiece to Thomas Henry Huxley's Man's Place in Nature. *"Parade" showing the similarity of structure among gibbon, orangutan, chimp, gorilla, and man, suggesting that the latter differs from the former (the anthropoid apes) in degree only.*

same stock to diverge in character. . . . That they have diverged greatly is *obvious* [e.a.], from the manner in which species of all kinds can be classed under genera . . . families . . . suborders, and so forth."[1] This leap of faith (that anything diverged from anything) is hardly scientific. "That one species could turn into a completely different species," is, said England's Richard Milton, "an intoxicating draft from the tankard of speculation."[2]

The question then is: Has evolution (which is the same as speciation) actually taken place? Loren Eiseley, chairman of University of Pennsylvania's Department of Anthropology after Carleton Coon, once marveled at how "a Devonian fish managed to end as a two-legged character with a straw hat."[3] And he believes this; it is not just one of his poems. Some zoologists, though, along with dear Benjamin Disraeli, can't see it: "[T]he archerfish is a very successful construction that has *always* [e.a.] existed in its present form. . . . This Intelligent Designer appears to have an endless number of novel ideas. . . . His creatures can adapt to their environments to a certain extent, but they did not evolve from other species and will never develop into new species."[4]

In the year 2000, the National Science Board found that roughly half of Americans reject the concept that humans developed from earlier species of animals. Some scientists doubt that the world is old enough for an organ like the human brain to have evolved from protozoa. How, then, does evolution (based on the concept of speciation, also know as transmutation) manage to remain the prevailing theory? Here, then, is a hypothetical debate between POE (professor of evolution) and A (adversary) on the question of speciation, the doctrine of one kind of animal, over time, turning into another kind of animal, the overall process known as phylogenesis.

WALKING WHALES
AND OTHER CHIMERAS

POE: The transmutation of species has been scientifically established for more than one hundred years. There are thousands of examples, such as the reptile's scales evolving into the mammal's insulating coat of fur—probably under the influence of climate cooling. Robert Broom, evidently one of your heroes, did a fine job showing how South African reptiles led to the first mammals.

A: OK, here's my position: No animal ever *became* human, nor did man *develop* organically at any time in the past. The French paleontologists, for one, refused to buy into the development of later forms out of earlier ones. "Evolution is a fairy-tale for grown-ups," quoth Jean Rostand. Even Lyell and Hooker, Darwin's closest allies, held to the permanence (immutability) of species.

Let a sign be upon the earth, so that man in his darkness may not believe that one animal changes and becomes another. . . . [When] different animals bring forth a new living animal . . . unlike either its mother or father . . . the new product [is] barren. . . . And this shall be testimony . . . that each and all the living were created after their own kind only.

OAHSPE, BOOK OF JEHOVIH 5:10–12

A: The penalty imposed by nature for an attempt to introduce confusion of species is barrenness, such as sterile mules, for in the mismatch the generative organs are thrown out of balance.

Figure 4.2. Cartoon by Marvin E. Herring.

POE: Barren, that's right. Darwin, from these facts, inferred that the result of a cross (of different human races) is inferior vitality and lessened fertility, possibly even premature death.

A: Wait a second. With *Homo*, we are not talking about different *species,* but only different *races* (subspecies). And with crossings of different races comes, not inferiority, but hybrid vigor, known as heterosis, which actually *increases* fertility and resistance to disease. Mixed offspring tend to inherit the strong traits of their parents (getting bigger and stronger) especially in the first few generations. Such hybridization, when you come down to it, precludes all need for human evolution!

POE: Nevertheless, we do indeed witness evolutionary changes in, say, the Neanderthal transition to the modern type. Excavations at the Israeli rock shelter of Skhul, for example, showed a population of Neanderthaloids trending in the modern direction: reduced dentition, a more vertical forehead, and the first sign of a genuine chin—all showing the gradual process of evolution toward true man. There are many such intermediate forms, all proving evolution.

A: For the moment I'll skip over Neanderthal's ineligibility as our ancestor, but only consider this: If species are *forever* changing by small degrees to become some other species, why do we have today only clearly defined ones?

POE: Remember selection? Survival of the fittest? The unmodified ones, or half-baked ones, if you will, simply die out. That's why.

A: Uh, let's get our terms straight, OK? What I am calling mixing of races (say, of Neanderthal and mod), you are calling evolution, that is, continuous change or transmutation or speciation, which you say are proven by transitional forms, like those intermediates at Skhul. You call it continuity. But all I see in the animal world is discontinuity, which is to say, deep divisions in the order of nature, a natural gulf between species that cannot be spanned. Why are there no transitional fossils between the major phyla? Even Darwin's most loyal devotees have admitted that his attempts to come to grips with nature's discontinuities were muddled and incoherent.[5]

POE: Concerning this gulf or deep division or discontinuity that you mention, well, the intermediates are missing simply because intervals of speciation may be quite brief. We are lucky to find the ones we have. Still and all, there *are* cases, such as the walking whales: You can still see vestigial legs and pelvis bones in today's whales. One of the best-documented examples of large-scale evolutionary change is *Pakicetus,* the 47-million-year-old ancestors of whales who were walking creatures. *Pakicetus*'s stubby forelimbs helped it steer through water, although its spine indicates it swam, like whales, by moving its lower back up and down. *Pakicetus,* discovered in 1994 in Pakistan, had hind legs, apparently something like a seal. Its skull, moreover, had whalelike morphology, altogether suggesting an amphibious stage between land and sea; they walked and swam.

A: Who says those flippers were ever any larger? Who says anything is vestigial? Today there are some snakelike lizards who have tiny, perfectly useless legs. Is that vestigial? Those tiny "feet" on whales, are you sure they are shrunken hind limbs?

POE: Sure, even snakes lost their legs

A: I doubt that!

POE: Well, take *Basilosaurus,* then, another ancestor of the whale, as Darwin thought, with semiaquatic otters or sea hogs as the in-between types. The hind limb of *Basilosaurus* has been reconstructed from fragments. It seems that over time, these creatures stayed more and more in the water until—

A: Please. Just picture it: If you convert a land quadruped into a whale while it is still on land, this imagined transitional beast could not use his hind legs and would be obliged to keep them permanently stretched out backward and drag himself about using his forelegs. I ask you, how could whales possibly have evolved from terrestrial animals by small genetic changes, if the advantage that is involved actually requires the *full development* of those features? A land mammal in process of becoming a whale would fall between two stools—fitted neither for life on land or sea.

POE: Clumsy perhaps, lumbering about a bit like a sea lion, but it could still get around.

A: Put it this way: The whale's skin, musculature, lungs, nose, and hemoglobin are so fitted to its seaborne life and so different from terrestrials, they could not conceivably have evolved from landlubbers by any pileup of genetic changes.

POE: It was a question of diet, you see, of survival; their ancestors, these four-legged mammals, turned to the sea for its resources, adapting to hunt in the oceans. Having been scavengers living near the sea, about 55 mya, they first ate dead fish along the shore, then began to chase prey in the shallows, wading deeper and deeper and—

A: Say, that *Basilosaurus* looks more like an eel. But look at the whale, its shape is that of a fish. Your theory rests on rampant guesswork: naming the ancestor of the whale either a deer, a pig, a wolf, or a bear!! But the recipe for whales is not contained in the genetic makeup of pigs, wolves, or bears.

POE: It remembers being a landlubber, though; when we see a beached whale, we are witnessing its residual instinct to return to land. I should also mention fossil types that phyletically link reptiles to birds, *Archaeopteryx,* for example—a reptilian protobird with *teeth.*

A: Last time I checked, *Archaeopteryx* was just an early bird about the size of a crow, skeletally like a swan.

POE: The resemblance between birds and predatory dinosaurs is undeniable. The theropod *Ornitholestes,* for example, had very birdlike feet. Just look at the morphology—it is halfway between a reptile and a bird: hips, pelvis, and legs of dinos are quite similar to birds. Anyone can see that such reptiles evolved to the avians.

A: Well, then, you might as well call the penguin an intermediate between a bird and a fish, if you are calling *Archaeopteryx* a phylogenetic link between dinosaur and bird. But birds already lived in the age of reptiles, like that fossil found in Colorado in 1977—a true bird that lived *at the same time* as *Archaeopteryx*. The latter, of course, had excellent wings, was a genuine bird with modern feathers—

POE: —but a long bony trailing tail just like a reptile and reptilian features of the skull and pelvis.

A: Even today some birds have bony tails. Besides, that Colorado bird

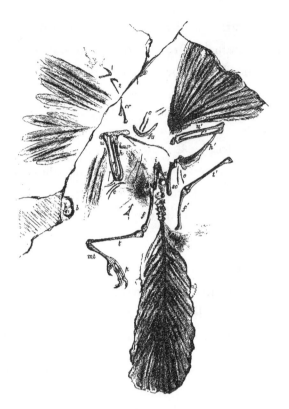

Figure 4.3. This drawing represents the first Archaeopteryx *found in 1861.*

had the telltale "wishbone" like all birds, and there are other primitive birds in addition to *Archaeopteryx* that had teeth. Even *today's* newborn birds have teeth to break the shell. In Chile they found a huge prehistoric seabird whose beak was full of spiky protrusions of bone. And why do we get fossils of modern birds in the *same rocks* as *Archaeopteryx*? Indeed, birds are found in a horizon below it. Professor Ernst Mayr said: "The particular bipedal dinosaurs that are most birdlike occurred . . . some 70 to 100 mya, while *Archaeopteryx,* the oldest known fossil bird, lived 145 mya . . . no birdlike dinosaurs are known from that period. . . . It is quite inconceivable how they could have possibly shifted to flight." In fact, bird and bat wings, when they appear in the record, are already developed, with no evidence of metamorphosing from any anterior type. Darwin himself brooded that his theory would fall to pieces if serial modifications could not be demonstrated for such complex structures as wings.

> *No amount of variation within . . . the [reptilian]*
> *Crocodilia would allow for the origin of the birds. For in*
> *the latter we find a number of innovations of which there*
> *are no trace in any crocodile, and if there was, then it*
> *would not be a crocodile. . . . A transitional stage between*
> *[the feather] and the reptilian scale is hardly imaginable.*
>
> SOREN LOVTRUP, *DARWINISM:*
> *THE REFUTATION OF A MYTH*

> *The great reptilian life . . . had no descendants.*
>
> JAMES CHURCHWARD, *THE SECOND BOOK*
> *OF THE COSMIC FORCES OF MU*

POE: Now there *is* evidence that birds evolved from dinos: flying pterosaurs in China about the size of a falcon, in the genus respectfully named *Darwinopterus, must* be transitional. Also the theropod *Sinosauropteryx* was indeed a feathered dinosaur.

A: Show me a fossil with scales developing the properties of fur or feathers.

POE: Several Cretaceous reptiles in China had incipient feathery fuzz; many of the coelurosaurs were feathered.

A: That Chinese dinosaur—they say it had protofeathers, but from the description (hairy filaments, shaggy bristles), it sounds more like a rough coat of fur. After all, dinos had a four-chambered heart, like mammals, and they may have been warm blooded.

POE: Well, how about *Mononykus,* then, the bird dinosaur with teeth and a long tail and a keeled breastbone; its clawed forearms substitute for wings. In Patagonia, paleontologists have found perhaps the most birdlike dino of all, *Unenlagia comahuensis*. All right, it did not fly, but it did hold its forelimbs in a winglike manner. Prebirds probably evolved flight from tree-climbing reptiles who slide down to the ground. Even some mammals glide, membranes stretched between their limbs, probably showing us how bats got their start. We can also see a continuum in very small animals that tend to float gently in the air. Even today there are creatures that beautifully illustrate every stage of evolutionary continuum, say, the flying squirrel, or frogs with big webs between their toes that glide, or tree snakes with flattened bodies that catch the air, or lizards with flaps along their bodies—

A: Did a reptile lay an egg and a bird hatch from it? Who is to say those frogs ever sprouted wings or your flapped lizards ever took wing? The pectoral girdle and hind legs of the theropod dinos are way too small and weak to have served as a wing that could lift the creature to the sky. No factors are known that could have caused a sudden drastic growth of those extremities.

> *It is particularly remarkable that no forms with the wings*
> *at an intermediate stage of development have been found.*
> FRED HOYLE AND N. C. WICKRAMASINGHE,
> *EVOLUTION FROM SPACE*

Figure 4.4. Reptile egg hatches bird.
Cartoon by Marvin E. Herring.

PROMISSORY MATERIALISM

*For all these creatures the mystery is the same: why are
there no transitional fossils leading up to them?*

FRANCIS HITCHING, *THE NECK OF THE GIRAFFE*

A: Neither can anyone explain the hominid gap between three and
two million years ago, openly called the black hole of evolution.

POE: It is only a matter of time before we recover specimens to fill
that gap. Literally thousands of transitional forms are discovered every
year. Are you familiar with Canada's recently discovered Arctic fish of
the Devonian, the tiktaalik, which had finlike limbs and a neck structure? Tiktaalik very nicely documents the evolution of limbs: this flat-
headed fish lived in the shallows and showed rudimentary arms and
joints, clearly transitional. It fills a significant gap—between swimming
fish and walking animals. *Lepidosiren* is another outstanding transitional between fish and amphibian.

A: But with no sign of a protolimb.

POE: *Eusthenopteron* had a budding limb.

A: For the life of me, I cannot figure why fish would want to leave the
water.

POE: A lot of reasons—drought, predators, oxygen.

A: None of them could walk, though.

POE: Alas, the vagaries of the fossil record leave the story incomplete—
precisely because natural selection eliminated those imperfect forms!
Living organisms, moreover, rarely die in circumstances amenable to
fossilization. We simply have not yet found the more advanced ones
who could walk.

A: Contrary to your promissory materialism, paleontologist George

Gaylord Simpson said the record is unmanageably rich, and still, your intermediates are missing in action.

POE: Owing to tiny populations or too short a duration for fossils to have formed. Keep in mind that geologically brief intervals of speciation are likely to escape detection—and preservation—which for sedimentary beds is very episodic. You'd be surprised by how much was destroyed by glaciers, too.

A: Yes, I would. As far as I can tell, discontinuity between animal types is the rule, ubiquitous throughout the living kingdom; and even if perfect intermediates (continuities) were found, it would not prove the evolutionary transitional model, which, we both know, was rejected by the leading biologists of the nineteenth century—Georges Cuvier and Louis Agassiz, Francois Jules Pictet and Heinrich Georg Bronn, Richard Owen, William Harvey the botanist, Andrew Murray, the entomologist, and even John Henslow, Darwin's mentor. Neither did Faraday and Maxwell, the most illustrious physicists of that day, buy into Darwinism. They simply saw no evidence for a sequential order in nature.

POE: Au contraire, we see a nice sequence in the primate order itself, with the beginning of communication symbols and speech in the great apes.

A: Just because animals have certain communication faculties does not mean they are precursors to human speech. Have you ever found an actual sequence from the African apes to man? Of course not.

POE: Well, dry conditions in Africa and acid soils of the forest, as you know, are not conducive to fossilization, so the best we can hope for is that more fossils will be found over the next few years, which will fill the present gaps and demonstrate a true, orderly, step-by-step sequence. Only when sufficiently abundant material is available for comparative study will the question be decided.

A: Which was OK to say in Darwin's time, but can we get away with it today? Let's face it, we never hit pay dirt and we never will. Some zoologists have declared that more than 80 percent of the varieties of mammals, reptiles, and amphibians have already been found—but not their transitional forms.

> *The fossils go missing in all the important places.*
> FRANCIS HITCHING, *THE NECK OF THE GIRAFFE*

POE: We simply have not yet assembled the record in enough detail. Many fossil-bearing strata lay on plates that have been subducted by tectonics. There are certain cases, too, where daughter species could have evolved elsewhere, or were very thin on the ground.

A: That's thin, all right. Even Darwin worried that the record would not support his theory: "Why is not every geological formation . . . full of such intermediate links?" Remember, his most formidable opponents were not clergymen, but fossil experts. Today there are numerous animal behaviorists who cannot abide step-by-step evolution, if only because the clumsy intermediate stages could not give the animal any evolutionary "advantage." Behavioral systems, it has so often been pointed out, are useful for an animal *only at their full level of development.* Previous steps would not have furnished any benefit at all—more likely a hindrance. Try to picture an elephant whose trunk does not reach the ground.

POE: Nevertheless, there are many useful gradations: animals with half a wing, a quarter of a wing, and so forth and what about the ear? Can't you imagine sensitive skin detectors being transformed into protoears in a step-by-step process? Every intermediate stage must have assisted survival to some degree.

A: That puts me in mind of Dr. Stephen Jay Gould's comment on the dung beetle, which, looking so much like dung, avoids predators. Did this evolve gradually? Gould's reply: "Can there be any advantage in looking 5 percent like a turd?" Ultimately the question is: Do your half wings and so forth really represent *intermediate* forms? This is sheer supposition. My guess is that Darwin, were he alive today, would most likely reject the theory of evolution on this basis: The fossil record today should be brimming with invertebrates possessing partially developed backbones, fish with little legs, reptiles with primitive wings.

> *Naturalists are chasing a phantom in their search for some*
> *material gradation among created beings.*
>
> Louis Agassiz, *Methods of Study in Natural History*

> *It might have been expected . . . that in those cases where*
> *the geological record is more or less complete . . . we would*
> *find closely graduated varieties of species. . . . Yet we do*
> *not find such a graduated series.*
>
> Gertrude Himmelfarb, *Darwin and the*
> *Darwinian Revolution*

POE: Ah, but you *do* find a state bridging aquatic invertebrates and vertebrates, in the hagfish, for example, which is capable of taking in food like an invertebrate. All right, I know it's the *human* record that interests you: so think of Laetoli Man—a transitional hominid somewhere between archaic and modern anatomy. Laetoli apparently had no divergence of big toe in those famous footprints, making it clearly transitional—

A: Wait a second. Transitional? How could they be transitional if they were *contemporaries* with their "ancestors," the australopiths? These Laetoli features are not modifications at all, only the result of blends—hybrids. That shortness of Laetoli stature (they stood only 130 cm high) came from the little people, Ihin genes, same with their modern foot.

POE: Well, they are a bit too human to belong to *Australopithecus afarensis* (although some say they do fit Lucy). There are many opinions; no one can be sure. But if I may jump ahead in time, the considerable anatomical variability among Neanderthals indicates evolutionary progression—which is to say, transitional forms evolving to the modern type.

A: To vary and evolve are two entirely different things. Species "do not evolve," said Pierre-Paul Grasse in *Traite de Zoologie,* who, I might add, also argued that selection acts to *conserve* the genotype, rather than transform it or modify it or transmute it. Variation is not at all a sign of evolutionary progress, quite the contrary, variation does not *allow* for evolution beyond the species limit. Let's go back to those Israeli specimens at Skhul, OK? Now the Neanderthal people normally look pretty much alike, right? But at the Mt. Carmel caves, there is tremendous variation, a sign of race mixing. Not evolution. Those different races overlapped in time and space; they "exchanged genes," as it is politely said.

POE: Or they simply represent a highly variable population.

A: Variable? No, not a variable population, but *more than one race or strain.*

HIGGLEDY-PIGGLEDY

POE: Honestly, I find the discussion of speciation a bit troublesome. The only good hominid evidence is Neanderthal's gradual separation from other populations, the process showing intermediate forms.

A: Or showing exogamous mating, which produced the chaos of races in the Stone Age, none of which are intermediate types, and many of which are indeed anachronisms, proving evolution impossible, because reversals don't really happen. But retrobreeding does. Just think of all the embarrassing reversals in the record, reversals in skull thickness and brow size. Devolution? I don't think so. Only mixing with lower types: back breeding. Pure and simple.

Neanderthal, who was the only hominid known in Darwin's time, looked like a good candidate for human evolution (as Professors Schaaffhausen, Hrdlicka, Weidenreich, Coon, and Brace believed). But Neanderthal man was not only too recent a creature to have been our all-purpose ancestor, he was very much the *contemporary*—not the antecedent—of the modern Cro-Magnon people in China, Denmark, Portugal, and Israel. What's more, Neanderthal mated with these Cro-Magnon neighbors, producing hybrid children of every imaginable blend.

Not our antecessor after all, Neanderthal man was then consigned to the dustbin of evolution—what else could be done with him?—and labeled an irrelevant side branch, dead end, extinct country cousin, whatever. He was not a forerunner of modern man, though it was still argued (to keep evolution alive) that the two share a "common ancestor." But even this face-saving strategy, making the two types divergent offshoots of a shared parent, hardly solves the riddle. If Neanderthal and mods merely branched off from the same stem, how did the former get so "uncouth and repellent" and the latter so elegant and worldly—and in no time at all! Scientists, after all, say it takes up to five million years to speciate,* in which case we must admit there was not nearly enough time either for Au to turn into *H. erectus* or *H.erectus* to turn into Neanderthal—or for Neanderthal to turn into anything else. Even the "cousin" idea for Neanderthal took a beating at stratified sites where diggers found archaic *H. sapiens* at levels *below* Neanderthal—as in France, Israel, South Africa, Borneo, Australia.† A similar out sequence was also observed at Steinheim, Swanscombe, Ehringsdorf, Krapina, and other sites where the more gracile Neanderthal actually preceded the more extreme, chimpanzee-like one.

---

*Goodman, *Genesis,* 198.

†Johanson and Edey, *Lucy,* 349.

POE: We do find a general sequence of improvement in Africa from *Au. afarensis* to *H. habilis* to *H. erectus* to *H. sapiens* to *H. sapiens sapiens*.

A: *Au. afarensis?* Even Johanson, who figured Lucy was ancestral to later australopiths, admits that these descendants actually "became more and more exaggerated . . . [with] bigger back teeth, a dish-faced profile, a very large jaw."[6] But consider this: Even when we do find a reasonable progressive sequence, it is not at all clear whether one form *led* *to* the more advanced one. Has any form ever been found that is transitional between the quadrupedal primates and the bipedal ones? No; in every case the divisions between the classes, such as reptiles and mammals, are absolutely clear, with nothing indicating one type of organism converting gradually into another.

POE: Oh, I can think of many instances of transitional forms and behaviors, many intermediate lineages—between, say, the four-toed extinct horse and the single-toed modern one. Or even the chimpanzee, our closest relative among the primates, who spends a lot of time walking upright. Gibbons also walk on their hind legs.

A: Big deal. Birds are bipedal, too, for Pete's sake. So were dinosaurs and the cursorial lizards who ran about on two feet. It's like Dr. John B. Newbrough's commentary on evolution in 1882, the year of Darwin's death:

> He chased the origin of man a little further back, and there left him. He failed utterly to grapple with the cause of different species. No real connecting link has ever been found between fishes and amphibians, amphibians and reptiles, reptiles and mammals, or monkeys and men. Darwin changed the word "creation" into "evolution" and there left it. The origin of life he left where it had always been. The doctrine of one species of animal being changed into another is easily squelched.

Figure 4.5. Dr. John B. Newbrough.

Protests, upon release of Darwin's *Origins*, included England's David Livingstone, Thomas Carlyle, John Stuart Mill, and America's John B. Newbrough. Mathematician and physicist Lord William Thomson Kelvin disbelieved Darwin's theory to the end of his life. Charles Lyell, a close associate of Darwin, also held out, not fully convinced of mutability or descent from "brutes": "grand speculation" he called it (in *Life, Letters and Journals, II*). Sir John Herschel dismissed Darwin's natural selection (which works at random) as the "law of higgledy-piggledy."* Louis Agassiz, the founder of American paleontology, who led the study of natural history in nineteenth-century America, viewed all species as created by the Almighty Intellect, ruling out the possibility of recent species descending from earlier ones. Richard Owen, biologist, paleontologist, and Britain's most influential anatomist, who coined the word *dinosaur* in 1842, did not believe man evolved from a monkey. William Whewell (the celebrated philosopher of science who coined the word *scientist*), said that we have a choice: accept the doctrine of transmutation or frankly recognize "the miraculous nature of creation." He himself chose the second.

* Mayr, *What Evolution Is*, 140, 199.

THE ISOLATION CARD

A: All mammals share a number of specific, defining, and diagnostic characteristics—hair or fur, mammaries, a diaphragm; no other class possesses these. All mammals have characteristics that are absolutely unique to mammals. All felines have characteristics that are absolutely unique to felines. All birds have characteristics that are absolutely unique to birds. The fishes of the Paleozoic Age are in no respect the ancestors of the reptiles of the Secondary Age, nor has it ever been shown that man "descends" from lower primates.

POE: But we do see microevolution, new species arising from preexisting ones, involving a gradual accumulation of small genetic changes. This was Darwin's "insensibly fine gradations."

A: Yes, the success of his theory is limited to microevolution—minutiae, like color changes in moths. But extrapolating from micro to macro, as did Darwin, is a hearty jump. If there is progression, why do breeders often encounter reversion to original traits? Back to square one.

POE: That's artificial breeding; but in nature, the tendency to random variation was seen by Darwin as having unlimited potential, given enough generations, with no significant tendency to go back.

> *Varieties have strict limits . . . beyond which each species*
> *cannot pass . . . in any number of generations.*
> Charles Lyell, *Principles of Geology*

A: Are there no limits? Darwin's critical error was assuming there were no limits to variations; bears, he surmised, could, given enough time, turn into whales. But change does not go on indefinitely. The species barrier, in the words of Arthur Koestler, entails "a restricted mutation spectrum."[7]

Even Ernst Mayr, one of the leading twentieth-century voices of biological evolution, had to warn his colleagues that "every genotype seems to have limits to its capacity for change." This he called genetic homeostasis, effectively a natural barrier beyond which selective breeding cannot pass. Indeed, there are "severe limits . . . No bird can ever evolve into a mammal, nor a beetle into a butterfly . . . [because] a given species can tolerate only a limited amount of variation."[8]

POE: Given nature's tendency to continued progression, varieties move in very slight degrees, farther and farther from the original type, and there appears no reason to assign any definite limits—

A: To modifications, perhaps, but what about transmutation into a totally different animal? I don't think so. This is why breeders say, if you get too far from the essential sheep, things break down, and eventually sheepishness reasserts itself. Beyond a certain point, further change is impossible.

POE: But in nature, over millions of years . . .

A: You get *horizontal* changes. But *macro*evolution (vertical), no. How will your advocates of continuity explain away the gaps, the missing hundreds of types shading imperceptibly one into the next? Paleontologists openly admit that the transitional forms required by Darwinism have never been found. Here lies cognitive dissonance. Split mind. Sustaining the paradigm—even though it has been checkmated.

POE: It is only because the process is poorly understood. One must take into account small populations and geological isolation—two indispensable features of evolution. It was isolation that led to morphological change in *Drosophila* (Hawaiian fruit flies) and also in the Galapagos finches. It is precisely such small isolated populations that give us the mechanism of change, because in large central populations, the gene flow (of any upgraded mutation) would get lost, diluted, swamped. But if that population is small and remote, the new trait can take hold. These peripheral isolates are the crux of speciation. This is how it happens. After a long period of isolation, the group diverges so far from the ancestral type that it becomes a new species, particularly when selection is strong enough to maintain genetic differences on the two sides of the mountain, desert, canyon, river, whatever. A demographic "bottleneck" sees development of new alleles under long periods of separation: the rise of a new population from a few individuals. The bottleneck is then followed by population growth.

A: As I understand it, such small isolated groups tend to *die out*. Moreover, such long-term inbreeding as you describe is not very likely to give any evolutionary advance.

POE: Let me offer Neanderthal as an example of peripheral development during isolation, which can then explain the coarse features of classic Neanderthal: arrival of the Wurm glaciation drove them into regions where they adapted to more Arctic conditions. Neanderthals were stuck in Europe during this long glacial period, separated from contact with others until the Upper Paleolithic. And this resulted in their unique, rugged, characteristics: bulbous nose, hatchet face, and so on. As a further example, arid conditions in Africa (brought on by an ice age beginning 186,000 years ago) may have forced humans into isolated pockets that contain enough water for survival. Then, with the return of lusher conditions 120,000 years ago, *H. sapiens* finally appeared—explaining the frustrating gap between archaic robust *H. sapiens* and more modern ones. Genetically isolated, the inhabitants would have developed the distinctive traits we view as modern—good chin, high forehead, and so forth.

A: How is it, then, that animal species often arise in places where isolation is impossible, as in certain lake systems. I dare say, the more isolating, the more they stay the same—rather than evolve. Take the long-isolated Australian Aborigines as an example—a people remarkably uniform in physical traits. I'd also like to mention that the Paleolithic landscape offers little support to any notion of separation, showing few signs of these cul-de-sacs. What can you point to on the map to prove the isolation mechanism? Such populations, in France, for example, lived close together in small areas, without any signs of a geographical barrier. In fact, I think the opposite is the case: plenty of crowding, overlap, and interbreeding.

POE: But you can't get speciation without isolation. Without it, novelties would simply be swamped. The necessary variations would be lost in crossbreeding—unless preserved by isolation.

A: Your system of isolation seems to assume plenty of Lebensraum and meager population levels—which I believe is deeply flawed. Your reasoning probably also contradicts evolution's fierce competition for resources (i.e., crowding).

POE: Only consider the Andes barrier, which clearly produced different animal species on the mountain's eastern and western slopes; ditto on the two sides of the Amazon river. And look at hobbit: those folks on that island (Flores) were tiny—obviously subject to isolation and shrinking. Island dwarfing. The phenomenon is well known in the animal kingdom.

A: But not in the primates; even Peter Brown—of hobbit fame—owns that *Homo floresiensis* might well have arrived on the island *already* small bodied. So much for island dwarfing. Other Negrito groups have been isolated for tens of thousands of years and have not changed one jot or tittle or become a different race. No matter how long the pygmies and Polynesians live isolated from one another, in separate geographic locales, they are *still the same species—Homo sapiens*! And what about the opposite case, where *sudden* big changes are in evidence? How does that work with Darwin's "insensibly fine gradations" over immense spans of time?

LEAPS AND BOUNDS

POE: Yes, in contrast to slow and gradual evolution, we know there is saltation, which is to say, abrupt jumps or rapid branching of species.

Lamarck thought the evolution of a species could occur within a few generations or even one. But he was incorrect.

A: So at least we can agree that the fossil record does show the sudden appearance of life-forms—such as the true mammals in the Mesozoic. Baron Georges Cuvier, the great French paleontologist, while working in the deposits of the Paris Basin, saw *sudden* changes in geology and fauna; organisms would appear suddenly, remain unchanged for long periods, then disappear in a flash. Even dinosaur and mastodon appeared suddenly. Evolution? I don't think so.

POE: Genetic recombination is one source of quick work, especially when large populations interact with much smaller ones. Plus DNA research now allows for quantum jumps in the record. But when paleontological science says sudden, you must realize the term is relative. It may actually mean thousands if not tens of thousands of years, even hundreds of thousands.

> *Genetic recombination can give rise to variations that are within the range for each species: a finch with a beak a little bigger than before, or a cow that yields more milk. . . . What it does not mean is that genetic variation is capable of explaining . . . entirely novel characteristics. It does not explain the appearance of a wing where before there was only an arm.*
>
> RICHARD MILTON,
> *SHATTERING THE MYTHS OF DARWINISM*

A: Why then were no ancestors ever found for the creatures of the famous Cambrian explosion 500 million years ago? The strata under the Cambrian are almost empty of animal fossils; we just can't find the forerunners of starfish, octopus, sea urchin, clams, or snails, only bacteria, worms, and primitive organisms like algae. The story is the same for flowering plants and chordates (the precursor of vertebrates), which appeared suddenly, not to mention sudden new species at the Triassic-Jurassic boundary.

POE: Perhaps the mutation rate was upped by something like cosmic rays or a vast change in world ecology. But as for the absence of pre-Cambrian fossils, that is because they were extremely tiny and soft bodied, no skeletons.

A: All right, but what about such important and abrupt *human* milestones as *Homo erectus,* Cro-Magnon man, and Neanderthal? Didn't *H. erectus* appear suddenly because hybridized? Isn't it the same for Neanderthal? Sudden, because a hybrid? Keith, who said Europe's Neanderthals appeared suddenly, also saw in Australia a "sudden transition from one type of mankind to another," not due to any transformation, but because a new type of man simply appeared.

> *Homo erectus seems to have appeared almost abruptly.*
> WILLIAM HOWELLS, *GETTING HERE*

And why the quickening of pace as the stages of evolution proceed? Does it make sense that simple species took millions of years to evolve but *H. sapiens*, the most advanced, managed in only a few thousand years? Doesn't that bother you?

POE: No, we must only suppose that in some cases evolution has proceeded at a faster pace for reasons yet to be discovered.

A: Well, what about the rapid jump from australopith's tiny brain (500 cc) to *H. erectus*'s 800 to 1,100 cc? This is evolutionarily inconceivable.

POE: *Homo habilis* is the missing link here; he had a distinctly bigger brain than *Australopithecus*. But this does not necessarily mean it happened gradually. Evolution in the *Homo* lineage may not have been gradual at all. Within a relatively brief period, man acquired the modern brain, pointing to surprisingly rapid development.

A: Sir Arthur Keith asked: How could the extraordinarily complex human brain have expanded in the relatively brief course of the Pleistocene? How, he asked, could the brain of *Pithecanthropus* (*Homo erectus*) "have evolved into the modern human form? I cannot credit such a rapid rate of evolution."[9] Bottom line: the way to account for this is crossbreeding with a more advanced type; after all, there is no increase in *H. erectus* cranial capacity (after the initial 50 percent increase from Au) during the next half million years: it stayed around 1,000 cc, until they were upgraded by crossbreeding with the mods in their midst.

POE: Have you seen the reports on *Au. sedipa*? These early hominids cast serious doubt on the long-standing belief that brains evolved gradually; rather, *Au. sedipa*'s brain must have increased in size—or rather shape—relatively quickly. That happened again, circa 600,000 years ago, another spurt in brain growth, a "punctuational" event. Think also of the breakthrough in Africa about 50,000 years ago: some fortuitous mutation in the genes spurred cultural advances, some radical reorganization, a neurological change that allowed man to develop culture, sophisticated behavior—some sort of intellectually advantageous mutation promoting the fully modern mind.

A: Genes didn't make the eighteenth-century enlightenment any more than they made the nineteenth-century Industrial Revolution or the twentieth-century age of electronics. There is nothing biological to it.

POE: Well, it could have lain fallow—*something* must have occurred in at least one human lineage, dovetailing nicely with the out-of-Africa phenomenon.

A: There is no trace whatsoever in any fossils of such neurological development. And why should man, the most complex and advanced of living beings, have evolved in the *least* amount of time? Why did *H. sapiens* develop so rapidly when major advances in simple plant life have taken millions of years?

POE: We learned from snails that evolution needn't operate at a "snail's pace." We find leaps in body plan among these mollusks; reared in petri dishes and exposed to platinum, many developed without external shells. This simulates genetic mutations that turned off shell production in ancient snails, paving the way for slugs. It's not always small incremental steps, you see. This must have happened many times in evolution, which is capable of proceeding by rapid episodes of change, after long periods of stasis. And the change period is so brief that intermediate forms are rarely preserved. The very restricted gene flow in an isolated population can cause rapid speciation. Maybe this is how the first real *Homo* came about.

A: I doubt it: the *H. habilis* to *H. erectus* transition involves too extreme a change, especially in size, too great an evolutionary leap. The very small body size of OH 62 (*H. habilis*) versus the near giantism of some *H. erectus* truly challenges evolution.*

> *The leopard doesn't change its spots.*
>
> PROVERB

AN ABOMINABLE MYSTERY

POE: In such instances, we could be looking at mutation rates much accelerated by increased environmental radiation. Ionizing radiation, however small, may hasten the evolutionary process. Radiation has produced giants, at least in the plant world. There is also the possibility of relatively rapid progress triggered by strong competition between different populations under severe environmental pressures. This is called punctuated equilibrium by Gould and Eldredge—punk eek—a burst of evolution. George Gaylord Simpson, a leading vertebrate paleontologist, called it quantum evolution. Same thing.

---

*It might be possible to explain their size and strength by hybrid vigor, as in Ngandong, OH9, Turkana, Weidenreich's "goliaths," and von Koenigswald's *Meganthropus*; see also discussion of cannibalistic giants in chapter 11.

A: Punk eek was supported by studies of 500 myr *trilobites* (primitive wood lice) and snails! I'm talking about human beings here. Not lice.

POE: Yet Gould thought *most* evolutionary change occurs quickly, an abrupt splitting from ancestral stocks.

A: I don't think punk eek explains anything. It's just a way of rephrasing the embarrassing gaps—which only hybridization (in my opinion) can resolve. And please explain to me why from 300,000 (the supposed end of *H. erectus* time) to 100,000 BP (beginning of mod man), no *H. erectus* fossils appear in the record—just where we should find them "transitioning." Where are your intermediates? Where did *H. sapiens* (Cro-Magnon) *come from* 50,000 years ago? Not from *H. erectus*—that's too great a jump—and not from Neanderthal either. He just came out of nowhere. The *H. sapiens* explosion is so "dazzling," thought Richard Leakey, that "we might be blind to the reality behind it."[10] Indeed, culture, tools, industries, art, religion—all appearing abruptly on the scene! Sewing needles, fish hooks, lamps, rope, homes, figurines, musical instruments, engraving, decorations, jewelry, trade—all without any sign of a precursor. I'm talking about the Aurignacian horizon that appeared in Italy, France, and Spain (El Castillo), about 39,000 years ago. Are you telling me that while it took 100,000 years to produce the chinless, thick-necked, grunting Neanderthal, it only took a short time to come up with elegant Cro-Magnon man? Whence came our sharp chin, vaulted forehead, thin skull, smaller teeth, jaw, and eye socket? Did *H. erectus* (or anyone else) magically turn into *H. sapiens* after hundreds of thousands of years of stasis? It was the same with Neanderthal and with *Au. afarensis*—basically unchanged from first to last.

POE: When homeostasis figures in the record, it reflects a very low rate of mutation, combined with rather minor selection pressures—yes, like *Au. afarensis* who survived relatively unchanged for 900,000 years. Genetic homeostasis favors the steady state.

A: That's circular. Besides, it's the same with the pongids: modern gorillas, orangs, and chimpanzees spring out of nowhere. They have no yesterday. Speciation—and along with it, Darwinism—is dead in the water.

POE: Ah, but rapid branching can account for speciation. The smaller, more inbred the group, the quicker mutations become fixed, speeding up the evolutionary process. Mendelian inheritance shows the sudden appearance of "sports" and monsters; anything can happen, really.

A: Such as separate creation? By a creator? Whereby the fossil record would and indeed does show creatures appearing, fully formed, as do all thirty-two mammal orders as well as invertebrates (of well-differentiated types), which appeared abruptly in the Cambrian.

POE: Let's leave a creator out of this for now, all right?

A: OK, but fish swim into the fossil record seemingly from nowhere, suddenly and fully formed—in a most un-Darwinian way, with only *gaps* where their ancestors are supposed to be. Darwin rightly called this an "abominable mystery"—after finding the flowering plants made to order, already much diversified when they first appeared in the Cretaceous, with nary a trace of their predecessors.

POE: Intermediates, you realize, might be hard to find due to a missing layer of sedimentation or other sporadic changes in the environment. Moreover, some sort of macromutation could account for sudden change—and gaps. The problem disappears altogether if we accept the saltation model, perhaps combined with migrations—away from those missing ancestors—or new people who came in from elsewhere. Let us not forget that the gaps usually reflect isolation—people living apart in restricted geographical areas over long-enough periods of time.

A: This saltation is megaevolution or macromutation, which is genetically improbable, some say impossible. Why not then adopt the simplest solution: crossbreeding. Or even spiritual influence capable of producing sudden change?

AUTHOR'S INTERLUDE

Be most searching, O man; for thou shalt find, in this day and generation, the legends and history of Apollo in all the divisions of the earth ... [for it was] in the time of Apollo that man in his present form was brought into being on the earth.

OAHSPE, THE LORDS' THIRD BOOK 3:10–12 AND
BOOK OF APOLLO 5:6–7

Mind and consciousness can directly influence physical processes without the intervention of material mechanisms.

CALLUM E. COOPER, *TELEPHONE CALLS FROM THE DEAD*

As I understand it, the time of Apollo, ca 18 kya, saw births "overshadowed"—and a new race was born: "the women were entranced in the sacred whirling dance and the impression on the soul shaped the unborn child."[11] It is this factor, unseen but potent, which Michael Cremo describes as "the transference of impressions from mothers to embryos."[12] This transference was dubbed psychogenesis by William Fix, who suggested that a "mother could have given birth to a being . . . governed by a spiritual agency. If hundreds of thousands or millions of spirits came down to earth together and acted in the same way at the same time, a whole generation of fledgling human beings could have been produced at once. . . . The human body can indeed be altered by mental influence."[13]

The angels of heaven ... by virtue of their presence, strove to make man wise and upright.... Upon the earth the number of such angels was millions.

OAHSPE, BOOK OF JEHOVIH 7:5

Thus was laid down the foundation for *H. sapiens sapiens*, the race called Ghans (appendix B), in the time of Apollo. The sacred dance raised the vibration, explains cosmogonist Peter Hartgens, so that the loo'is (angels of generation) could better do their job. "The loo'is have been preparing these matters . . . and shall [go] in among mortals, finding the most comely-formed men, women and young children. And when they have chosen them . . . they shall be quickened by signs and miracles."[14]

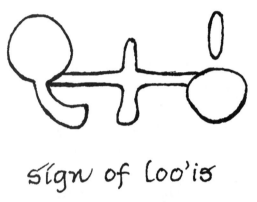

sign of loo'is

Figure 4.6. The sign of loo'is. Their office was to inspire certain men and women to wed, in such a way that their offspring would rise higher than the parents in wisdom, love, and power.

Go to the root of the matter, O My beloved. . . . Follow them! Double the number of loo'is; leave no young man alone; leave no young woman alone. Keep watch over them day and night; give them visions and dreams of Apollo. For, I am concerted to remold the forms of the earth-born. . . . And the women look on, receiving the spirit of the matter in their souls, the act of which entails on their offspring that which is desired by the Gods. . . . Unseen by mortals, the Lords stir up the whole world. In one generation, behold, a new race is born. . . . And the Lords inspired kings and queens to erect *images* [e.a.] in the temples, and the images were given a name signifying Harmony, Symmetry and Music.

OAHSPE, BOOK OF APOLLO 5:6

Images

The legions of Apollo inspired men to make images of stone and wood, so the Druks could learn from them.* It was at this time that refined figurines were introduced, to inspire, to serve as impressing models of ideal form. Images were placed in the oracle houses for the angels to come and "teach the holy doctrines."† And in the time of worship, the angels came and possessed the idols and spake therefrom, with audible voices spake before mortals.‡ Thus are Cro-Magnon's figurines recognized by anthropologists as representing "a complex range of spiritual beliefs." For man had received the commandment to remember the God of harmony, symmetry, and music, and to build images of him in all the divisions of Earth.§ And people were taught, under inspiration of angels, how to make the images, for there were few mortals sufficiently perfect to serve as models. "And the images shall have short arms and long legs." Thus do we find shortened arms on some of the Venus figures, reflecting the ideal of the well-proportioned modern races. In Japan, the Jomon *haniwa* are small figurines depicting little people (Ihins) with noble faces, just as the earliest stone figures in Europe (near Hamburg) show quite modern features.¶ Mesopotamian figurines with tapered waists stand in contrast to the no-waist, barrel-chested, long-armed tribes of Druks and barbarians.

They mold the inhabitants of the earth as clay is molded in a potter's hand. He comes to the young mother's dream, and shapes her unborn, with limbs like a racer. . . . Out of the idolatry of My Son, Apollo, I will beautify the inhabitants of the earth.

OAHSPE, BOOK OF APOLLO 5:2–11 AND 14:5–6

---

*Oahspe, The Lords' Third Book 3:11.

†Oahspe, First Book of the First Lords 2:16.

‡Oahspe, First Book of the First Lords 1:8.

§Oahspe, Book of Apollo 4:6, 5:7, 13:3.

¶Kolosimo, *Timeless Earth,* 5.

"And man depended on the angels for all things helpful to his understanding."[15] He was instructed by his unseen mentors in arts and morals, science and industry, navigation and agriculture. The angels watched over man, "teaching him oft times unknowing to himself in all good works and industry."[16] They taught them "how to make fire by striking flint stones."[17] Learning from angels, we might surmise, eliminates the need for mutations or long and slow evolution of ideas.

The Inca shamans of Peru averred that their forebears were taught building techniques by the light beings (*apus*), as did the Amazonian Tukano people, who also learned from their unseen teachers (the *abe mango*) how to make pottery and woven cloth, and how to cook food: "by the angels was he taught and guided in all things . . . knowledge of the sun and moon and stars . . . how to find the times and seasons of the earth."[18] "Apollo told us when to plant and when to reap."[19]

> *Perhaps tutelary spirits taught us to farm and gave us the*
> *domestic breeds that we needed.*
>
> BEN STOECKER,
> THE ATLANTIS CONSPIRACY

It will be said by some that there is a law of evolution whereby man riseth from a lower to a higher state as the earth groweth older. But I say unto you, there is no such law. Save but by the labor of thy Lord and thy God, through their angels, man riseth not upward, but he goeth the other way.

OAHSPE, THE LORDS' FOURTH BOOK 4:8–9

A: According to the Malaysian little people, it was the *hala'* (angel mentor) who helped early humans, teaching them rules of mating and all proper behavior, for man of himself was helpless and needed to be shown what to do. It is on record that "man was naked and not ashamed, neither knew he the sin of incest," so the angels were sent down,[20] and they went among the Druks and taught the law of incest.* The Batek say the earliest human beings married their brothers and sisters because they did not yet know that it was prohibited. Don't you think this departure from incest could change the genetic features of a people in a short time?

POE: Unfortunately there are no fossils that leave a record for that sort of thing. Even less is there any evidence whatsoever of so-called angel interference.

GOOD-SHAPED MEN

A: Well, what about the marvelous cave paintings of the Bushmen and Cro-Magnon? Several authors have developed the theme that Paleolithic cave art is mostly about communing with spirit helpers; with the famous paintings of shamans, those sites are generally regarded as shrines and mystery schools. Even today many caves, according to natives, are as "cathedrals," holy or magical conclaves. "The spirits oft appeared, teaching openly their several doctrines" in the temples—many of which were *caves*.[21] Otherworldly inspiration was the legacy of the sacred little people (Ihins) who were the first to practice spirit communion and to teach others how to receive instruction from the invisible beings. Among the Andamanese, one of the oldest tribes on Earth, a man becomes a shaman by contact with spirits, either in dream or by "dying" and coming back. (The main Andaman tribe are the Onge people. Ong means "spiritual light"; see chapter 7.)

---

*Many more intriguing passages of this sort, for the curious reader, in the Oahspe Bible: Book of Aph 13.12, The Lords' First Book 1.61, First Book of the First Lords 3.3, The Lords' Second Book 1.4, The Lords' Third Book 2.6, The Lords' Fourth Book 1.5 and 2.5, and First Book of God 5.

How can anyone believe that one species morphs into a different species? Have breeders ever turned a cat into a dog? Such changes are a scientific illusion, a falsehood so big that we, dumbfounded, believe it. Anthropologists say the brain of australopithecus did not increase in size or become more human for a million years. The *H. erectus* race changed remarkably little, without any brain increase during their entire tenure on Earth and with the same hand axes throughout; later assemblages are as crude as those in the beginning. At the stratified site of Choukoutien in China, we see no steady increase in brain capacity or shape, from the bottom to the top of the excavation: *Sinanthropus* remained the same over the entire period examined. Neither is there any evolutionary change among Neanderthals. Certainly no brain size increase among them, or *since* them for that matter. And after Cro-Magnon appeared without warning, they didn't change much afterward either. Where is the evolution in all this?

POE: Point taken, many groups were retarded in evolution, but then came the rapid development of the brain in the late Pleistocene. The cave paintings and other creative work that you allude to, appearing forty or fifty thousand years ago in Europe, only suggest there was some intellectually advantageous gene mutation around that time that spurred rapid cognitive and cultural advances.[22] If not, then we still have no adequate explanation for our big brain. Perhaps Mr. Wallace (Alfred Russel Wallace) was right, that the brain had a surprisingly rapid development. It's also possible that Dr. Tilly Edinger of Harvard was right, in that our brain has not evolved at all from early man but somehow developed at a later stage.

SO LITTLE HAS CHANGED

A: I hasten to add that this stasis we are talking about also holds for numerous animal species that looked exactly the same millions of years ago as they do today—the snapping turtle appeared in the distant past with the same form and habits it has today, as did drag-onflies, beetles, sharks, pelicans, rats, beavers, wolves, aardvarks, porcupines, and sturgeons. The tuatara, a lizardlike "living fossil" on islands off New Zealand, has shown little morphological evolution for nearly 200 million years, like the turtle, opossum, nautilus, horseshoe crab, and alligator—all virtually unchanged for millions of years.

POE: These are a bit of a time anomaly, probably because their environments have not changed, signaling an evolutionary backwater. They are persistent types, or in some cases, the rate of mutation is exceptionally low. Some of the large birds, for instance, have evolved rather slowly. Other species become stagnant, simply reach a standstill, having attained their evolutionary limit.

A: Thus accounting for the ant of 150 million years ago, which is almost the same as today, likewise, the starfish, platypus, oyster, and bacteria—unchanged for hundreds of millions of years. The cockroach hardly changed since the Carboniferous.

POE: Well, cockroaches obviously became a successful species several hundred million years ago and just stopped evolving.

A: Even Louis Leakey, undoubtedly an evolutionist, thought that man, rhino, flamingo, and so on have remained relatively unchanged since their first appearance. Ancient African deposits found by his daughter-in-law, paleoanthropologist Meave Leakey, held thousands of tracks identical to those of today's animals: antelopes, hares, giraffes, hyenas, horses, pigs.

POE: Please bear in mind that a trait may not evolve at all if it serves its purpose well—in which case you can have stasis for millions of years and then a great spurt of evolution.

A: Stasis—spurt—stasis—spurt. Pretty bipolar, if you ask me.

POE: If a species is well adapted to its ecological niche, the pressure to change is nil—they simply remain the same for a very long period of time indeed. Even the human body form has not altered appreciably in 100,000 years. Species may simply present as static in their range—a kind of genetic inertia, especially in large interbreeding populations. Conditions are not always propitious for evolutionary innovation.

A: *Not always?* Didn't Dr. Gould establish that stasis is actually the *rule?* Most species exhibit no evolutionary change during their time on Earth.

> *Stasis typified much of prehistory.*
> ROGER LEWIN, *THE ORIGIN OF MODERN HUMANS*

POE: Clearly there are some species in the fossil record that remain unchanged for millions of years, having reached their evolutionary plateau in a stable ecological niche. But then sudden appearances might follow, under acute selective pressure.

A: Right, no change, that is, until they *mixed*—starting in the very beginning with Asu/Ihin blends. Take Gona, Ethiopia, for example—with carefully manufactured tools made by *Au. garhi*. Paleontologists say this happened 2 million years after the first human ancestor began to walk upright. Until *garhi,* nothing much had changed. Crossbreeding was the "change."

A HORSE IS A HORSE IS A HORSE

POE: Neanderthals changed because they *specialized* during the isolation created by Ice Age conditions. Early Neanderthal had been of a generalized type, without the severe traits of later classic Neanderthal. Zinj was also generalized, without any of those specialized traits seen in the animals: stingers, spurs, spines, scales, talons, tusks.

A: "Generalized" strikes me as a purely circular argument invented to cover a negative meaning only—the *absence* of traits that scientists *assume* developed afterward by "selective pressures." It is meaningless otherwise. You call classic Neanderthal specialized because of his strikingly coarse features, and progressive Neanderthal generalized because most of his features are vaguely modern (except ridges and thick cranium). Richard Leakey calls *H. erectus*'s massive brow a more advanced feature, specialized, even though this so-called advance is *away* from the modern type! This is nonsense, a semantic blanket, covering anachronisms, which, taken at face value, simply force us to abandon Darwinian evolution.

POE: The species Neanderthal is a collateral development that went nowhere, not really primitive per se. Their features are simply the consequence of special physiological adaptations to northern cold. They became too specialized in this regard, then died out with the postglacial warmth, just as Solo River Man and Rhodesian Man paid the penalty of overspecialization, and went extinct. In fact, all specialized creatures are dead ends, really, like the great apes who became specialized for forest living and are now dying out.

A: I have a problem with this. Natural selection, the very pillar of evolution, entails animals specializing in various ways to meet the demands of the environment. How then can you call specialization the road to extinction? Those adaptations are supposed to be the road to *survival*! Your ad hoc mechanisms are contradicting each other! Didn't Le Gros Clark (back in 1955) establish "the astonishing *conservatism* of morphological elements," such as the 3-million-year stasis of leopards and impalas?

POE: Conservatism? My dear woman, this outdated immutability of species or conservatism was discredited more than 150 years ago.

A: Right, by Band-Aids and jargon: As soon as a feature doesn't fit with the expected model of evolutionary sequence, it is called a specialization or divergent modification or a collateral branch of the *Homo* tree. They even call Ardi a side branch! And that is our true ancestor, the first man—Asu! And since Negritos in *Europe* also came as a surprise, they were labeled either "specialized" or "pathologic" or "aberrant offshoot," like the Grimaldi Woman, whose prognathism was labeled a pathologic condition: she had suffered presumably from a phenomenon of ortho-dontics, her face hence more protruding. But prognathism is a mark of those early races that inhabited every portion of the globe and whose genes were retained in the Grimaldi type.

POE: The point is that the more generalized an animal, the more it is amenable to, and capable of, making new adaptations; they are not so narrowly conformed or restricted to a given environment as are more specialized creatures. Take mankind: Our species is peculiarly unspecialized, blessed with infinite creativity and marvelous adaptabil-ity, going all the way back to Lucy's bunch, who were great general-ists. Look at the huge amounts of climate change they survived in the face of overwhelming challenges. They had that plasticity that we label "generalized."

> *[There is] no evidence whatsoever that the earlier members of any long-continued group were more generalized in structure than the later ones.*
>
> THOMAS HENRY HUXLEY, "PALEONTOLOGY AND THE DOCTRINE OF EVOLUTION,"
> *CRITIQUES AND ADDRESSES*

A: I really wonder who is to say which animal is specialized and which is not—or which is simple and which is complex. Hindsight makes it so, but it is pure assumption that simpler forms became specialized (grew horns, stingers, and so on) by progressive morphological changes. Doesn't biochemistry tell us that even the most rudimentary forms of life are already exceedingly complex and sophisticated?

> *Fossil residues of ancient life-forms . . . do not reveal a simple beginning.*
>
> FRED HOYLE AND N. C. WICKRAMASINGHE,
> *EVOLUTION FROM SPACE*

Simple structures . . . have staggering complexity.
. . . Darwin never imagined the exquisitely profound
complexity that exists even at the most basic levels of life.
MICHAEL BEHE, *DARWIN'S BLACK BOX*

POE: Evolution from a primordial common ancestor has unfolded over an immense period of time and has followed a step-by-step sequence from primitive to advanced, complex, forms of life. Animal life, for example, followed plant life.

A: But complex plant life and animal life appeared *together* in the Cambrian. The Cambrian explosion, the biological Big Bang, saw the sudden appearance of highly organized marine fauna, the first shells appearing *fully developed*.

POE: Do you doubt that organisms progress from simpler to complex? There are thousands of examples. Evolution does not necessarily imply or require progress, just change—even though things in fact have gotten more organized and complex.

A: Fact? Or opinion? Sir Richard Owen, statesman, anatomist, and zoologist, "the British Cuvier" and headman at the British Museum in the mid-1900s, said this is not the case among reptiles. Dinosaurs, for example, were actually more complex than today's lowly reptiles.* Dinos were the apex of reptilian life, and (if you don't mind my mentioning this), according to Sir Richard, they did not evolve from some simpler reptile but had been created by God. To him, the *coexistence* of very simple and very complex life-forms *today* refutes natural selection; there was, he said, not a single proven instance of transmutation of species. If simple organisms were always "evolving" into more complex ones, why are the most numerous creatures in nature today of the simplest form?

---

*In fact, the Scottish geologist Hugh Miller argued that in each of the geological epochs, it was the higher form that *preceded* the lower.

Figure 4.7. Thomas Henry Huxley and his cartoon of "Eohomo" (dawn man) riding the dawn horse, Eohippus. By showing the graded evolution of the horse, Huxley thought he verified Darwin's predictions. Others say the early horse is the size of a fox and may not be a horse at all.

POE: We must touch on complexity theory here—complex adaptive systems, in which the individual elements interact, adapting their behavior to changing conditions. Out of the process—and over time—emerge complexity and diversity. Horse evolution, as an example, marks a path from multitoed horses (the dog-sized Eohippus) to today's single-toed horse, the species known as Hipparion providing the intermediate type, the progression moving from small to large, from simple grinding teeth to complicated cusps.

A: *Eohippus,* I am afraid, has been found in the same strata as the modern horse. You cannot say horse evolution was progressive. They grew taller, then shorter, then taller again. No single branch led from smaller to taller; some had three toes, others had just one. Paleontologists tell us there have been numerous regressions in the number of toes. Even today a modern horse with three toes is occasionally born. In any case, from multiple toes to hoof (single working toe) is merely a modification—which in no way made the animal either more or less complex. Fewer toes, longer legs, larger molars—is this complexity or just change? Besides, the so-called horse sequence uses specimens from different places!

> *Most of them [intermediates] are simply varieties . . .*
> *artificially arranged in a certain order to demonstrate*
> *Darwinism at work, and then rearranged every time a*
> *new discovery casts doubt upon the arrangement.*
>
> FRANCIS HITCHING, *THE NECK OF THE GIRAFFE*

POE: Horse evolution still is not quite worked out, but we *can* demonstrate that the seed-eating and vegetation-eating finches indeed evolved from simple insect-eating finches: after a drought, larger-beaked finches were selected, for they were able to eat tough seeds, hence survive; whereas in a rainy year, with softer seeds, the smaller beaked finches were more likely to prosper.

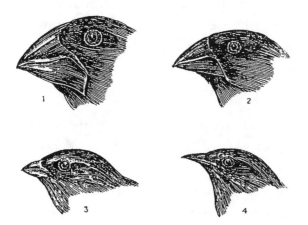

Figure 4.8. Galapagos finches, illustrated in Darwin's Journal of Researches, *show differences in beak structure. It is then argued that isolated populations derived from a single parent species, which diverged on different islands, producing new species, each adapted to a different environment. But who is to say these different beak shapes were not in the original phenotype from the beginning?*

A: The seed-eating finches are not more evolved than the insect-eating ones, or any higher on some imagined evolutionary ladder. Besides, the shape of the finches' beaks are not all that different. Neither does the concept of higher and lower apply to the Galapagos finches or perhaps anywhere else. When biologist Lynn Margulis was asked to comment on "speciation happening" among the Galapagos finches, she replied: "They saw lots of variation within a species, changes over time. But they never found any new species—never."[23]

POE: Ah, but if we observe, in other birds, that a slight increase in the wing flaps can make the difference between life and death, then we can see evolution, survival of the fittest, in action. Human societies, too, pass through stages of increasing complexity—steps in advancing evolution.

A: What has been termed steps in evolution are mere modifications, slight adjustments that seem to suit creatures to their environment, without in any way making animals more complex.

POE: So where have the more elaborately organized forms of life come from? They must have evolved from simpler forms—no alternative interpretation has ever been offered. Man is living proof of this, certainly the culminating peak of primate evolution—the last to become differentiated from the common ancestor of all.

A: Actually, ammonites were succeeded by simpler forms. When a simple form appears *earlier* than a complex one, it does not mean it is a phyletic precursor, just that the time was ripe for that sort of organism. If the fossil record shows a history of life from simple to complex, this reflects the ages of Earth—at first able to sustain only seaweeds and algae. If complexity came later, Earth was ready for it, but this does not prove one *came from the other* or from a "common ancestor."

POE: Unfortunately, our knowledge of the development of multicellular organisms is very limited, because they seldom leave a fossil.

A: The one-celled organisms that are supposed to be at the top of the evolutionary tree—why are they still around? Can you tell me why even the first fish were already highly complex vertebrates? At the lowest level of geologic strata, the fossil record consists of quite complex creatures—such as the Cambrian trilobites. Life does not start with only a few rudimentary species. It begins with many life-forms. Darwin wrote in *Origins* that if any complex organ existed that could not possibly have been formed by slight modifications over time, "my theory would absolutely break down."

POE: We now know a lot more about living forms than they did in Darwin's day. Still, there is not a single case of a complex organ that could not have been formed by numerous successive slight modifications.

A: Even though the simplest of all living things, bacterial cells, are exceedingly complex, with thousands of design pieces, far more complicated than any machine built by man. I'm not sure any living system can be labeled primitive or ancestral, simple or complex.

POE: The earliest membrane-bound cells were extremely simple. The first self-replicating organisms were not up to the complexity of today's DNA and proteins. Bacteria was probably more primitive than eukaryote cells; first came archaea, then bacteria, then eukaryotes.

> *It came as a great surprise in biology that a handful of fungi cells contain [sic] as much DNA as a man, and that the salamander contains a great deal more.*
>
> FRED HOYLE AND N. C. WICKRAMASINGHE,
> *EVOLUTION FROM SPACE*

A: Explain, then, why the skull of a mammal or a bird, as Professor Mayr pointed out, "is not nearly as complex as that of their early fish ancestors." Even the amount of DNA does not increase up the evolutionary scale. Indeed, toads have more DNA than mammals and "the organism with the most DNA . . . is the lily."[24] It is also a fact that tomatoes have 7,000 more genes than humans and rice almost twice as many as humans.

> *If evolution has taken place . . . it ought to be relatively easy to assemble not merely a handful but hundreds of species arranged in lineal descent. Schoolchildren should be able to do this on an afternoon's nature study trip to the local quarry, but even the world's foremost paleontologists have failed to do so with the whole Earth to choose from and the resources of the world's greatest universities at their disposal.*
>
> RICHARD MILTON,
> *SHATTERING THE MYTHS OF DARWINISM*

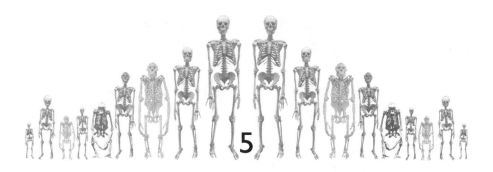

5

THE MATING GAME

Crossbreeding from Day One

As for those genealogies which you have recounted to us . . .
they are no better than the tales of children.

PLATO, TIMAEUS

The glut of inconsistencies and unsolved mysteries in the evolutionary family tree simply vanishes with an understanding of crossbreeding, one of the oldest habits of mankind. It is because man is and has
always been a mixer that we so regularly find a striking combination of
primitive and advanced traits in hominids. These mosaics are screaming
hybridization, *Homo hybridis*, *Homo cohabitensis*. But are paleontologists listening?

We are all a complete mixture.

BRYAN SYKES, THE SEVEN DAUGHTERS OF EVE

Students and scholars alike are forever struggling with the *relation-*
ship between the different fossil men. Well, that's the right word. They
had relationship, relations. Unthinkable? Not really. The hybridized

"ascent of man"—which easily replaces the theory of evolution—is the other shoe about to drop on the scientific world.

> *Man is perhaps the most promiscuous animal . . . in the*
> *matter of indiscriminate interbreeding.*
>
> EARNEST HOOTON, *UP FROM THE APE*

IS NEANDERTHAL OUR ANCESTOR?

C. Loring Brace, representing America's "pro-Neanderthal" school, and the Smithsonian's Ales Hrdlicka, seized on Israel's Mt. Carmel evidence to establish Neanderthal as forerunner of *Homo sapiens*. Other American scientists such as Franz Weidenreich and Milford Wolpoff also believed this to be so. Their European counterparts would have none of it. Neanderthal was merely some irrelevant distant cousin, for traces of mods *earlier* than Neanderthal came into existence, instantly disqualifying him as an ancestor.

Earnest Hooton also pointed out that the time line was much too brief: How could "he have changed thus rapidly into modern man . . . in so brief a space of time"? Besides, Neanderthal's brain was too large to be a phylogenetic link between low-order hominids and modern man—"a less credible miracle," Hooton thought, "than the changing of water into wine." Rather "radical race mixture [would be] . . . exactly the sort of phenomena that are shown in the skeletal series from the caves of Skuhl and Tabun in Palestine."[1] Hooton thought only the progressive Neanderthals stood a chance of evolving to moderns. (Two Neanderthal types have been discovered: a more robust specimen with a brutal continuous torus or visor called classic Neanderthal, which I call Neanderthal 2 [N#2], and a somewhat more refined, progressive type, which I designate Neanderthal 1 [N#1]. The more gracile Neanderthal 1 should, according to evolution, follow Neanderthal 2, but it actually came before.)

Wilfrid E. Le Gros Clark, for his part, thought the skeletal differences and "pattern of growth" in classic Neanderthals represented an isolated group not in the *Homo* line. Swiss workers (after making skull comparisons), concluded that Neanderthals belong to a separate species and did not give rise to living Europeans. More recently (2008) Germany-based geneticists, despite finding "no difference between Neanderthals and modern humans," proceeded to argue that "Neanderthals were a separate species."[2]

Mod traits have been found in Ethiopia *before* Neanderthals vanished from Europe and Asia. Neanderthal could not possibly have evolved into *H. sapiens*, as seen also at Dordogne, where a Cro-Magnon skeleton was found just above Neanderthal ones—a few thousand years is not enough time. Elsewhere in France mods at Combe Capelle and Mentone were again much too close in time to Neanderthals to have evolved from them: just as 35 kyr St. Cesaire Neanderthals could not have evolved in 3,000 years to 32 kyr Cro-Magnons.

The same situation applies to Grimaldi Man whose existence "in close association with the much more primitive Neanderthal form shows that *Homo neanderthalensis* could not be the ancestor of *Homo sapiens,* since both species were contemporary."[3] All told there is no real transition in Europe from Neanderthal to Cro-Magnon, the latter appearing not by evolution but quite suddenly.

Neanderthal, others argued, was too specialized to become anything else, no less modern man; and how could he be the ancestor of us all if his bones were basically confined to southern Europe (and adjacent western Asia)? Or if "fully modern men . . . lived in sub-Saharan Africa, Australia, and other areas of the world while the Neanderthals still inhabited Europe."[4] And he could not be our ancestor, if only because *H. sapiens replaced* or absorbed him in Europe rather than evolved from him (see chapter 12).

Even Thomas Henry Huxley, who would have loved to, could not see Neanderthal as an ancestor, because he was too much like us; though according to others, such as England's Chris Stringer, Neanderthal was too *different* from us (especially in skull shape) to be our ancestor. (Go figure.)

But a way out of these difficulties was found—just create a common ancestor (out of thin air). Svante Pääbo of the Max Planck research team in Leipzig, declared: "We shared a common ancestor with the Neanderthal" before diverging from them, and evolving apart. But Neanderthal is simply, very simply, a retrobred race resulting from Ihuan (Cro-Magnon) "indiscretions." He is not "pre-modern," but actually post-modern, offspring of the back-breeding Cro-Magnons. Neanderthal owes his existence, not to any separating or splitting, but to a *coming together* of Ihuan and Druk. Marcellin Boule came very close to the truth when he suggested that Neanderthal "was a degenerate species."

Let's start with Neanderthal's cohabitations and work our way back to the "affairs" of earlier times.

KISSING COUSINS: HIGHLIGHTS OF NEANDERTHAL DISCOVERIES

1829   About two years before the young naturalist Charles Darwin sets sail on HMS *Beagle,* the very first Neanderthal skull is found in a cave in Belgium, that of a child. Within twenty years, further remains of a proto-Neanderthal woman and child are uncovered in Gibraltar, followed a few years later (in 1856) by the Dusseldorf/Feldhofer find, the German Neander Valley famously lending its name to the specimen christened *Homo neanderthalensis* in 1864. The limbs of this creature were bowed, the nose large, the brow bulging. Some saw it as merely a pathological modern, otherwise no different (except for robustness) than the rest of the human race. Indeed, that is his classification today: *Homo sapiens neanderthalensis.*

1908   With the discovery and examination of France's La Chapelle-aux-Saints skull, French paleontologist Boule indignantly removes Neanderthal from the human family. The head, he

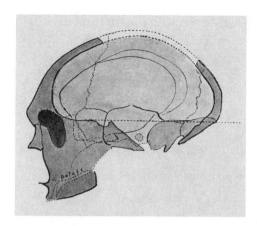
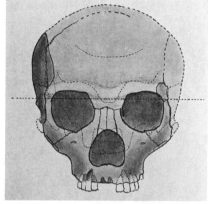

Figure 5.1. Two views of Gibraltar Man's skull.

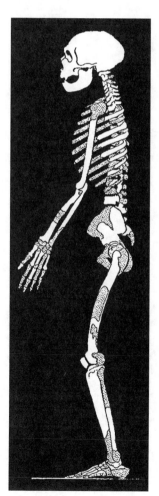
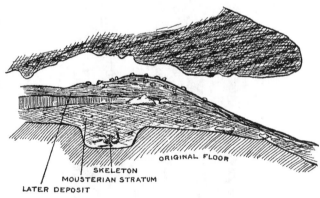

Figure 5.2. La Chapelle-aux-Saints skeleton and strata.

thought, had too much in common with the chimpanzee—its cranium long and low, its brow prodigious. The "structural inferiority" of the toothless specimen too much offended his Gallic pride to be classed with men—despite an awesome brain of 1,625 cc, well over the modern average! The obvious answer, that a dose of Cro-Magnon (Ihuan) genes supplied the large brain, was not considered. At Shanidar, Iraq, Neanderthal's cranial capacity was again well over 1,600 cc, outstripping our own, today's average somewhere around 1,400.

Figure 5.3. Marcellin Boule (1861–1942). To Boule, the idea of Neanderthal as our forerunner was "absurd."

Figure 5.4. Neanderthal reconstruction, published by Marcellin Boule.

Figure 5.5. This reconstruction, published in L'Illustration *in 1909, based on the work of Marcellin Boule certainly exaggerated the animalistic appearance of Neanderthal.*

Figure 5.6. A more civilized portrait of Neanderthal from Illustrated London News, *1910.*

1911   To Arthur Keith, there was no reason to suppose that if Neanderthal and *Homo sapiens* mated, "they would not produce fertile offspring."[5]

1931   Hooton stated simply, concerning Neanderthal and Cro-Magnon: "They surely bred."

1937   Back to France, where the Fontéchevade specimen is discovered, a Neanderthal evidently of the Mousterian period. However, the layer *beneath* held skull fragments of a *more modern* type, with lighter build and milder browridge; this at last was "the earliest Frenchman." Yet the reversed sequence toppled the smooth evolutionary paradigm of gradual *improvement* over time. (See appendix E on reversals.)

1950   Howells declares in *Mankind So Far* that "the Neanderthal brain was most positively and definitely not smaller than our own; and this is a rather bitter pill: it appears to have been perhaps a little larger."

1955   Le Gros Clark, esteemed Oxford professor of anatomy, in his book *The Antecedents of Man* tries to settle the mounting dilemma of reversals. To resolve the difficulty between Neanderthal 1 and 2, Le Gros Clark reasoned that the later Neanderthal 2 must have been isolated for a long time (due to climate conditions), which led to the differentiation. The older Neanderthal 1, mainly outside Europe, had a fairly well-developed forehead (absent in 2) and less massive jaws. The skull approximated that of *H. sapiens,* and the limb bones were also quite modern, being lighter and straighter than Neanderthal 2. In this case, Le Gros Clark astutely derived this man from an *H. sapiens* type with regressive changes in the skull, which he actually attributed to retrobreeding with a lower type.

1970s   Something called the spectrum hypotheses, introduced in this decade, admits much more interaction (sexual mixing) between different groups than we previously thought;

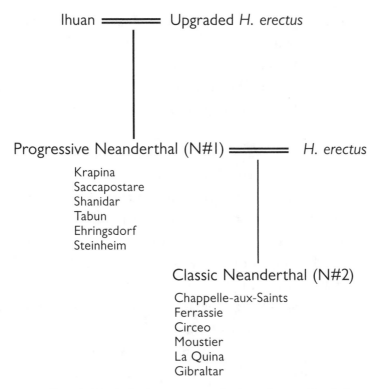

Ihuan ════════ Upgraded *H. erectus*

Progressive Neanderthal (N#1) ════════ *H. erectus*

Krapina
Saccapostare
Shanidar
Tabun
Ehringsdorf
Steinheim

Classic Neanderthal (N#2)

Chappelle-aux-Saints
Ferrassie
Circeo
Moustier
La Quina
Gibraltar

Figure 5.7. Author's conception of crosses that produced the two different kinds of Neanderthals.

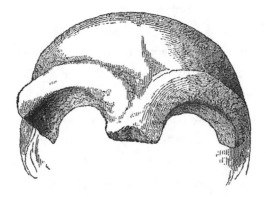

Figure 5.8. Torus. How can we call this brow "apelike" when orangutan is an ape, and it does not have a prominent torus; neither did some robust Au, such as Kromdraai, perhaps due to an admixture of AMH (Ihin) genes.

indeed, the concept anticipates today's theorists who are talking more and more of "gene exchanges," though guardedly, catching up with Louis Leakey who, decades ago, had asked outright: "Is it not possible that they are all the result of crossbreeding between *Homo sapiens* and *Homo erectus*?"[6]

1986   A twelve-year dig at Vindija Cave, Croatia, now completed, reveals that these Croat Neanderthals had sophisticated tools. Their skeletal structure was also a bit more delicate and modern than expected, which paleontologist Jacov Radovcic and others interpreted correctly as the result of interbreeding between Neanderthals and mods. Sure, many individuals had the ferocious browridge, sloping forehead, and powerful jaw of classic Neanderthal, but others had refined faces, cranial vaults, and limbs (matching similar finds in Italy). These Vindija people were dubbed "modernized Neanderthals"—a lovely euphemism for serious fraternizing. Normally Neanderthals among themselves look very much alike, though at Vindija there was considerable variation. How could this be explained? Further muddying the picture, the earliest mods at this location have numerous Neanderthal traits.

1988   "The status of Neanderthals has changed more times than the mini-skirt has gone in and out of fashion," quipped Roger Lewin in his 1988 book *In the Age of Mankind*. And now, on the slopes of Mt. Carmel, Israel, a mixed burial (studied actually since the 1930s) resurfaces to shake the family tree. Not too different from the Croatian sequence, Israel's Qafzeh Cave revealed an early *H. sapiens* mixed in with others who looked considerably less modern. These Palestinian caves confirmed the coexistence of *H. sapiens* and Neanderthals. And with every variety of intergradation, *crossbreeding* ("signs of hybridization"[7] at Skhul and Qafzeh) came into view.

The fusion of types found in these Israeli caves (Tabun, Skhul, and Qafzeh) shows modern and Neanderthal traits freely admixed. Most have the trademark (heavy) torus, but variability of features is extensive, many specimens markedly modern in cranial vault, occiput, face size, and limbs. Yet those who posited the hybridization of these two different races are now silenced by a semantic ploy—labeling them Neanderthal*oid*, and arguing for a "transitional" stage of evolution.

Transitional? For goodness sake, these people were all *living together!* The fact, moreover, that the AMHs at Qafzeh were using Neanderthal tools makes us suspect that the main result of these crossings was *retrogressive,* that is, the back breeding of AMH-Ihuans into Neanderthal stock (which thereby inherited the Ihuan big brains). But to make them look "transitional," paleontologists adjusted the date from the third interglacial to fourth; using C14, Tabun was dated 60 kyr and Skhul 35 kyr, making the Skhul population intermediate (transitional between Neanderthal and mod), thus "eliminate[ing] all need for theories involving hybridization, with their attendant difficulties."[8] Despite these maneuvers, Ashley Montagu saw here a clear "intermixture between a modernlike form and Neanderthal. . . . The Mount Carmel population was the product of that intermixture. There is every reason to believe that similar intermixture (hybridization) occurred between populations *throughout the long prehistory of man.*"[9] [e.a.]

Making matters worse for the evolutionist, at Mt. Carmel, as in France in 1937, *H. sapiens* was *older* than Neanderthal. Indeed, a similar out sequence was observed at Steinheim, Swanscombe, Ehringsdorf, Krapina, and other sites, where the more gracile Neanderthal (Neanderthal 1) *preceded* the more extreme, "chimpanzee-like" one (Neanderthal 2).

1992   Speaking of chimpanzees, in his book released this year, *The Third Chimpanzee,* Jared Diamond confidently declares (despite Carmel and Vindija), "No skeletons that could reasonably be considered Neanderthal-Cro-Magnon hybrids are known." Yet that same year, fossils in Spain (Atapuerca) prove to be a farrago of *H. erectus,* Neanderthal, *and H. sapiens* traits! "There could have been a big party of different hominids living in Europe together," offers archaeologist Clive Gamble.[10] And now there is further evidence from other parts of Europe that Neanderthal and early mods (Cro-Magnon) partied together: at Mladec, Czechoslovakia; Steinheim, Germany; and Muierii, Romania.

1996   This year, Donald Johanson and Blake Edgar publish their lavish book *From Lucy to Language,* informing us that Neanderthal and Cro-Magnon "were too biologically distinct to have shared anything more than culture."[11] Others, though, were indeed floating hybridization to explain the softened features of St. Cesaire's Neanderthals, especially since excellent Cro-Magnon tools were found there as well as at another French Neanderthal site: Arcy-sur-Cure. Further confusing the issue, Johanson and Edgar state: "Even if it [meaning: interbreeding] did happen rarely, it would not mean that Neanderthals and modern humans were a single species"—a downright contrarian position, defying the very definition of species, as I will soon make clear.

1997   Hominid Paleogenomics takes the stage by storm, as biologists are now equipped to analyze the Neanderthal genome in earnest: wherein the Max Planck (a group of reknowned scientific research institutes) people supposedly deal "a powerful blow" to the interbreeding idea. If it happened at all, "it was too rare to leave a trace of Neanderthal DNA in the cells of living people."[12] Also in 1997, Chris Stringer of London's Natural History Museum agrees that Neanderthal are "probably" a different species. Not a good bet, as we will soon see.

1998 For, now a small and stocky breed of 24 kyr people has been discovered at Abrigo do Lagar Velho, Portugal; it is a child's skeleton, presenting a striking blend of mod and Neanderthal traits—thousands of years *after* the latter supposedly went extinct. While the child's teeth and chin are modern enough, the jaw is heavy and short, the bones thick, forcing some to reluctantly concede that there could have been "significant interbreeding" between Neanderthal and Cro-Magnon here. Others, though, hold to "the complete absence of any Neanderthal mitochondrial DNA in modern Europe. . . . [I]f this interbreeding were a frequent occurrence, then surely we would see the evidence in the modern mitochondrial gene pool, and we just don't."[13] This comment is almost immediately contradicted by new updates that say 4 percent of the DNA in Europeans and Asians appears to be derived from Neanderthals, due to hybridization. Paleoanthropologist Milford Wolpoff concurs: Neanderthal DNA falls within the modern type.

1999 Following up on the controversial Portugal discovery, paleo-anthropologist Erik Trinkaus of Washington University, St. Louis, points out that Portugal's 24,000-year-old boy is a blend of mod and Neanderthal, but hedges: "the boy could simply be the love child from a single prehistoric one-night stand." Ian Tattersall of the American Museum of Natural History, also playing down the rising specter of crossbreeding, remarks: "It's just a chunky modern kid. There's nothing special about it."[14] After all, too *much* interbreeding, and the house of evolution all falls down.

2000 Tattersall goes on to cite DNA evidence confirming that the human race does not bear any biological connection to the Neanderthal "species."

2002–3   But a new trend is afoot, and now a dig at Pestera cu Oase (Cave of Bones) in Romania comes up with more hybrids, *H. sapiens* sporting unmistakable Neanderthal features: huge long noses, great molars, occipital buns. Dated anywhere from 30 to 40 kya, the mandible is robust but combines archaic and mod features. The specimens are clearly of mixed type. Four years later (in 2007) when digs in Czechoslovakia come up with similar results, the finds once again whisper of repeated dalliance between modern humans (Cro-Magnon Ihuans) and Neanderthals (more than a one-night stand). Nevertheless, Richard Klein, teamed up with Blake Edgar in *The Dawn of Human Culture,* assures us that "if Cro-Magnon and Neanderthals interbred, it was probably on a very small scale."[15] The same year sees Stephen Oppenheimer, in *The Real Eve,* asserting that "no genetic traces of them [Neanderthals] remain in living humans."[16] But all our naysayers will soon be eating crow.

2006   In November, Lawrence Berkeley National Laboratory, emboldened by recent progress in paleogenomics (using 38,000-year-old DNA), issues a press release asserting that Neanderthal and ancient mods were not likely to have interbred. Even though they coexisted, say these scientists, there is no evidence of significant crossing. But that same year, paleoanthropologist Erik Trinkaus reenters the fray, reminding us of definite hybrids discovered in southwest Romania (in 1952), comparable to the 40,000-year-old Romanian cave skull (found in 2002), which has been classed as basically modern, though the forehead is extremely long and flat, and the molars very large. Trinkaus, as before, manages to cover both possibilities by stating that interbreeding would not be a "surprise." But very shortly after, as the evidence mounts, the professor loosens up: "Why not? Humans are not known to be choosy. Sex happens."[17] Indeed, a separate study that year finds a haplotype in these 40,000-year-old genes that may have been passed on during an "encounter" with Neanderthals.

2007   *Discover* magazine runs a piece that mentions that "our ancestors may have gotten up to 25 percent of their DNA from Neanderthals."[18]

2008   Despite the theoretical crisis caused by all this unevolutionary nookie, genome researchers at Max Planck Institute for Evolutionary Anthropology in Leipzig doggedly maintain that Neanderthal mtDNA falls outside the range of human variation, indicating an entirely separate species—even though Neanderthal and *H. sapiens* share about 99.5 percent of DNA! Cautiously, however, the team asserts that this doesn't mean crossbreeding would have been impossible. Hybridization, though now out of the bag, must still be made to play a minor role.* Thus do these scientists, now working on 70 percent of 40,000-year-old Croatian DNA, say that so far the project has unearthed no sign of DNA transfer between the two hominid lineages.

Nature magazine is quick to cover the story, parroting that Neanderthals *probably* did not interbreed with their "human" neighbors. *Scientific American* publishes similar findings but does mention the alternative multiregional model of occasional interbreeding, which Wolpoff has been saying all along, citing, for example, the especially modern jaw structure (mandibular nerve) of Neanderthals.

Meanwhile, the Max Planck people, evidently sniffing

---

*Why? Because genetics is responsible for the popular out-of-Africa (OOA) theory (see chapter 11) that *brooks no interbreeding* between these races, but opts instead for replacement of the inferior by the superior one. One of the major league OOA popularizers, Chris Stringer, is quoted in *National Geographic* this year: "Most Neanderthals and modern humans probably lived most of their lives without seeing each other." (Quoted in Hall, "Last of the Neanderthals," 54.)

a change in the wind, are now quoted as saying "some gene exchange *might* [e.a.] have occurred, but the result was later found to have been a false signal because of sample contamination."[19] Other publications in 2008 backed up this stalwart conclusion, such as Ker Than in *National Geographic*.[20] Also, in their 2008 book, *Human Origins*, Rob DeSalle and Ian Tattersall stand by the agenda: "the two kinds of hominid did not mingle. . . . Neanderthals and moderns represented two distinct species, and thus would not have effectively interbred."[21]

2009   Leading a Max Plank team, Pääbo, having successfully recovered 60 percent of the Neanderthal genome, affirms his earlier assessment of no "significant" evidence of crossbreeding.

2010   "DNA has even shown that a *few* [e.a.] Neandertals interbred with our ancestors," reports now allow. The thing is, we now have data on the *coloring* of these specimens and lo and behold, some Neanderthals had red hair and pale skin. Well, of course they did; all the races, past and present, have some Ihin traits—be it fair complexion, mod brain, slender build, or other AMH characteristics. But Douglas Palmer's book, *Origins: Human Evolution Revealed,* published in 2010, backs up the naysayers, agreeing that one finds "no genetic mixing between Neanderthals and modern humans."

Pääbo, blowing hot and cold, first stresses that there is no sign of admixture. Using sequences of nucleotides, he notes that Neanderthal mtDNA is "readily distinguishable" from mods. But things are changing fast and by May 2010 the Pääbo-led Max Planck team changes its conclusion, acknowledging that interbreeding probably occurred: the "team's second conclusion—[huh?]—is that there was probably interbreeding between Neanderthals and modern humans. . . . Neanderthals mated with some modern humans, after all."[22] Non-African moderns, it is now discovered, have 1 to 4 percent Neanderthal genes. Nevertheless they are still telling us that Neanderthal DNA does not seem to have played an *important* role in human evolution, although interbreeding "would not be greatly surprising, given that the [two] species overlapped from 44,000 . . . to 30,000 years ago."[23]

By October of 2010, an article titled "Human Family Tree Gets Bushy, Grows Roots" appears in *Discover* magazine relating "one stunner: early humans mated with Neanderthals" according to genetic studies (using DNA fragments from Neanderthal bones). Pääbo can now trace that interbreeding back 60,000 years to the Middle East. Now he says "humans and Neanderthals must have mated at some point."

2011   *Discover* magazine, in January, lures readers with "Stone-Age Romeos and Juliets," the piece confirming that "Neanderthals and modern humans probably had interbred . . . at least once." Only once? Now, earnest paleoanthropologists, poised like concerned parents of wayward teens, promise to "pin down how much interbreeding occurred." Then, in May, *Discover* returns to the subject: "Maybe we all carry a bit of Neanderthal in our DNA," although the differences presumably "are great enough to rule out significant interbreeding." So paleogenetic DNA sequencing has now established that AMHs "interbred with *Homo neanderthalensis* . . . in the Middle East. . . . We shared the planet with our cousins."[24] But not just in the Middle East . . .

Finally, as if to make it all OK and respectable that our ancestors fooled around with lowly Neanderthals: "First modern humans protected themselves against disease after leaving Africa by interbreeding with Neanderthals. . . . Early humans picked up genes which protected them . . . and eventually helped them to populate the planet."[25] (So nookie with the cavemen wasn't all that bad.)

DIVERGED OR MERGED?

Since Darwin's day, many missing links were found—creatures that seem to link Au to *H. erectus, H. erectus* to Neanderthal, and so on. So what was the problem? Something in the morphology of these intermediate types, the links, always precluded a smooth sequence from one to the other. Always something that doesn't fit. It is, some have commented, as if these mosaics were all made of spare parts. Each fossil seems to combine the laundry list of morphological traits in a different way, ruining the flow from one to the other. Lee Berger's *Au. sedipa*, for example, is "a jumble of parts." A spot note in *Smithsonian* titled "Body Parts" refers to our "idiosyncratic histories" and the idea that *H. erectus* is made out of "different parts. . . . It is amazing what evolution has made out of bits and pieces."[26] But is it evolution that jumbles spare parts, or simply crossbreeding?

"The nookie factor," the whimsical term coined by my friend, archaeologist Chris Hardaker, gives us a much better idea of these spare parts: a picture of melting pot Earth, mix-and-match humanity with the urge to merge since day one, which is the only honest solution to the so-called transitional and mosaic types that abound in the fossil record. The hybrid model of man's lineage does not look for *an* ancestor but for (at least) *two* ancestors. In the words of Franz Weidenreich: Most populations have *several* ancestors.

Paleontologists have, to date, assembled many fossil men, both primitive and more advanced, but can reach no agreement on sequence, how these fossils relate to each other, most critically how or if *one is ancestral to the other.* But let's streamline the whole shooting match: Think of the endless quibbling, even ungentlemanly rancor, over hominid classification—deciding *which* taxon specimens belong to—when mixing, plain and flat, is the magic bullet, Occam's razor. Parsimony *par excellence.*

Think of how things will slip into place if early (second race, Ihin) is recognized as the source of *all* small, gracile, and big-brained specimens. The late nineteenth-century German biologist, geneticist, and zoologist

August Weismann laid out a fundamental principle, the law of parsi-mony (though he didn't follow it himself), demanding that extraneous forces not be invoked. Weismann thought Darwinism leaned too much on unknown factors, hypothetical forces. Account for phenomena that can be readily explained by *known* causes, Weismann argued. Well now we know . . . about hybridization.

Parsimony also means the fewer *assumptions* made, the better the theory. It is a kind of thrifty, minimalist thing: the more a *single* axiom or fact can account for a wide range of phenomena, the truer it rings. This way, too, we remove the paradox of polygenesis: the problem being— How could races all over the world have coincidentally evolved in the same way? This difficulty evaporates once we recognize the Asu-Ihin-Druk blends in *all parts* of the world. It has been asked, for example: If a single mutation of a gene could change the length of limbs, why is this "random" mutation the *same* in far different places? Obviously, there is something amiss in our searchlight. Nothing ever "changed" the length of limbs; gene flow (race mixture) alone brought about such differences.

Milford Wolpoff in *Race and Human Evolution* incisively points out that the same DNA variations "are found over and over again, in the most widely separated populations." How is that possible? How could they all evolve the same way? The answer is profoundly simple: They didn't. The same basic types *intermixed*; they did not evolve.

Follow these hybrids dispassionately and we also dissolve the decades-old debate of why bipedalism appears to have preceded big brain. Lucy, for example, was bipedal but pea-brained; Taung Child also was bipedal with the "brain of an ape." And why was Taung's foramen magnum and dentition more advanced than his jaw and face? Mixing is the answer. All the mixed traits in these australopiths (see table 1.1 on page 41) devolve quite simply on their mixed parentage.

Differences between *H. rudolfensis* and *H. habilis* (the two were contemporaries in Africa) have been ascribed to "rapid evolution." Nonsense. It is simply a matter of two races blending, consorting—and in a matter of generations producing a new type. Why call the time

Figure 5.9. A shot at hybridization. A grapefruit chastising an orange for cheating on her husband, having two children—a tangerine and an apple. Cartoon by Marvin E. Herring.

between *H. habilis* and *H. erectus* "rapid hominid evolution," when it is plain that mixing gets even quicker results—with no evolution in sight? Besides, it was too big a change in *size*/stature for *H. habilis* (a bit over three feet) to evolve into creatures like tall and lanky Turkana Boy. Did *H. erectus evolve* at all? Not really. They appeared suddenly. Neither did they "lead" to higher forms:

> *Nowhere can it be demonstrated that men of the Homo erectus grade did evolve into modern populations.*
>
> J. B. BIRDSELL, *HUMAN EVOLUTION: AN INTRODUCTION TO THE NEW PHYSICAL ANTHROPOLOGY*

Zoologist T. Dobzhansky had no problem accepting hybridization among human populations throughout the Pleistocene. Crossbreeding readily explains the so-called rapid evolution of certain fossil men, the inexplicable quantum leaps in the record (evolution's big bangs or explosions). From the Au lineage a new "species" rapidly became much larger, brain size increased dramatically. No, this is not evolution. This is something else.

In the plant kingdom it was found that primroses made similar jumps, not by Darwinian mutations, but simply by hybridization, producing traits the parents did not have. Breeders have obtained plants and animals so surprisingly different, they would be immediately classed as a different species entirely if found in the wild. Hooton, discussing the appearance of very tall and raw-boned men, wrote "I have no doubt . . . [they] owe their increased size and ruggedness . . . to heterosis or hybrid vigor. . . . [I]n a world of men who have been migratory and promiscuous for scores of thousands of years, the origin of new types through hybridization . . . and recombinations . . . is a much more likely phenomenon than pure line ancient forms. . . . Hybrid vigor . . . means an increase in size and strength. . . . [M]ost of the tallest human groups are of mixed racial origin."[27]

ROOM OF MIRRORS

By general definition, a species is a breeding group that does not or rather *cannot* productively mate with any other group. Even if they do mate and there are offspring, the latter are infertile. Sometimes, though, it is difficult to know where one species leaves off and another begins. When it comes to wolves and coyotes, for instance, it is hard to say quite where one "species" stops and another starts, because wolves and coyotes *can* successfully breed. They should not, therefore, be called different species.

You could say zoology, like paleoanthropology, has become species happy. Even among the thirteen supposedly distinct species of Darwin's Galapagos finches (see figure 4.8 on page 187), some do interbreed; just as there are two supposedly different species of Mexican howler monkeys that still mate with each other. Even lions and tigers have interbred (in a zoo): if the father is a lion, the cub is called a liger; if the sire is a tiger, it is a tigon.

What about the human species? William Howells held that humanity has always been a single species (therefore interfertile), even though separated geographically. T. Dobzhansky was another who protested the unnecessary multiplication of species and genera for hominids; let them all be one species and one genus *Homo*. Though correct, this is today a minority opinion.

For many years it was argued that Neanderthal was a different species from us; *H. erectus* also was considered, by some, not only a different species but also a separate *genus* from *Homo*. Richard Leakey, wisely, thought that *H. erectus* and *H. sapiens* were not only the same genus but also the same species, *H. erectus* but an early version of *H. sapiens*. His father, Louis, toward the end of his long career, even asked if *crossings* could not in fact account for the multifarious *H. sapiens–H. erectus* composites so puzzling to analysts. Nor did Coon or Weidenreich rule out the interfertility of these two types of men: AMH and Druk.

As an example of the oxymoronic confusion bedeviling these matters are statements like: "Neanderthals and modern humans are separate species, but do not rule out some interbreeding."[28] If they *could* interbreed, and did interbreed (with fertile offspring), we have no business calling them different species; they would be merely different races (subspecies). Also violating the definition are statements like: "Once we shared the planet with other human species, competing with them and interbreeding with them."[29] Come again? I thought different species don't interbreed by definition!

The very concept of species has been debated for decades, with no final agreement among biologists on what species are. Are they simply a gene pool, meaning closely related populations who reproduce with one another? In the event that they *don't* interbreed (like different "species" of fruit flies that look exactly alike), it doesn't mean that they *can't*. Separate gene pools may not be such a good criterion; the Bushmen, for example, comprise a separate gene pool from Eskimo-Inuit and don't exchange genes with them, but still both are of the same species. We should follow Ernst Mayr in this regard, defining species according to their interbreeding *potential*.

Weidenreich, noting the "irresistible trend to interbreed" among humans, recognized that all people were cross-fertile, that they can, do, and *did,* crossbreed, this ability proving we are all one species. Indeed, one might even "eliminate all the generic names"[30] of fossil man,* as far as he was concerned. Although his insight goes back seventy-plus years, I believe it is the correct one, yet it is disparagingly labeled "already in disrepute"[31] by some of today's leading evolutionists.

Officially, *H. erectus* and *H. sapiens* couldn't interbreed because they are (taxonomically) classed as different species. But they did breed together, producing viable, fertile offspring, such as *Homo floresiensis,* the small-bodied fossils found in Palau, and Broken Hill Man (see chapter 11), where a combination of *H. erectus* and *H. sapiens* traits are all too apparent. The Palauans of Oceania were the same size as small Au, yet some of these Micronesian fossils are as recent as 1,400 years old, and certain craniofacial features are in keeping with *H. sapiens.* As for Zimbabwe's Broken Hill Man, he possessed a large brain and gracile limbs, though his face and skull resembled Neanderthal and *H. erectus.* Some then called it Neanderthaloid, but with such mixed features, scholars argued back and forth whether to classify it as *H. sapiens* or Neanderthal. Europe's *H. heidelbergensis* was another who represented a marvelous alloy of *H. erectus,* Neanderthal, and *H. sapiens* traits. In fact, *H. erectus* almost everywhere has been found with *H. sapiens* traits sprinkled in, such as in Europe where a mosaic of features abounds—not to mention *H. erectus*–mod blends in the Far East, at Choukoutien and Ngandong. And in Africa: mixed in one creature, *Homo ergaster* (Skull KNM-ER 3733) are massive (archaic) features along with thin bones and modern nasal structure. Students of the African hominids have noted "the puzzling variability in early *Homo.*"[32] Of course there's variation! It is not puzzling; that's what happens when races mix. No, variability is not "hiding in the genes"; nor is it DNA "mutations" that generate variability. All we are seeing are the mix-and-match results of crossbreeding. We have taken infinitesimal measurements with calipers and run endless computer programs, and argued about classification till the cows come home. But there is and

---

*For example: *Australopithecus* is called a different genus than *Homo,* and *Ardipithecus* is called a different genus than *Australopithecus.*

has always been only one species of man on Earth, different kinds for sure, but all assorted members of the family of man.

In an e-mail, archaeologist Christopher Hardaker told me they edited out the following portion from his 2007 book *The First American:* "By definition, a people cannot successfully mate and have viable offspring with another species. . . . There might be nookie but no babies. . . . If babies, then same species. . . . How does '*H. erectus sapiens*' sound? If everyone from *H. erectus* forward could 'do it' with everyone else, it would certainly give an added dimension to the name *erectus*. . . . You are warned on the first day of Phys Anth 101 that the world of human evolution is divided into splitters and lumpers . . . [but] it is a room of mirrors."

A NEW RACE WAS BORN

If you want to get *really* confused, make a close study of who's who in the human fossil family. Paleontologists keep changing their mind as to who belongs in the genus *Homo*. The earliest candidates, Africa's australopithecines, were a bewildering hodgepodge of human and sub-human. East Africa's Olduvai and Turkana finds were so "varied and baffling" that debate raged for decades over their proper classification.[33] All that fussing and fighting, when they were, after all, nothing but hybrids—as were their parents and grandparents.

At sites like Torralba, Spain, and Olorgesailie, Kenya, large numbers of primitive and sophisticated tools (flake and core-bifaces, respectively) occur together in open-air sites. And when *H. erectus* tools were found mixed in with more primitive ones at Olduvai Bed II, some analysts were forthright enough to say it looked like a case of "interbreeding. . . . The two gene pools are completely mixed. . . . This sort of inter-action must have happened many times during the course of human evolution."[34]

How will Darwinian phylogenesis stand up to the lack of evidence for *H. erectus* evolving to *H. sapiens*? According to Ian Tattersall, a method called cladistic analysis does not allow it; according to Birdsell, brain size difference does not allow it. And if so, there is no evolution either, for with Neanderthal eliminated as our ancestor that leaves only *H. erectus* to fill the slot. Neither was Louis Leakey convinced that the pithecanthropines led to *H. sapiens,* his opinion based on morphology of the skull and *H. erectus*'s huge visor or torus, so tremendous it strains even the most credulous imagination to see in it the precursor of man. Meanwhile Marvin L. Lubenow has decried the wholesale ignoring of the Selenka-Trinil expedition, which found that mods and pithecan-thropines lived at the very same time in Java, the latter playing no part in human evolution. "It is an amazing conspiracy of silence," comments Lubenow. [35]

In Hindu scripture it is written: "the first race Asu [Ardi] tempted the white people"* In other words, the Ihins broke the command-ment of endogamy (in-marriage) which enjoined: "Neither shall ye permit the Ihins to dwell with Asu, lest his seed go down in dark-ness."† But the little people strayed out of the Garden of Paradise and "began to dwell with the Asuans, and there was born into the world a new race, the third race of man, called Druk [a.k.a. Cain a.k.a. *Homo erectus*], and they had not the light of the father in them, neither could they be inspired with heavenly things."‡ And the Druks were the first people to go to war, hence Cain's "mark of blood" in the Old Testament.

Those of the lesser light were called Cain (Druk), because their trust was more in corporeal than in spiritual things; and those with the higher light were called Faithists (led by Ihins) because they perceived that Wisdom shaped all things and ruled to the ultimate glory of the All One.§

*Oahspe, The Lords' Fifth Book, Ch 3, 1:14.

†Oahspe, First Book of the First Lords 1:10.

‡Brice Johnson, ed., *Scroll of the Great Spirit*, 26:21.

§Oahspe, The Lords' First Book 1:19.

Despite overwhelming evidence to the contrary, *H. erectus* is still widely considered our immediate ancestor, but he is nothing more than the result of Ihin-Asu "interaction." If neither Au nor *H. erectus* is ancestral to us, then what? It's the nookie factor, for the Ihins repeatedly blended with the autochthonous races. Wherever short stature, smooth features, gracile build, or unpredictably modern crania are observed in otherwise archaic forms, Ihin genes are at play.

LUMPERS AND SPLITTERS

Darwin mentioned theorists who envision anywhere from two to sixty different species of man. Well, the more they dig, the more mixes they find, erroneously calling them species—and gloating over the ever-expanding variety of human and prehuman species. Yet critics do ask: "Is it possible that the scientists, who have given new species names to every early *Homo* find with significant differences, have made our family tree more complicated than it really is?"[36]

Indeed, the "lumpers" (including Thomas Henry Huxley, T. Dobzhansky, George Gaylord Simpson, Franz Weidenreich, and Ernst Mayr) all have argued against this unnecessary proliferation of species. Let all hominids be one species. It's the splitters who gave us 700 species of *Drosophila* and 13 species of Galapagos finches, when they may just be varieties or races (subspecies); according to Mayr, there is extensive hybridization among six of those "species" of finches. "[You] might as well have called them all one species," says another analyst.[37] Fact is, Darwin himself thought the finches were merely varieties, until the ornithologist John Gould convinced him they were separate species!

But that's birds; what about man? The fundamental design of the human cranium is the same from the earliest pithecanthropines right up to the modern races. These features, they say, vary among recent races as much as between different fossil types; therefore, we might just as well split up recent man into several species—which would be a falsehood. (If a bone man were to compare the morphology of an African Khoikhoi and a Swede, without knowing the source of the specimens, he might well designate them as different species.)

Multiple human species—no; multiple *races*—yes.

The fossil people are trifling with us: a slightly different shaped jaw—and presto!—a new species is born. This was the scenario when Lucy's discoverers assigned her to a different species than *Au. africanus,* who had slightly more specialized teeth. But they were not different species, only different proportions of Asu and Ihin genetic makeup. (C. Loring Brace thought the teeth of the two not very different.) Splitters continue to confuse the issue by seizing on minor differences, boldly naming a new species based on one or two fragments.

Richard Milton, in *Shattering the Myths of Darwinism,* reminds us that a few years after Louis Leakey "insisted that his discovery [of Zinj] was entirely novel," it was found to be just another robust Au. Also in East Africa, Ethiopia's *Ardipithecus kadabba,* although known only from fragments, was made the "probable ancestor" of *Ar. ramidus.* Originally considered a subspecies (race only), the older Ardi was now elevated to *Ar. kadabba,* a new species, on the basis of bone scraps and a few teeth with a slightly different wear pattern than *Ar. ramidus.* Are different teeth really enough to crank out a brand-new species?

The splitters are a determined lot, telling us that "the notion that multiple human species are the norm . . . has only got stronger with a series of major scientific discoveries. Since 1994, four *new species* [e.a.] of hominid have been added to the human family tree."[38] But rather than species, these are simply different *races.*

Built into anthro-speak is a slippery, disorienting use of this term species, *where only races apply.* Even Stephen J. Gould, the late-twentieth-century darling of paleontology, couldn't get it right: "Before commerce and migration mixed us up, each race is a separate biological species."[39] This is a contradiction in terms! We ought to remove Neanderthal from our species, Gould also argued—in agreement with Tattersall, a consummate splitter, who urges that Neanderthal be restored to separate species status, ignoring the now incontestable evidence that they *interbred* productively with mods. And how about this: J. B. S. Haldane, the esteemed English biochemist and founder of the new synthesis (neo-Darwinism) said "new species may arise by hybridization."[40] Well, in that case we don't need evolution at all. England's Douglas Palmer counts at least twenty species among our ancestors, again using "species," where really only race (subspecies) applies. In fact many of these "species" are not even distinct races— just variations, strains, crosses, mongrels.

A recent discovery in Africa, honored with a new species name, *Australopithecus sedipa*, was presented to the world with all the fanfare of an exciting new discovery who will "rewrite long-standing theories." Dated to about two million years, it was initially hailed as our direct ancestor. However, critics say *Au. sedipa* actually came after *H. habilis* and hence is too young to be our ancestor. All the while *Au. sedipa* screams hybrid—nothing but an upgraded Au. Found in 2008 near Johannesburg, this creature is made of "spare parts," said to be an "odd blend . . . head-to-foot combination of features of *Australopithecus* and the human genus, *Homo*."[41] And, yes, *it is just that.* Testifying to Ihin genes are: reorganized front brain, projecting nose, smaller teeth, humanlike hips and pelvis, longer legs, and precision grip.

In 2010, when they found "distinct" mtDNA in the Siberian Denisova remains (it was only part of a pinky finger), it instantly became a new species. Assigning new species to every new fossil, sometimes named after the finder or the sponsor, is like having a star named after you. After all, there is little hope of hitting the news with mere varieties or races; only new species will do the trick. One paleontologist decried "the rashest statements in the face of evidence . . . generalizing on single observations. . . . Journalism, my dear boy, journalism pure and simple."[42] With Denisova, the glory went to the geneticists, the finger fragment supposedly "marking an entirely new group of ancestral humans."[43]

All these new species create "a terrible paleontological mess. . . .

Everyone who had a fossil come into their hands . . . wanted it to be something new . . . for their purposes of self-aggrandizement. . . . Some were unhappy when their prize species were lumped together with those that others had discovered."[44]

Weidenreich, one of the best minds in the game ("that masterly student of ancient man," according to Hooton[45]), was right to warn that raising the differences between racial groups to the rank of species is artificial, an illusion, a "taxonomic trick," exaggerating dissimilarities to make your find important. Like Weidenreich, Dixon and Hooton explained the confusing varieties of man simply, parsimoniously, by way of amalgamation. Harvard supported this for a while, though the debate fired up again: Were races actually species (splitters) or just variants (lumpers)?; today you get as many opinions as professors. Interfertility, for example, is quibbled with and not always considered a perfect criterion for species inclusion. (Significantly, it is biologists—not taxonomists—who are satisfied with using interfertility as the best available guide to define species.)

Darwin muddied the picture with floundering logic, arguing that even if all human races are interfertile, still this is not a safe criterion for single species. These matters, he said, are governed by "highly complex laws . . . a naturalist might feel himself fully justified in ranking the races of man as distinct species, for . . . they are distinguished by many differences in structure. . . . The enormous range of man . . . is a great anomaly in the class of mammals, if mankind be viewed as a single species. . . . The mutual fertility of all the races has not as yet been fully proved."[46] It took the twentieth century—with its uptick in racial blending—to prove it.

[No] animals very different will breed together.
CHARLES DARWIN, *NOTEBOOK C*

It is a fact, too, that some species that look quite distinct interbreed regularly.

Frankly, Darwin's ideas on fertility and sterility of hybrids came mostly from his study of flowers. Though he denied it for *man,* he did note the increased fertility of crossbred *flowers.*

CUTE IN THEIR OWN WAY

Whenever the populations of early man met they did exactly what modern populations do, they interbred.

ASHLEY MONTAGU,
MAN: HIS FIRST TWO MILLION YEARS

Perhaps there is some ingrained repugnance at the idea of elegant AMHs mating with clumsy cavemen, even though the Indo-Europeans, our forebears, "spread across the world . . . interbreeding with peoples of the most diverse hues and features."[47] Tattersall, a great splitter, thinks we're much too fine to be lumped together with the early hominids: "Our own living species, *Homo sapiens,* is as distinctive an entity as exists on the face of Earth, and should be dignified as such instead of being adulterated with every reasonably large-brained hominid fossil that happened to come along."[48]

In fact, I suspect a hidden factor behind the slow acceptance of amalgamation as the true ascent of man: xenophobic bigotry. Awarding species status to mere race differences may be part of a lingering deep-seated racialism that views these differences as irreconcilable and unassimilable, the same spirit of apartheid that invented and used the word miscegenation to denote interracial marriage.

Tattersall goes on to bolster his claim with the argument that "Neanderthals . . . and *Homo sapiens* would probably not have recognized each other as very desirable mating partners."[49] "Few Cro-Magnons may have wanted to mate with Neanderthals," agrees Jared Diamond, who supposes that "the differences may have been a mutual turnoff . . . [H]ybridization occurred rarely if ever . . . [one] doubts that living people of European descent carry any Neanderthal genes."[50] Cro-

Magnon, say others, regarded uncouth Neanderthal as "little more than animals."[51]

Despite these parochial opinions, the fact remains that "race pride or the differences of religion and customs have never prevented the Anglo-Saxons from crossing with the lowest of savages," as Armand de Quatrefages made bold to assert in *The Pygmies.* Let's not allow our puritanical or other views to blind us to these liaisons. Highbrow and lowbrow have somehow always bridged the gap; folks from opposite sides of the track, so to speak, have always been drawn together across the divide. "One thing human beings don't do," observed prehistorian/ anthropologist/archaeologist Robert Braidwood, "and never have done, is to mate for purity."[52]

Franz Weidenreich, in *Apes, Giants, and Man,* also stressed "the tendency of man to interbreed without any regard to existing racial difference. This is so today . . . and there is no reason to believe that man was more exclusive in this respect in still earlier times." And who knows, some of these hybrids might have been quite fetching. Hey, if high cheekbones and full lips were in then, as they are now, no telling what these mosaics looked like. Writers are beginning to ask if the Neanderthals "were the ugly brutes often portrayed by popular science, or were they cute in their own way?"[53] (Some had freckles.)

Figure 5.10. Mixed Fijian. In the Fiji archipelago Polynesians and Negritos crossed in all degrees.

Historically, some of the most backward people have mated with more sophisticated tribes: Alpheus H. Verrill reported that "the living Indians of these districts [Nicoya, Chiriqui, Terriba, in South America] are the result of mixtures of the cultured races and the

TABLE 5.1. SIGNS OF CROSSBREEDING: MIXED MORPHOLOGY AMONG EUROPEAN FOSSIL MEN

Where/What	Modern Traits	Archaic Traits
Croatia/Krapina Man*	Frontal region, Beetle brow, sloping forehead, great jaw, large face and orbits, barrel chest	1,300–1,500 cc, brachycephalic, limbs mod, light build
France/Arago Man	"A fascinating mosaic of old and new features"[†]	
France/Fontéchevade Man	High forehead, 1,400 cc, light build	Heavy browridges
France/La Chapelle Man	Big brain: 1,625 cc	Heavy browridges, long and low cranium
Germany/Heidelberg Man	Modern dentition	Massive mandible, broad jaw, no chin
Greece/Petralona Man	"Hints of modernity"	"Distinctly primitive"[‡]
Germany/Steinheim Man	Well-developed forehead	Heavy browridges
Hungary/Vertesszollos Man	Some modern features, 1,500 cc	Thick and broad cranium
Portugal/skeleton	Modern chin and teeth	Thick bones, short legs, hefty jaw Romania/skull Modern skull Long and flat forehead, large molars
Spain/Atapuerca Man	Modern chin	Heidelbergensis in type
UK/Swanscombe Man	Big brain: 1,325 cc	Thick skull and bones, primitive features

*Going against the flow, Klaatsch of Breslau was the first to suggest that Krapina's remains represent two races who coexisted and cohabited, rather than any sort of evolutionary transitional.
[†]Lewin, Mankind, 127.
[‡]Hooton, Ape, 127.

more savage tribes."[54] In Brazil, Percy Harrison Fawcett thought the Morcegos (see chapter 12) and their kindred the "most bestial and degenerate savages in existence," yet some of them intermarried with the more noble Tupis and Caribs, and from this amalgamation sprang two different types, the Botocudos ("Neanderthaloid in type," according to Czech anthropologist Ales Hrdlicka) and the extremely varied Aymaras. This region also had milk-white Indians, their skin

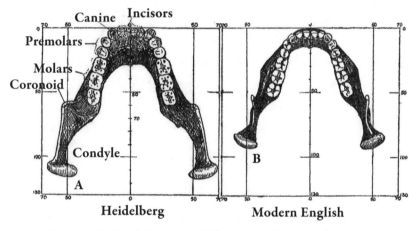

Figure 5.11. Heidelberg mandible compared to a modern one.

Figure 5.12. Heidelberg site.

covered with short down, who saw best in moonlight; those moon-eyed types are of extraordinarily composite heritage.

Betwixt and between: The pattern remains the same as we go farther back in time—earlier hominids were profound blends as well: *H. habilis* (found in the same general area as Au) shows foot morphologically quite mixed, the ankle in particular. While he was erect and bipedal, with modern femur and dentition and thin skull, nevertheless his stride was shambling, his arms long, legs short, and hands curved. Although *H. habilis*'s Ihin genes gave him a notably short stature, his small brain betrays the legacy of Au. Those early men of the African savanna came in a great variety of shapes and sizes, some scholars (noting the great range of types) even doubting that there *was* a separate species of *H. habilis*; Mayr, for example, calls *H. habilis* an Au.

LAST DITCH: DIFFERENTIAL EVOLUTION

We need to take these mixtures more seriously, before tossing them in the "variability" bin or squeezing them into imagined evolutionary sequences or inventing such chimeras as differential, rapid, or mosaic evolution. Differential evolution is the hypothesis that all aspects of an organism do not evolve together, but each at its own rate. African hominids in particular needed this theory; consider Kenya's Black Skull, for example: "parts of this skull were evolving faster than other parts."[55] How does that grab you? In the case of the australopiths at Olduvai with a foot more modern than the hand, or *Au. garhi* with relatively mod legs but long (archaic) forearms, supposedly the legs "evolved" quicker than their forearms. And in the case of Laetoli, the feet supposedly evolved before any other body part.

Or take Turkana Boy whose well-preserved skeleton from East Africa ("a mosaic of *erectus* and *sapiens* features") was clearly modern "from the neck down" (the rib cage almost identical to modern man's), whereas his skull was markedly primitive—with hardly any forehead or chin, and the frontal lobes rather small (880 cc). Now, all this indicated, at least to Brian Fagan, that "different parts of the human body evolved at different rates."[56] *Evolved?* I don't think so.[57] If face/head/ brain evolved *last* (as the evolutionists tell us, and as surmised in the case of Turkana Boy), then why did Taung Child of South Africa, an upgraded Au, have a high, round forehead, smooth browridge, delicate cheekbones, and flat (not prognathous) face? Taung, the first *Australopithecus africanus* ever discovered, was modern *from the neck up,* just the opposite of Fagan's differential evolution of Turkana Boy.

Must we resort to *ex post facto* theories like differential evolution (or modular structure of the genotype) to explain (away) these mosaics, or to explain how the primitive features of Neanderthal compute with his large brain? The Neanderthal skull is more modern than its femur (the same combination found in Swanscombe Man), which is to say, these people were modern from *neck up.* The same is true for that troublesome Portuguese skeleton, with the perfectly modern chin and teeth that contrast with his thick bones and short legs (of the classic Neanderthal type), reversing the so-called differential evolution of body parts conjured by Fagan for Turkana Boy. And so, at least with the Portuguese find, archaeologists grudgingly conceded "significant interbreeding." We don't *need* differential evolution or any other principle to explain these patent hybrids.

But the theories are piled high and deep. Turkana Boy's combination of modern body surmounted by a primitive head was so "improbable," an observer might "wonder if . . . [he] was a visitor from an alternative universe or perhaps the product of some strange genetic experiment."[58] How silly and inane. Java Man, like Turkana Boy, had a primitive skull combined with a fully upright body. Noting the modern femur of this *H. erectus,* anthropologist Jeffrey Schwartz ventured "our ancestors had become human first from the waist down."[59] But if so, why do Au and *H. habilis* have humanlike dentition? Ad hoc and inconsistent, a differential rate of evolution is held nonetheless as a "principle," Turkana Boy nicely confirming that different parts of the body evolved at different rates. Why else would *H. erectus* appear with modern limbs but primitive jaw, dentition, and cranium? S. J. Gould saw evolution "proceeding at different rates in different structures. . . . This must be true or evolution couldn't have happened."[60] Well, maybe it *didn't* happen.

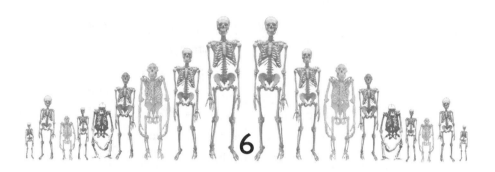

THE TIMEKEEPERS

The Uncertainties of Scientific Dating

In every case, the story [of man] seemed to hinge on the age of things.

JAMES SHREEVE, *THE NEANDERTAL ENIGMA*

Current geological dating is arbitrary at best, and wildly inaccurate at worst.

DAVID HATCHER CHILDRESS,
LOST CITIES OF NORTH AND CENTRAL AMERICA

I pay not the least attention to the generally received chronologies.

GODFREY HIGGINS, *ANACALYPSIS*

INFLATIONARY TIMES

Does mankind (or Earth itself, for that matter) have as long a history as we've been taught? Millions and billions of years? I have come to doubt it. One century ago, geologists like George Frederick Wright placed the first forms of life at only 24 mya and the age of man as no older than 100 kyr—in contrast to the five *million* currently assigned to him. (The subject of Earth's age is touched on again in chapter 10).

Historically, it was Darwin's ponderously gradual and long evolution that kicked off deep time, for his theory *demanded* vast eons for organisms to evolve. Yet, it is a plain fact that with hybridization of the races, new types can happen quickly, in a matter of generations. Darwin, we know, had postulated accumulated and minute adaptations building up over huge amounts of time; after all, "millions on millions of generations" are needed for speciation to happen.[1] Darwin's detractors, however, twitted his conveniently "indefinite time . . . an unlimited number of generations for the transitions to take place."[2]

I want to stress that the inflated dates handed to us for life on Earth bear the heavy imprint of evolutionism. And we know that long dating stands at the very opposite extreme (the opposite fallacy) of young Earth, the biblical age of Adam said to have occurred a meager 6 or 10 thousand years ago, which is just as suspect as the dates given us by today's deep-time enthusiasts. I think the truth actually lies somewhere in between. Sure, dating by the bone people is reliable enough up to, say, 40 or 50 kya (the carbon-14 limit), but beyond that time, things go hog wild, off the charts. I would sooner accept their dates *relatively* (reflecting sequences), than as an absolute time count.

> *Absolute dating . . . is still subject to a very large margin of error.*
>
> GEORGE GAYLORD SIMPSON,
> *THE MAJOR FEATURES OF EVOLUTION*

ADDING ZEROS

"Not a trace of unquestionable evidence of man's existence has been found in strata older than the Pleistocene," said William J. Sollas, Oxford geologist.[3] Before the 1950s, the Pleistocene was believed to extend back some 700 kyr; today, that's tripled to at least 2 million years.

> *Man today has no more conception of how many years ago*
> *the Pleistocene commenced, or the length in years of any*
> *geological time, era, or period, than the ancient fossil on*
> *my library table.*
>
> JAMES CHURCHWARD, *THE LOST CONTINENT OF MU*

Moving man (hominids) beyond the Pleistocene, back to the Tertiary (up to 6 mya) has been de rigueur since Louis Leakey's time (as per Zinj) and the rise of potassium-argon dating (K-Ar method) in the 1960s. This is how the bone people keep score, and one thing is for sure—*the older the better.* Given the academic race to outstrip the *earliest* dated man, most dates are suspiciously overestimated.

Figure 6.1. Louis Leakey
as a young man.

And it is dominoes. Almost overnight, *H. erectus* of Java was made to jump from 500 kyr to 1 million years old, Louis Leakey's headliner Olduvai finds pushing everything back even further. (A cute consequence: In 1957, Ashley Montagu's book was titled, *Man: His First Million Years.* Reprinted in 1969, the revised title was: *Man: His First Two Million Years.*)

Just one century ago, earliest man's age was set (reasonably) at around 100,000 kya. I believe this figure is very close to correct (see more on this in chapter 10). Marcellin Boule, in 1923, had man no older than 125 kyr; at this time the Paleolithic was thought to have begun only 50 kya, but in 1929 that was changed to 400 kya.[4] What is going on here? Comparing these dates we soon find that man's age has been increased in the past century by a factor of twenty! In the 1940s and 1950s, *Australopithecus* (who was the earliest hominid at that time) was dated to no more than 500 kyr. Then by the 1970s the first hominids were "at least a million years old."[5] And with American anthropologist F. Clark Howell's finds in Ethiopia (just after the K-Ar method was brought out), that figure instantly doubled to 2 myr.

Nevertheless, Le Gros Clark (in 1955) noted that "all the australopithecine deposits in South Africa are not very widely separated in geological time."[6] But then after Leakey and Howell broke the Tertiary barrier, their new expanded dates became the gold standard, and everything must be fitted around this benchmark—even surpass it.

Now mtDNA says the mod type of man arose perhaps 130 kya: Ethiopia's Omo I, originally dated at less than 50 kyr, is now offered as possibly the oldest *Homo sapiens* anywhere at 130 kyr.[7] But this figure will not stay put; watch it change to 200,000 (to fit the African Eve theory). What is the oldest reliable date for the modern type? Back in 1955, Le Gros Clark dated AMHs to a modest 30,000 years ago. But since the 1990s, given the push for the out-of-Africa paradigm, date changes seem more politics, more agenda, than impartial science.

Long dating had a good head start, for it was advanced from the beginning by Darwinists in order to supply a plausible length of time for man to develop his great brain. British naturalist Alfred Russel Wallace (as we will see in chapters 7 and 9) thought at least 15 million years of evolution were needed to produce the human brain (or "the symphonies of Beethoven"). Evolution theory needs very substantial time also to get back to the less specialized pongids of the Miocene, from which both apes and humans presumably evolved (especially since the more recent Pliocene apes are not too different than today's editions).

WORLD'S GREATEST FOSSIL: *NE PLUS ULTRA*

Let me ask you this: How many times have you read the phrase *we now know*—persuading the innocent reader to put his or her faith in the most *recent* findings of science? Let's try to remember, though, that hominid discoveries in Africa have caused the family tree to be rewritten several times. Which one is right? "[A]ll the evolutionary stories I learned as a student . . . have now been 'debunked,'" owned Derek V. Ager.[8] What are the chances of today's "breakthroughs," with all their splashy fanfare, holding up to tomorrow's surprises?

Let's not be naive. Latest studies are not necessarily the best. It is nothing but the misleading rule of progress (forever onward and upward) that Darwinism itself inculcates. For example, American paleoanthropologist John Hawks in his weblog imperiously dismisses earlier work, criticizing authors who use "completely obsolete anthropological information" more than fifty years out of date. Objection! More often than we may suppose, the older theorists (using their brains and immense fount of knowledge) more closely skirted the truth than many of their modern replacements (using supercomputers and the latest vogue in their field).

And how would Charles Darwin fit in Hawk's reproof? Isn't Darwin 150 years out of date? I have come to suspect the vaunted claims that new or recent studies are automatically better. Franz Weidenreich, as an example, was on the money *75 years ago,* when he rejected an out-of-Africa or any other single-origin theory of man, showing instead that *H. erectus* was followed by mods *everywhere* (not just Africa). As occasionally noted in these chapters, I frankly find the work of some of the self-made old-timers—like Weidenreich, Hooton, Dixon, and Coon—more impressive, reliable, and realistic than the smoke and drone of today's generation of skin-deep theorists and sycophants.

But this is how it's done: New finds are trumpeted as ground-breaking fossils, rewriting human history, seriously revolutionizing everything, and threatening to deal a major blow to previous concepts. Headline-seeking superlatives like "utterly dazzling find" and "oldest human ancestor" sound more like a marketing phenomenon than sober science. Feeding off the glamour of deep time, for example, Lee Berger's *Australopithecus sedipa,* dated to a whopping 5.5 myr, "changes the game."[9] Don Johanson well knew the value of "the world's oldest humans . . . to dangle before foundations and private funders."[10]

Discovered (but in fact only rediscovered) in 2003, hobbit Flo made her flashy debut as the hominid that "overturns everything"[11] (utterly disregarding earlier hobbits found in the 1950s). It was the same with Kenya's Black Skull, which was touted as "overturn[ing] all previous notions of the course of early hominid evolution," the swashbuckling posture answered by head-shaking critics who aver it was exactly "what you would expect to find . . . it isn't even a new fossil" (the very same type had been found back in 1967 by French diggers).[12]

Combine the funds-seeking diggers with the sensation-seeking press and boom—each new specimen is an electrifying challenge to orthodoxy. The script: "Our assumptions were wrong, but we now know . . ." deceptively creates the impression of a humble, open-minded, self-correcting science.

How shall I account for the difference betwixt thy arguments now and the other time?

OAHSPE, BOOK OF FRAGAPATTI 9:17

In reality, such finds are blowback—exposing the bogus theories and pseudobreakthroughs of yesterday's news. One disgusted observer commented facetiously: "Every new fossil hominid specimen is the most important ever found . . . and is completely different from all previous ones, and probably a new genus, and therefore deserve a new name."[13]

I fear that much of this date changing, this time inflation, is a form of damage control, saving evolution from the wrecking ball of hard facts. It is difficult to get a footing here. Whose dates shall we believe? Some say Neanderthal and *H. sapiens* split off from an (imagined) common ancestor 300 kya, but others say that happened 800 kya.[14] Was there an unstated reason for setting Neanderthal farther back in time? Yes, there was: Greater age is needed for Neanderthal, to account for huge differences in physiology between his lineage and all others. In the continuing chaos of dates for Neanderthals, analysts have dated them anywhere from 150 to 353 kya. Still others give Neanderthal no more than 100,000 years on this Earth.[15]

Robert Braidwood dated the earliest Neanderthal to 85,000 years.[16] There is no undisputed documentation of Neanderthal, some say, before 75 kya. Neanderthal is basically man of the Mousterian age, most clustering around 40 to 50 kyr. He does not show up earlier than 50 kya in most parts of Europe, Russia, Siberia, and Iran[17]; and his tenure on Earth can be no more than 50,000 years altogether.[18] His remains are found in largely organic state (not fossilized), indicating relatively recent time.

CONTROVERSY, CONTAMINATION, AND CONFUSION

These wide and indeterminate date ranges are called scatter. Vertesszollos Man, a problematic fellow for evolutionists, with awfully early AMH features, has been dated (scattered) anywhere from 100 kyr to 700 kyr! Petralona Man, another early European troublemaker, is dated anywhere from 70 to 700 kya[19]—the 70 kyr tag fits the modern aspects of his morphology, while the older date, 700 kyr, fits the primitive ones. Why even bother with such indeterminate ranges? In the case of Ihroud, Morocco Man, one method gave him 40 kyr and another method gave him 125 kyr—the older date favored, for it helped him fit the bill as a transitional species in the out-of-Africa theory. Dates changed for Africa's Kabwe Man, too, scattered from 11 kya to 40 kya to 300 or 400 kya[20]—also to better fit the out-of-Africa model.

It's a rollercoaster ride: Java's Ngandong Man, found in 1931, has been dated anywhere from 77 to 170 to 300 kya. But consider the largely ignored report by zoologist Emil Selenka and zoologist and paleontologist Margarethe Selenka concerning Java's deposits of *Homo erectus*. Here violent eruptions of a nearby volcano and subsequent flooding changed the landscape dramatically: the Solo River actually changed its course. Those beds might have been less than 1,000 years old, for the degree of fossilization was the "result of the chemical nature of the volcanic material, not the result of vast age."[21]* Keith, a century ago, had cautioned that "the mineralized condition of bones does not signify much . . . [if] impregnated with iron and silica from the stratum in which they were embedded."[22]

Modern man's age has been scattered anywhere from 30 kya up to 500 kya. But don't make any AMH *too* old, as that would put mods too early in the record, throwing off the whole evolutionary staircase, onward and upward. Let's see how this works: The AMH Keilor skull from Australia was thought to be of very early (third interglacial) age, but being AMH, it was automatically elevated to "definitely postglacial times,"[23] possibly as young as 15 kyr. Before the long-dating craze began a half century ago, if they found an AMH too early in the record (as indeed they are), they simply bumped up the date to make him more recent. Nowadays they're more inclined to leave the old date as is and say the whole shooting match began earlier than we thought—in the Pliocene. Deep time. But man did not exist in the Pliocene (see discussion in chapter 10).

How do you like the scatter for China's Choukoutien *Homo erectus,* variously dated anywhere from 200 kyr to 750 kyr? Good thing I used a pencil (with a stout eraser) during my note gathering for these dates. How does one know which guess is correct? They will shelve any hominid at mid-Pleistocene (say, 800 kya) *automatically* if the morphology is

---

*When Mt. St. Helens blew in Washington state in 1980, with hot ash and gas surging at hurricane speeds, it only took *hours* to lay down significant sediments; thick layers do not necessarily take a long time to form.

H. erectus–like. This is called dating by morphology; for example, "We must regard a small brain cavity . . . as an indication of antiquity."[24] But beware: That was before the "shocking" discovery of "too young" *H. erectus* types like Flores's hobbit, with her pea brain of 380 cc, who lived as recently as 12 kya. Forget antiquity.

TABLE 6.1. EXAMPLES OF DEEP-TIME GAFFES

Item	Date Originally Assigned	Subsequent Date Assigned
Skeletons, United States	38–70,000 BP(AAR method)	7,000 BP*
Neanderthal, Croatia	130,000 BP	ca 30,000 BP[†]
Fontéchevade Man, France	800,000 BP	100,000, 70,000, or 40,000 BP
Galley Hill Man, UK	mid-Pleistocene, ca 1 myr	Neolithic, less than 10,000 BP[‡]
Hamburg skull, Germany	36,000 BP (carbon dated)	ca 14,000 BP
KBS tuff, Africa	220 myr (K-Ar method)	2 to 6 mya
Krapina Man, Croatia	Third interglacial	Upper Paleolithic
Moab Man, Utah	65 mya	Native American bones from the past few centuries
Olduvai Bed II	500 kyr (dated by Louis Leakey)	17,000 BP[§]
Skull 1470, Africa	230 mya (volcanic rock test)	1.8 myr
Solo River *H. erectus*, Java	900,000 BP	30,000 BP[¶]

*Childress, *Lost Cities of North and Central America*, 12: For example, Paleo-Indian skulls in California are dated 70 kyr by aspartic acid racemization (AAR), which is sensitive to moisture. Amino acid dating got only 7 kyr.

[†]Coon, *The Living Races of Man*, 52.

[‡]The Galley Hill specimen was contemporary with Heidelberg Man who was supposedly 300 kyr; Le Gros Clark, *The Fossil Evidence for Human Evolution*, 56–58.

[§]Coon, *The Living Races of Man*, 52, 70.

[¶]Howells, *Getting Here*, 107.

THE DIFFICULTIES OF ACCURATE DATING

Radiometric Dating

Radiocarbon dating (C-14) was introduced in the late 1940s and works all right for relatively recent material only: "The men who run the tests would report that they cannot date with accuracy beyond 3,000 years."[25] C-14 analysis has actually given shells of *living* mollusks an age of 2,300 years. One of the problems is this—the decay rate of C-14 may not have been so constant in the past, particularly if you factor in the great abundance of vegetation on the landscape in earlier times. The globe was once more tropical with luxuriant plant life; without taking this into account, you get older ages from C-14 readings, which may also result from the proportion of C-14 isotopes in the atmosphere, which has not always been the same. There is also the problem of contamination by natural coal, slanting carbon dates in the too-old direction. The real question is: Does organic material take on C-14 at a constant rate? Open to debate, the C-14 margin of error, some say, is as high as 80 percent (up to 10 kya).[26*]

Potassium-argon dating (K-Ar), another radiometric method, enabled geologists and archaeologists to date very old materials. First developed in the late 1940s, it wasn't used in an archaeological setting until 1959. This new method began to give astoundingly old dates. It was an instant media sensation, attracting headlines—and funding. Come up with a fossil specimen *older* than the record holder—and you're a player. A 165,000-year-old shellfish dinner, for example *"pushes back* [e.a.] the earliest known seafood meal by 40,000 years"[27]—how about that! When Richard Leakey dated his Lake Rudolph Skull 1470 to 2.8 myr, he bragged that "our past has now been pushed back at least 10,000 centuries [by this] extraordinarily important [find] . . . the old-est skull of early man yet discovered."[28] Competition among diggers,

---

*Even ice-core dates, with that ring of hard science, may be less than dependable; Robert Schoch, in *Forgotten Civilization,* mentioned one such date changed from 14 kyr to 78 kyr.

even "science wars," have become a game of breaking records, like Ardi who "smashed the 4 myr barrier."[29]

In K-Ar dating, potassium, a leading parameter, is a radioactive element with a supposedly steady decay rate: apparently it takes almost 3 myr for 0.1 percent of "parent" K-40 to turn into "daughter" Ar-40. It is not the fossil itself that is dated but the potassium-bearing deposits in which it is found. However, collecting undisturbed and uncontaminated rock samples of indisputable association with the fossils is more iffy than we are led to believe. Too, we are relying on the constant rate of decay from potassium into argon: even though the margin of error is claimed to be less than 1 percent,[30] decay rates are only statistical averages.

The story of K-Ar began in 1959 with *Zinjanthropus boisei* at East Africa's Olduvai Gorge; at this time the University of California had just brought out the K-Ar method. Louis Leakey's first guesstimate for Zinj's age had been 600 kyr, which itself was looked on as extravagant, even outrageous, compared to a mere 10 kyr, presumed by other workers.[31] But K-Ar now dated Zinj almost 2 myr. Quite a leap—from five to seven digits! An overnight sensation.

Let's take a closer look: Radiometric analysis by the K-Ar method is used specifically to date lava and other volcanic rocks—despite the fact that volcanic ash is apt to be infused with much older crystals. Ash samples that are badly weathered are also useless. Consider this: The K-Ar method performs best on things 2 to 3 myr; in fact, K-Ar *cannot* date younger materials, that is, rocks younger than 100 kyr. Its capability extends to deposits only if they are *older than 500 kyr,* whereby it has more than tripled man's age and inaugurated the wayward trend toward deep-time or long dating, back-dating genus *Homo* into the *millions* of years.

Given that the K-Ar method refers to the age of volcanic soils, *not to the bones themselves,* it is a poor method if the bones are not securely in place, and many if not most fossils are, indeed, surface finds, kicked up from who knows where. Since argon 40 is a gas, it easily migrates in and out; potassium is also mobile. K-Ar in certain minerals will therefore show bogus age if argon gets trapped inside; indeed, argon could have already been in the rocks before they solidified. How can we be sure of the ratio of parent:daughter elements in the original sample? The delusory K-Ar method has yielded dates ranging everywhere from 160 million to 3 billion years old for the *same material*—rocks actually formed fewer than 200 years ago!

First man Ardi was dated wildly at 23 myr by testing the basalt (lava deposit) closest to the fossils. They realized, of course, that was way too old for any hominid, so they sampled some other basalt a ways off and "selected" those samples that gave a comfortable age of 4.4 mya—because that would be about right in the current scheme of evolutionary dating. We find a similar scenario in connection with one very controversial *Homo habilis*–like specimen, a skull, for which early testing again gave a too-old date of 230 mya! But humans weren't around then, of course, so they kept sampling volcanic rocks in the area, until settling on one that gave an age of 2.8 myr. Why? Because *this* age aligned with previous published studies and was assumed to be about right.

Radiometric dating, as David Pitman, Australia's brilliant young cosmogonist, has explained (in a personal communication), makes the blithe assumption that

> no (or very little) new material has been added to the earth since its initial formation. . . . But the planet has undergone periods when vast amounts of new material, both organic and inorganic, have been thrown down to the earth. Apparently, this is where our fossil fuels came from (i.e., they are not fossils at all!). . . . And the age you are measuring is *greater* than the length of time that the material has been on the earth. . . . It probably spent a whole bunch of time rolling around in space before it was caught in the earth's vortex and dumped on the surface. The accuracy of radiometric dates depends on these conditions. . . . It is impossible to know which particular measurements will be accurate without first establishing the vortexian history of the fossil location.

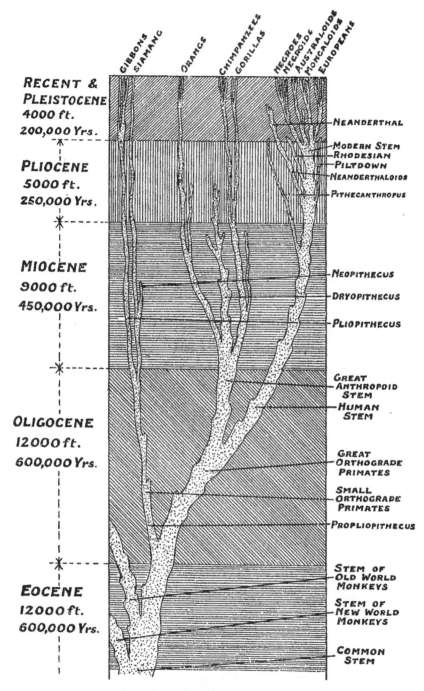

Figure 6.2. Genealogical tree (1929) shows estimated depth of deposits (left-hand column).

Our Earth is composed of material that gathered as it formed, whirled into shape by its powerful vortexian (field) currents*; the tendency to accumulate outside (extraterrestrial) material increases the farther back we go. Nonterrestrial material could have been in space for millions of years before being pulled into the earth-forming vortex. Without factoring in the influx of very old interstellar particles, radiometric dates are distorted in the too-old direction. Are we measuring Earth's age—or stuff that fell to the planet from dissolved stars and spheres?

As Pitman has observed,

> decay rates are likely to change over time . . . [and are] highly subject to the change in atmospherean currents. Think of the meteoris belts where solid elements like iron are apparently held in solution and condense out to create meteorites. . . . At this altitude, iron would be radioactive because it would be capable of decaying into pure energy. These belts, their location and intensity, are controlled by the atmospherean currents, and these currents change over time. Therefore the decay rate of a particular sample will depend on its vortexian environment, and this environment changes greatly over time. . . . The decay rate [is] mostly tending to slow down over time. . . . Using it for dating becomes very very shaky the more distant into the past you go. . . . Yep, we just need to take a bunch of zeros off those dates! . . . Heavy elements will tend to decay faster; the degree of how much faster will get greater the further into the past we go (when the vortex's rotation is proportionally faster). Thus there will be a direct relationship between the speed of the earth's rotation and the rate of radioactive decay—and thus the degree of inaccuracy of radiometric dating.

Indeed, some areas of the deepest ocean are highly radioactive, suggesting a much higher rate of radioactive decay in times past.[32] The quantity of cosmic rays penetrating the atmosphere may also affect

---

*More on vortex at my website www.earthvortex.com.

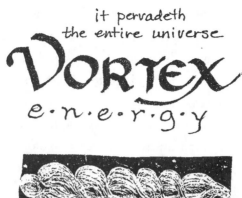

Figure 6.3. Vortex energy: the rotary force behind all motion; the force that driveth all things. The force is movement toward center, from external to internal, symbolized by the whirlwind.

the amount of C-14, research indicating that past surges of radiocarbon affected all parts of the globe. It is thought that the amount of rays varies as Earth passes through magnetic clouds; the strength of the magnetic field (which is ever decreasing) also affects the amount of rays that reach us. Baby Earth registered a fabulously turbo-charged atmosphere in ancient rocks, showing a hundred times the expected magnetism. The rocks are faithful recorders of the enormous amplitude of our planet's original force field, that force in turn having led to bigger earthquakes and greater volcanic activity—leaving lava beds almost a mile deep in the western United States. Lava flows in the Cretaceous covered land the size of India: today the Deccan Traps still extend over 200 thousand square miles, a mile and a half thick in some places.

From Molecular Dating to Genetic Analysis

When evolution turned "molecular" in the 1960s, we were told that this new science was based on proteins evolving at a known and constant rate, even though different proteins have different mutation rates. Anyway, this molecular revolution was followed, in the 1970s by the sequencing of DNA. In the same way that the DNA breakthroughs of the late twentieth century revolutionized the identification, the "fingerprint," of individuals (especially criminals), it also transformed the paleoanthropological arts, the celebrity status of DNA even threatening to rob the thunder of the good old fossil hunters.

Genetics (Mendelian), having once saved Darwinism (when it fell out of favor around the turn of the twentieth century), now made another go at it. Recent advances in molecular biology and genetics, it is now bruited about, can determine the history of species and virtually replace morphology (referring to the skeletal anatomy recovered at digs). Genetic distance can be measured, telling us when species diverged (split off), all of which is then nicely plotted on an evolutionary tree.

But outside the English-speaking cabal, there are geneticists like Maciej Giertych (Polish Academy of Sciences) who find no biological data relevant to tree genetics. Giertych says he could pursue his entire career without ever mentioning evolution. Built on supposition, this new school of population genetics *assumes* (1) that there is a constant rate of mutations and (2) that species have arisen by splitting off (diverging) from a parent stock—the theory blithely *assuming* a common ancestor, *which it is supposed to confirm!* All this is null and void if there has been extensive *crossbreeding* in the human family.

> *Race mixture is a very ancient phenomenon.*
> EARNEST HOOTON, *UP FROM THE APE*

Geneticists using mtDNA contend they can assess the amount of time that separates two species by measuring the accumulated mutations, the method called, appealingly, "the molecular clock." Offsetting this confidence is the fact that mtDNA is only a small fragment of our genetic heritage. And what if variations in mtDNA are due to factors other than mutation (like hybridization)? There is also the little problem of species that have *not* changed, mutated, or evolved in millions of years. Do we really have an "exact" (or relevant) science here? In an intriguing interview shortly before her death, biologist Lynn Margulis

Bunny Suits

Archaic DNA is easily contaminated with extraneous DNA, making the process of looking for ancient DNA extremely sensitive to contamination. Workers at an excavation site in Spain have taken to wearing "clean-room" bunny suits (jumpsuit, gloves, booties, plastic face mask) to protect samples from their *own* DNA. Handling, or even breathing on, a sample can easily throw off results.

Current work in RNA molecules has researchers claiming to have isolated collagen from a 68-myr *Tyrannosaurus rex*. Other scientists, however, suspect that the collagen comes from *bacteria* rather than dinosaurs, citing similar claims, for instance, of 80-myr dinosaur DNA, which actually came from a human. In any event, despite the claim to have found blood cells of some unfossilized dinosaur bone, it is doubtful that cells could have lasted any more than a few thousand years, no less 65 myr!

> *Dinosaurs have been gone too long for any genetic material to remain in fossils.*
>
> Bryan Walsh, "The Walking Dead,"
> *Time*, April 15, 2013

Most recovered Neanderthal bones have been extensively contaminated with modern human DNA. It is a painstaking process isolating these ancient genomes; as a rule, bacteria contaminates fossils. Recovering genetic material from such remains is fraught with difficulty, if only because DNA itself degrades rapidly. (Protein and DNA break down so fast, it is hard to believe they could survive more than 30 kyr or so.) Svante Pääbo's genome project was based on a leg bone from Croatia, and there were no bunny suits around when those Vindija Neanderthal bones were first collected; the remains sat in a drawer in Zagreb, the DNA unprotected for 30 years.

called the population geneticists who have "taken over evolutionary biology ... reductionists ad absurdum."[33]

As molecular genetics steps into anthropology's bailiwick, it boldly presumes to trump the paleontological arts. With DNA sequencing technology* upstaging morphological analysis (the bones) and claiming its increased rigor, "fresh evolutionary insights do not necessarily require any fossils at all."[34] Highly specialized biochemists (who are neither historians, archaeologists, anatomists, or paleoanthropologists) are then left to call the shots on the human record; and these population geneticists are saying that *H. sapiens* and Neanderthal did *not* interbreed.[35] This, we know can't be right.

Evolutionary geneticists Svante Pääbo and Johannes Krause of Germany's Max Planck Institute, using high-tech lab equipment and supercomputers, were the first to identify a novel Siberian hominid (Denisova) by using genetic analysis *alone, without* any real fossilized remains (the analysis was based on fragments of a pinky finger). Studying that precious pinky's mtDNA, scientists determined it was "twice as distant from us as Neanderthals." OK, but you cannot infer anything about the organism from the genes, which is to say, they have no idea what this creature looked like—its anatomy, context, ecology, or behavior. Nothing.

You don't discover a Lucy in the molecular lab, protested one anthropologist, others agreeing that genetics can never replace the hard work of the dig. Cold weather, significantly, helps preserve ancient DNA, but, hey, most early fossils are from tropical regions where conditions for DNA survival are poor. Neanderthal DNA, we learn, is only viable for 35 kyr; and protein will not last more than 100 kyr.[36] How then can this science tell us anything about creatures who roamed Earth (supposedly) millions of years ago?

---

*I would like to refer the interested reader to James Shreeve, *The Neandertal Enigma,* pages 74 to 78, where the flaws of dating by mtDNA are incisively drawn, including the refreshingly candid remarks of Milford Wolpoff.

> *Organic matter, metals and nonmetals . . . disintegrate*
> *within a few thousand years . . . a million years from*
> *now, the skeletons of Rodin, Renoir, Einstein. . . . . Will*
> *have turned to dust, along with the coffins in which they*
> *were buried.*
>
> ROBERT CHARROUX, *MAN'S UNKNOWN HISTORY*

> *Passing the buck . . . the evolutionists resolve all their*
> *problems by pushing them over to the geneticists.*
>
> GORDON RATTRAY TAYLOR,
> *THE GREAT EVOLUTION MYSTERY*

And so molecules (*and* radiometrics) are now setting the time frame for human evolution, the molecular clock supposedly measuring the passage of species time, even though mutation rates are not so fixed or reliable, as discovered in the work of molecular biologist John Cairns, who found bacteria were stress dependent, mutating at a faster rate when faced with an environmental challenge. Even the esteemed Ernst Mayr, considered perhaps the greatest twentieth-century evolutionary biologist, warned this eager new generation of scientists against the alleged constancy of rates.

The vaunted molecular clock uses *average* mutation rates, although there is "no theoretical reason why the accumulation of genetic change should be steady through time."[37] The rate of DNA mutations can fluctuate: "they fail to take account of the different rates at which mutations can accumulate."[38] Mutations can be brought on by cosmic rays, radiation, just as X-rays can produce mutations.[39] In excess, it "gives an array of freaks, not evolution." We well know that radiation can increase mutations, as the atomic age has made plain such horrible genetic effects.

Some scientists think supernova events threw colossal amounts of radiation to Earth. "The explosion of supernovae has subjected the earth at least fifteen times to showers of radiation strong enough to kill most forms. . . . These bouts of high radiation could well have been a significant factor in the evolution of life," says Lyall Watson in *Supernature*. Paleobiologist of Chicago's Field Museum, David Raup, in *The Nemesis Affair*, predicts that "as we near the plane of the galaxy, various aspects of our cosmic environment change. . . . Interstellar clouds of gas and dust, and the levels of certain kinds of radiation may increase . . . [which] might produce biological effects on earth."

And how much stock shall we place in genetic distance when chimpanzees are said to be *genetically* closer to human beings than to gorillas and we humans are 98.4 percent identical to chimps? Or when 95 percent of DNA is "junk DNA"?[40] Or when Neanderthal sequences come out as "somewhere between modern human and chimp sequences"? Paradoxically, one molecular biologist, John Marcus, after performing his own DNA distance comparisons, found "the Neandertal sequence is actually further away from . . . the chimpanzee sequences than the modern human sequences are." Humans closer to chimps than Neanderthals are? Marcus concluded: "the Neandertal is no more related to chimps than any of the humans. If anything, Neandertal is *less* related to chimps."[41] *Huh?*

Successfully intimidating with complicated formulas and daunting nomenclature, the work of the geneticist has been properly scientized by all these: molecular phylogenetics, allopatric populations, epistatic interactions, allelic combinations, prezygotic isolation, sympatric speciation, and amphiploids—none of which anyone outside the cabal understands and little of it in fact pertaining to *humans*—mostly to plants and fruit flies. Does the number of bristles on fruit flies tell us anything about human evolution?

Chemical and Faunal Dating

One chemical dating technique measures the amount of nitrogen lost and fluorine gained in bones buried in a deposit. But the amount of fluorine in the bones depends on the amount of fluorine in the soil; if rich in it, the bones there embedded can become rapidly saturated. Neither does fluorine, which yields highly variable results, accrue at a uniform rate; its absorption in volcanic areas tends to be quite erratic. Result: Bones of wildly different dates may have similar fluorine contents.

Fontéchevade Man has been loosely assigned a meaningless scatter of ages, anywhere from 40 to 800,000 years. But recovered from damp, clayey soil, it makes the latter date most unlikely. The nitrogen preserved in bone, a factor used in dating, varies greatly depending on the amount of clay in the soil. England's acclaimed Ipswich skeleton, beneath only four feet of clay, was assumed nevertheless to antedate the last glaciations, even though "the geological evidence of antiquity was actually quite inadequate," as noted by Le Gros Clark in *The Fossil Evidence for Human Evolution;* he further remarked that "even so, it was seized with avidity by those who were particularly anxious to bolster up arguments for the remote origin" of our species. (To Boule, these AMHs were "in reality barely prehistoric.")

Concerning Au dating, Le Gros Clark went on to warn that South African deposits are mostly in the form of breccias in which stratification (observable layers) is either absent or too poorly defined to permit geological dating. So an alternative was to base dating here on associated fauna—even though South African faunal correlations of those early periods were still not worked out. Some of these fossils (mammals) put Au in association with *Pliocene* fauna, even though a number of these mammals actually survived to a much later date.

Faunal dating uses animal bones to determine the age of sedimentary layers or fossils buried in those layers. This method is a touch circular: How is the geological age of rock determined? By its index fossils, meaning the fauna (animals) most characteristic of a particular stratum; but that fossil in turn is dated according to an assumed evolutionary sequence *determined by the rocks.* In other words, rocks are used to date fossils; fossils are used to date rocks. Trilobites, for example, have been dated about 550 myr because they are found in Cambrian layers; how do we know it is the Cambrian? Because of the presence of trilobites (index fossil). Faunal fossil markers (animal remains) may be mistakenly labeled, say, Tertiary, when they might actually be much more recent. As an example, the fauna used to long-date Krapina Man to 130 kyr actually survived to a much later time, 30 kyr. When such "index fossils" turn out to be long lived throughout several strata (or in fact still *extant*), you can throw out the whole ball game. (Some modern, extant, species previously thought to be extinct for millions of years are listed in chapter 12). Although *Australopithecus africanus* was dated by faunal assemblages, "African faunal sequences are not very precise indicators," warned William Howells; faunal dating is "notoriously capricious."[42]

Paleomagnetic Dating

Paleomagnetic dating is yet another method of determining age, this one according to magnetic field reversals: north pole becomes the south pole and vice versa. Such pole shifts are evidenced by the alternating zig or zag direction marked in ancient rocks: this is because volcanic basalt (the commonest rock on Earth) cools in the direction of the magnetic field. Solidified lava is then magnetized like the compass needle, facing North.

It was paleomagnetic dating that came up with that extraordinarily old date for *Au. sedipa*. But, trust me, the timing of these magnetic pole shifts is still indeterminate. Magnetic polar shifts are not the same as geographical pole shifts, the notion that the rotational axis of the Earth flips or turns over in space either due to crustal slippage or movement of the entire planet. Geographical pole shifts have been investigated in John W. White's *Pole Shift* and found by him to be pseudoscientific, without basis in fact.

I have researched these shifts and have never succeeded in finding any scientific consensus on the *rate* of these reversals. Again, it's a question of time. To give you a taste of the discrepancies: Some say a reversal occurs once every 5,000, 7,000, or 28,000 years; others say, every 100 kyr; others 200 kyr; still others 250, 500, 550, 780, or 1 myr. The Mammoth reversal supposedly lasted from 3.1 to 3.0 mya; the Gilbert reversal from 3.6 to 3.4 mya. Alternatively, a pole shift happened 26,500 years ago.[43] Or the Mungo event occurred some 35 kya; the Gothenburg event 13 kya; a "fully confirmed" field reversal 10,000 years ago[44]; or twelve pole flips have happened in the last 5 myr. Take your pick!

A GEOLOGICAL NIGHTMARE

Earth's crust and ancient beds, it is well known, may be distorted by faulting and folding and redeposited gravels, or material carried from elsewhere: newer stuff can get wedged in the gap of an older rock layer. When a hominid fossil is labeled intrusive, that means it has been accidentally reburied in some other deposits. Louis Leakey's 1-myr Oldoway (Olduvai) Man, for example, turned out to be a modern *H. sapiens* accidentally buried less than 20 kya in older deposits, which had been scrambled by faulting.[45] Workers know Olduvai Gorge is a place that has seen recurrent faulting and deformation. Even older than Olduvai Man, touted as the world's earliest *H. sapiens,* was Kanjera Man, another Leakey exploit in East Africa, which also turned out to be a modern human buried somehow in older sediments.

Neither is river plain a good stratigraphical context: sediments are so mixed up by flowing water. Many important fossils were recovered in river plains, Java's famous *H. erectus* man found right at the edge of a river. Alkaline washing, too, gives dates much older than they should be; some of the carbon-14 necessary for accurate dating is apparently removed by exposure to such treatment, thus deepening the age of the sample. In Africa, many of the key Kenyan fossils are from dry streambeds. Turkana Boy, found in a riverbed, was long dated to 1.6 mya. In Europe, an early dated Neanderthal type, Steinheim, was found in river gravels and dated to 300 kyr.

With many geologists frankly doubting that sedimentation rates are uniform, nineteenth-century uniformitarianism has been seriously challenged. So many factors can affect the transportation of Earth materials: speed of flow, roughness of channel, temperature, type of material carried, direction and volume of flow, depth, slope, and chemicals present. *And most fossil bones are indeed found in sedimentary rock.*

In places like Ethiopia where Tim White discovered what he trumpeted as "5.5-million-year-old hominids," the setting is nonetheless a "geological nightmare. You have a patchwork quilt of different aged sediments on the surface."[46] White's old partner in East Africa, Donald Johanson, said: "At Hadar the minerals in some of the volcanic ash layers we normally use to date fossil-bearing formations had been altered or contaminated by later geologic processes."[47]

Ethiopia's now-famous Ardi was discovered in a rainy region. Torrential downpours wash up traces of ancient stone and bone from *different* eras; remains as old as 5.7 myr can get mixed up with stuff as recent as 80 kyr. And that's just the point: We will return to these dates in chapter 10, where that figure, 80 kyr, is proposed as the oldest possible date for man. Extended pluvial periods in East Africa complicate the reading of deposits; here dating is surrounded by "clouds of dust."[48]

> *Red sandstone** ... *limestone, gravel conglomerates, and*
> *other formations extend over exceedingly wide geographical*
> *areas. The problem posed by these strata is that they suggest*
> *a blanketing of the planet from extraterrestrial sources.*
> *Sedimentary rocks are something more than we have*
> *been taught. . . . Our planet has been blanketed by much*
> *extraterrestrial matter in previous times.*
>
> VINE DELORIA JR., *RED EARTH, WHITE LIES*

Surface finds present their own share of problems. Hominid fossils found near the surface are much more likely to be contaminated with superfluous material of both different origin and different age. Java's Sangiran hominids almost all occur on the surface, making their real age uncertain. In Africa, *H. habilis* femurs were found right on the surface—which didn't stop Richard Leakey from dating them to 2.6 myr. Few of the locations where important paleoanthropological discoveries have been made can thus

---

*A rain of sandstone is mentioned in the Mexican Codex Chimalpopoca. A recent Colorado State University study suggests that *rapid layering* of sandstone is possible. An extraterrestrial origin of loess (silt) has also been proposed.

qualify—and that includes the whole australopithecine family of Africa, as well as *H. habilis* and *H. erectus,* which occur on the surface or in cave deposits, the latter also notoriously hard to gauge geologically.

Most hominid fossils gotten prior to World War II were taken from caves. Yet dating cave finds is iffy, for they tend to get "churned" up; recent material can get sucked down in "swallow holes." Glaciers also mixed up cave deposits. Significantly, most caves *older than 150 kyr* have collapsed or been flushed of deposits—yet Sterkfontein and Swartkrans and most other South African Au for that matter are from caves and insouciantly dated *ten times* older than that! "Time resolution of cave deposits is poor," admits Richard Leakey.[49] Often a jumble of material, such finds tend to be hopelessly scrambled. Most Neanderthals were found in caves, as were the hominids at Sima de los Huesos, Grimaldi, Krapina, Chapelle, Belgium, Dusseldorf, Moravia, Mt. Carmel, and Peking.

Volcanic regions are also notoriously problematic. On Java, those (surface) Sangiran hominids were found at the foot of a volcano; here, Trinil's *H. erectus,* in volcanic rock, is dated to the early Pleistocene or even Pliocene (more than 2 myr). Africa's Skull 1470 was also found in a volcanic context; the KBS tuff for 1470 gave a scatter from 500 kyr to 17 myr! While volcanic activity does enhance fossilization, the ash and pumice in such beds tend to become eroded and transported by streams and rivers.

The Olduvai Beds in Tanzania are also interspersed with volcanic material: the radiometric dating of these ash layers, encasing the deposits, gives Ardi an age of 4.4 myr. This date, however, has been questioned, for Ardi's region is difficult to date radiometrically. Also dated in the *millions* of years, East Africa's controversial Laetoli footprints are impressed in volcanic ash, just as Lucy herself was dated from lava flows; but lava flow may easily contain older rocks.

COSMIC CLIMATE

While evolutionists routinely turn to climate shifts and other supposed environmental challenges as the alleged trigger of natural selection and, therefore, of evolution itself—largely overlooked are the vagaries of *cosmic* change, the *cosmic* climate—and its inexorable effect on terrestrial geology. Not much is known about these nonterrestrial imponderables.

Scientists have spoken of the difficulty in dating objects that have had contact with electricity, which can distort the results of radiometric dating and of mutation rates. This has been (experimentally) shown in plant seeds and fish eggs. An electrical field of high concentration, or solar flares, can alter decay rates. In this regard, Stephen E. Robbins has warned: "When the atoms of the nucleus are excited, decay is much quicker, making things look vastly older. Cataclysms on a vast scale involve high energies that could easily alter radiometric clocks."[50]

> *Decay rates were much faster . . . in the past.*
> JONATHAN SARFATI, *REFUTING COMPROMISE*

Blanketing is another phenomenon, which is to say, extraneous material falling to Earth (extraterrestrial debris), resulting in contamination by older *cosmic* material gathered up in Earth's vortex. This interstellar dust (which NASA calls space clouds and which, according to the nebula theory, actually formed Earth) is *older* than Earth itself. As reported by the U.S. Geological Survey, each year tens of thousands of tons of interstellar dust still fall to Earth (some say 80 million pounds of space dust are added to Earth per annum). This cosmic dust influx, others estimate, could amount to anywhere from 10,000 to 700,000 tons per day. With so much space dust accumulated, could a 5-million-year-old creature even be dug up from the depths?

Consider some of the highly charged events in our planet's history, as witnessed by regions of Earth like the Gobi Desert with its irradiated sands, or by Earth scars and bombardments like the ancient tektites (glassy blobs) in Java, India, Australia, and France; or the strange black stones called *harras* in the Arabian desert; or vitrified areas in India and Libya, so like those in America's Death Valley and the fused stone ruins near the Gila River (remembered by the Hopi as fire from heaven). Or the melted stones at Brazil's Sete Cidades. Or Scotland's and Ireland's vitrified hill forts, calcinated by extreme heat. Or Turkey's and Iran's fused rocks that appear blasted from above (similar to the floor of New Mexico, where it was scarred by the first A-bomb tests).

Do these represent the brimstone of high antiquity? seventy-five thousand years ago when Earth's atmosphere had not quite settled down, "the gases of her low regions [were] purified to make more places for mortals. . . . Fire, brimstone,* iron and phosphorus fell upon the earth . . . and this shower reached into the five divisions of the earth."[51]

Some time around 22,000 years ago, "the earth . . . [was] dripping wet and cold in the ji'ayan eddies . . . [bringing] a spell of darkness."[52] And again, ca 8,000 years ago, "a'ji began to fall. The belt of meteoris gave up its stones, and showers of them rained down on the earth."[53] Aren't thick rock layers (showing little time between layers) consistent with the rapid deposition of such sky falls?

Figure 6.4. A'ji: degree of density needed to create a world; dark period on Earth sometimes accompanied by stone showers and other strange phenomena. A'ji signifies interstellar fields of greater densities and different properties.

Contrary to the uniformitarian view, so essential to evolutionism, early Earth and past processes cannot be judged altogether by present ones. "Extraterrestrial forces periodically disrupt the normal course of life. . . . The episodic course of natural history [redefines] the uniformitarianism of Hutton and Lyell . . . looking beyond the planet."[54] Thomas Henry

*Reminding us of the brimstone of red-hot salt and sulfur that covered the Vale of Siddim (Dead Sea area), represented in the biblical (Genesis 19) Sodom and Gomorrah.

Huxley, in fact, thought the remote period saw Earth passing through physical and chemical conditions "which it can no more see again than a man can recall his infancy."[55] His good friend Charles Darwin had to acknowledge Huxley's and Lord Kelvin's view that early Earth was subject to more rapid and violent changes. Indeed, Darwin deleted his section on steady sedimentation as a reliable chronometer (from the 1868 edition of *Origin*), after contemplating Lord Kelvin's work, including his, Kelvin's, *more recent* dating of Earth (see chapter 10).

Thousands of feet of sediments have buried the coral reefs; *but not always in steady increments*. Certain (polystrate) fossil trees seem to have been laid down rapidly; and there are signs of other episodes of super-sedimentation: consider the times of "Luts wherein there falls on a planet condensed earthy substances, such as clay, stones, ashes, and

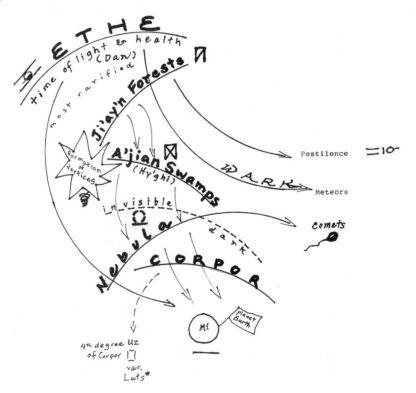

Figure 6.5. Luts: seasons of the firmament. Luts is the opposite of dan (light). It is also the brimstone of biblical fame. "Great cities [were] . . . covered up by falling nebulae" (Oahspe, Book of Sethantes 9.13).

disseminated molten metals, in such great quantities that it can be compared to snowstorms, piling up corporeal substance on the earth to a depth of many feet, and in drifts up to hundreds of feet. Luts was . . . a time of destruction."[56] Around 12 kya, "there came great darkness on the earth, with falling ashes and heat and fevers."[57]

The Finnish *Kalevala* recalls hailstones of iron that fell from the sky. Before our planet stabilized, long periods of darkness, luts, meteorites, stone showers, "star oil," and other destructive sky falls assailed Earth, lasting sometimes for hundreds of years. The quantity and quality of such precipitations to Earth have been greatly variable. The substance of a'ji and falling nebulae may be so fine, it is invisible to the eye; nevertheless, it is capable of building up in very short periods. "Mortals did not see the a'ji; but they saw their cities and temples sinking, as it were, into the ground; yet in truth they were not sinking, but were covered by the a'ji falling and condensing."[58]

BIG POPULATIONS AND EARLY CIVILIZATIONS

[A]ll the growing body of evidence for "art" before 40,000 years ago is simply dismissed and ignored.

PAUL BAHN, *NATURE* MAGAZINE

Ironically, the flip side of science's extravagant long dating is the stubborn refusal to recognize any sign of civilization earlier than, say, 6 or 7 kya. The same intellectual establishment that has *over*estimated man's time on Earth (by millions of years) turns around and *under*estimates man's achievements by at least 40,000 years.

Similarly, there is the fixed idea that human populations in the Pleistocene were small. This belief reflects academia's wholesale dismissal of lost races and extinct civilizations, a factor that, sooner or later, will blow up in their face.

A World Filled with People

Spencer Wells, a population geneticist and molecular anthropologist who collected DNA samples from (living) people all over the world to trace the roots of human history, states that his data reveal "as few as 2,000 people were alive some 70,000 years ago."[59] This outrageous claim is on a par perhaps with Brian Fagan's assertion that mankind became almost extinct 73,000 years ago due to a Sumatran volcanic eruption: "there were very few of them indeed."[60]

The geneticists' assumptions concerning human populations may be off target: The problem here is that each population has a separate demographic history, which invalidates the use of mtDNA to clock past events in any standard way. Population size alone can throw off its accuracy: for example, if a population grows more in one region than another, this can lead to greater (genetic) diversity; it would not, as claimed, be necessarily *older*—which circumstance queers the whole genetic distance premise.

> *Banish the thought that primitive populations were always sparse.*
>
> WILLIAM CORLISS, *THE UNEXPLAINED*

Small population? Why, man filled Earth, at least according to Genesis 1:28, 6:1, and 6:11, as well as chapters 10 and 11. Ironically, it was Malthus's very idea of *too-large* populations that inspired Darwin to postulate fierce competition and survival only of the fittest. Despite this, Darwin did indeed embrace small populations, if only because his model of speciation required small isolated groups in order for changes to take hold.

The world hath been peopled over many times and many times laid desolate.

OAHSPE, BOOK OF APOLLO 2:7

Peak populations followed by swift collapse have been well documented by British historian Arnold Toynbee and more recently by Jared Diamond. There were, to give one example, up to fourteen million people in Central Peten (classic Mayan) civilization, reduced to a mere thirty thousand by the sixteenth century. In tribal legend, too, the people of Melanesia and Vanuatu actually say the mortality of men (the origin of death) was due to overcrowding on Earth. Likewise does Vietnamese legend speak of primordial overcrowding that got so bad the poor lizard could not go about without someone stepping on its tail. Sources such as Babylonian and Persian creation cycles say the reason for the flood was overpopulation; just as the Greek Zeus planned a war to reduce population.

The Oahspe chronicles recount that the Ihins alone once "covered over the whole earth, more than a thousand million of them."[61] Worldwide population counts embedded in these scriptural histories may be summarized as follows.[62]

TABLE 6.2. WORLDWIDE POPULATION COUNTS

Years Ago	Population
70,000	Over 2 billion
67,000	Almost 4 billion
63,000	4.8 billion
60,000	6.4 billion
57,000	8 billion
53,000	9 billion
50,000	9.4 billion
41,000	10.8 billion

While William Whiston, an eighteenth-century Cambridge theorist, correctly, I think, estimated the preflood population at more than 8 billion, paleontologists today guess at only a few million people or possibly as few as 1.3 million humans ca 50 kya,[63] at which time there were supposedly only 15,000 to 20,000 Neanderthals in Europe and Eurasia: "tiny numbers of people,"[64] or as Erik Trinkaus put it: "There were very few people on the landscape."[65]

Around this time (Aurignacian), however, a great accumulation of Cro-Magnon "cultural debris"[66] speaks of quite large populations in Europe as well as a "very large population in Swaziland" (35 kya).[67] All the continents and islands of Earth were inhabited. There was no wilderness, for mortals were prolific, many of the mothers bringing forth two score sons and daughters, and from two to four at a birth.*

> *The inhabitants of the Earth were before the Flood vastly more numerous than the present Earth.*
>
> THOMAS BURNET, *SACRED THEORY OF THE EARTH*

There were also great cities in all the divisions of Earth before the flood, but all were cast down. Later, after the flood, in the time of Apollo (ca 18 kya), the Ihins in America alone numbered 4 million souls; and at that time there were in the world hundreds of millions of Ihuans and ground people.[68] Yet, for this period, the Mesolithic, paleontologists put a mere 10 million people on Earth, or as few as 5 million.[69]

Twenty Thousand Years of Civilization

If human beings didn't really put together an advanced society until well into the Holocene, why does the Paleolithic give us cities tens of thousands of years ago? Freeze-frame 70 kya: "Mortals had given up the ... [wilderness] and come to live in villages and cities."[70] Oahspe tells us that around this time the first written language in the world (Semoin) was developed, with characters engraved and copies brought to all the cities in the world (see figure 6.6).[71]

---

*Synopsis of Sixteen Cycles 1.19, also Book of Divinity 13.16: the people of the Old World fell from holiness and population increased dramatically. Also: 8,000 years ago, in time of Fragapatti: darkness was on Earth for 166 years, and the Zarathustrians gave up celibacy by hundreds and thousands, and married and begot children in great numbers, many women giving birth to 20 and some to 25 children. And some men were the fathers of 70 children, and not a few even of 100.

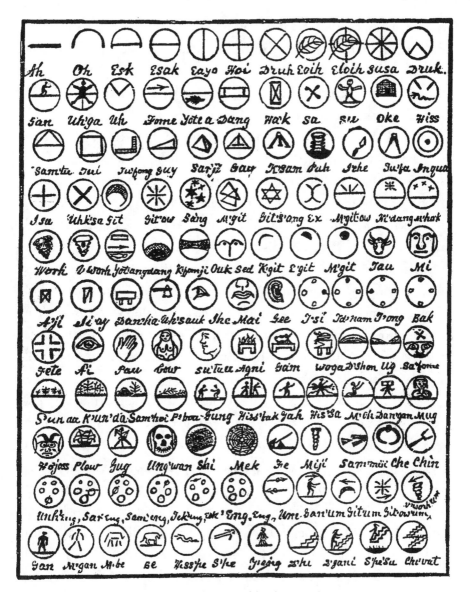

Figure 6.6. Semoin tablet from Oahspe.

Independent researchers have discovered that cultured AMHs *coexisted* with archaic types: "parallel to the hunter gatherer societies . . . a higher level of civilization also existed. . . . Humans more than 20,000 years ago had precise knowledge of major star coordinates. . . . The Egyptian calendar's starting date . . . is within 400 years of the oldest

Mayan date, recorded as 51,611 years ago on the Chincultic ceramic disk."[72]* Indeed, Martianus Capella stated that the Egyptians had secretly studied the stars for 40,000 years before revealing their knowledge; Diogenes Laertius also dated Egyptian astronomy to 48,863 years before Alexander the Great.

The Babylonian priest Kidinnu, about 15,000 years ago, was an astronomer who knew the facts relating to the yearly movement of the sun and moon, this science also predicting lunar eclipses with precision. North of Babylon (and possibly its mother culture), Gobekli Tepe, with its ancient observatory (the site also a ritual center, a sanctuary), is dated to 14,000 BP. This recently discovered star chamber in the highlands of Old Turkey, with finely carved reliefs on massive columns (five meters high), is *aligned north–south* (as is the much later Great Pyramid of Egypt), betraying a precise knowledge of geodesy 7,000 years *earlier* than the presumed beginning of exact science. Similar temples of the same age are being explored at Karahantepe, Sefertepe, and Hamzantepe, all antedating the Neolithic.[73]

The first period of civilization brought navigation, printed books, schools, astronomy, and agriculture.[74] Agriculture and other arts may go back as far as 45 kya: "Her people have tilled all the soil of the earth . . . feeding great centers with hundreds of thousands of inhabitants in all the five divisions of the earth."[75] Pottery fragments found in Belgium in Mousterian layers are dated 50 kyr, and here some of the Spy II specimens show AMH traits.[76] The ceramic arts (usually associated with sedentary agriculturalists) are in evidence in Czechoslovakia (Moravia) as well, at the 29 kyr site of Dolni Vestonice, a populace that also produced carvings, engravings, portraiture. "Explain Dolni Vestonice, and you explain humanity."[77] This extensive settlement, boasting dwellings up to fifty by thirty feet, was a thriving town long before the Holocene,

---

*The Maya developed writing, cities, canals, mining, architecture, carving, engineering, draining of swampland, irrigation, and navigation quite a bit earlier than supposed. The Puebla material at Valsequillo (see appendix F), in particular the Dorenberg skull, could well be 80 kyr. (VanLandingham, "Blocking Data, Part 2," 2–3.)

even though their ceramics (fired clay) and kilns are not supposed to be there until the agricultural revolution of the Neolithic, 20 kyr later. Archaeologists have also uncovered 28 kyr agriculture (evidence of cultivated grains and related tools) in the Solomon Islands.[78]

Twenty thousand years ago, in a short-lived golden age, there were thousands of cities with great canal works, but all were destroyed by the time of Apollo.[79] However, in time, they built up again and by 12 kya, thousands of cities thrived in the lands of Ham (Africa) and Guatama (America)[80] where it has been estimated that the great mounds along the lower Mississippi had "a population as numerous as . . . the Nile or the Euphrates. . . . Cities similar to those of ancient Mexico, of several hundred thousand souls, have existed in this country."[81] Man's existence in America, we realize, has been steadily pushed back from 6 kya to 10 to 20 to 30 and even 40 kya (see more on this in appendix F). Louis Leakey, in his New World foray, ignored conventional chronologies and dated American material culture to 48 kya.

Even if archaeologists say the first cities in the world came about 4500 BP (and strictly in the Old World), a lost horizon lies buried deep in the hidden earth: "The Ihins . . . built mounds of wood and earth . . . hundreds and thousands of cities and mounds built they."[82] Nor should we dismiss records of such lost cultures as known to the Hindus, Tibetans, Persians, Chinese, Polynesians, Maya, Zimbabweans—each with traditions of previous "worlds" and every one of them waved off, discredited, or simply ignored by the same experts who promote the fable of 5-myr man! Proceed cautiously within their precincts. Therein lies wizardry.

If our dating is wrong, "we might," mused Robert Schoch, "be forced to not only rethink our science, but to rethink our history as well. . . . The stakes could not be higher."[83]

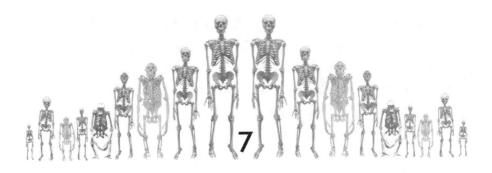

THE SPARK

Factoring In the Human Spirit

Two entities hast thou—that which is flesh, and that which is spirit.

OAHSPE, FIRST BOOK OF THE FIRST LORDS 3:17

*Alone of all the animals on earth man is twofold, mortal
through the body, immortal through the essential Man.*

THE DIVINE POIMANDRES OF HERMES TRISMEGISTUS

MY SPECIAL HERESY

There is a ghost in the man machine, unseen but potent. *Homo sapiens,*
a kind of man-god, has something of the transcendental in his makeup.
Anthropologists deploy words like *cognitive, symbolic, ritual, imagination,* or *abstract* as an acceptable substitute for outright soul and spirit,
which designates the incorporeal component of our being. Though
they do not believe it themselves, ethnographers have found, among
the ancient tribes, detailed eschatology and grasp of the afterlife. This
knowledge of immortal life is also inferred from burial sites with grave

goods and other tokens of the Ongoing (the life beyond). Were men of past ages deluded (as the archaeologists' subtext often implies), or were they on to something we have lost?

Should we look toward ape or angel, god or gibbon, for the secret of our beginnings? Hermes, Zoroaster, Homer, Aristotle, Plato, Pythagoras, and many other sages of the ancient day believed the human soul to be an emanation of the divine. It is not only recent religions or fundamentalists that think this way. The spirit part can be understood as something distinct from biological life; evolutionists, however, do not hold this vision: "There was nothing special about how *Homo sapiens* came to be."[1] In other words, *Homo darwinensis* is an animal, a talented, intelligent animal. Evolutionism talks of "our place in nature," ignoring our place in supernature, in the cosmos, in the invisible world. Entirely disengaged from any subtle or intangible factor, evolutionism labors under the one side of life—the physical, corporeal side. Only.

> *If all life is material, what is the mind and what is the soul?*
>
> KEITH THOMSON, *THE WATCH ON THE HEATH*

Charles Darwin and Karl Marx, double-handedly pulled off what Jacques Barzun called "the final separation between man and his soul." Darwin, Marx,* and Freud were among the most influential thinkers of the late nineteenth century and all atheists. But then came Einstein who, though mistakenly called an atheist, believed somehow and explored the great mystery. He knew that a science that ignores the intangible would never be complete. His vision of the all one cosmos required beauty and purpose, design, order, and majesty.

---

*Marx believed Darwin had provided a scientific basis for communism, but when he asked Darwin for permission to dedicate his book *Das Kapital* in Darwin's honor—or if he would write an introduction—Darwin declined the pleasure.

Figure 7.1. Charles Darwin. Outdoorsman in his youth (he loved shooting birds), country squire in his prime, independently wealthy, ungregarious, sickly, voracious reader, great naturalist, and mad scientist (ingenious botanical experiments), Darwin was able to work out his theories at his own tempo.

Although Darwin ended up with all the credit for evolution, his countryman, the naturalist Alfred Russel Wallace, was actually a codiscoverer. A complicated and well-documented set of circumstances put Wallace in the shade. But that's OK; he had his own destiny to fulfill, for all this was happening in the great age of spiritualism, in the mid-nineteenth century. The fact is, Wallace became a confirmed spiritualist by the 1860s, and when it came to man's true origin, his whimsy was that the first human beings were "plucked spiritually from nature." Whatever evolution was, it was "divinely directed"; human thought was the one thing that could not have been the result of natural selection (the subject of chapter 9). Some higher power had to be behind it. There was to Wallace an unknown reality. Up against Darwin, this was, as Wallace put it, "my special heresy." Since consciousness was to him "completely beyond all possibility of explanation by matter," there must be "an unseen universe—a world of spirit, to which the world of matter is altogether subordinate."[2]

Why wilt thou, O man, search forever in corpor [material world] for the cause of things? Behold, the unseen part of thyself ruleth over the seen.

OAHSPE, GOD'S BOOK OF BEN 2:9

This was not unlike Charles Lyell's position, which rejected evolution or natural selection as an explanation of mind. "I rather hail Wallace's suggestion," Lyell wrote (to Darwin's dismay) in 1869, "that there may be a Supreme Will and Power . . . guid[ing] the forces and laws of Nature."³

Figure 7.2. Alfred Russel Wallace (as quoted in Jeffrey Goodman's The Genesis Mystery) *thought "a superior intelligence had guided the development of man . . . and for a special purpose, just as man guides the development of many animal and vegetable forms.*

Though Wallace invoked "Great Mind" overall as something beyond material measurement ("disconnected from a physical brain"), the bone people steadfastly maintain there is no separate moral realm or ground of consciousness, and even if there is, it has nothing to do with the actual ascent of man. Yet Wallace stood his ground, insisting that the rapid development of the human brain demands a force of transcendent power at work. Today, evolutionists smugly call this

Wallace's "vague and goofy" ideas of human origins[4]; perhaps he "had simply gone mad."[5] Stephen Jay Gould sniffs at "Wallace's error . . . the fallacy of Wallace's argument . . . [which is] deeply wrong"[6]—lording over all the fools who believe the human spirit is a unique endowment.

But no one has yet to figure out why consciousness, if it is a *by-product* of evolution, "seems to spring into existence full blown."[7] Why is it that man's rise is so "explosively short" (Loren Eiseley) and all other animals' ploddingly slow and gradual?

St. George Mivart, biologist, zoologist, and acquaintance of both Darwin and Wallace, did not see human beings as part of physical evolution; the acquisition of a soul or mind must be by some spiritual agency, some higher power. How else could intellect or psyche have been gotten? Not through natural selection.

To sever consciousness from its source (albeit unseen) is, in my view, the most soul-deadening act of materialist science. "The existence," says a team of Hungarian ethologists, "of nerve cells in itself is not an explanation for consciousness. Consciousness is a characteristic of the soul."[8] And that's the problem: It pertains to something independent of matter; no wonder theorists wash their hands of consciousness, pegging it an "impenetrable mystery" (best left to "goofy" metaphysicians).

Science can tell us a zillion things about the brain but nothing much about consciousness. And when agnostics try to deal with it, it becomes a nonsensical string of words, a tangle of empty jargon: "Perhaps the extraordinary phenomenon of modern consciousness is . . . a chance combination of exaptations that were locked into place as an unexpected functioning whole by a final keystone acquisition that appeared . . . independently of any functional role. What that acquisition was . . . will have to await a more profound knowledge of the . . . human brain."[9] This is an example of *obscurum per obscuris*—explaining the unclear by means of the more unclear. Sentences like this make me swear out loud.

Prophet of materialism, Darwin was a child of privilege, though fashionably liberal and nonconformist; he was a cautious, thorough, methodical, diffident, and reclusive man. Although a blameless figure in most respects, he behaved rather badly to the spiritualists, calling their work "rubbish," their mediums "scandalous."[10] Committed to strict materialism (animal behavior was his particular niche), he made man simply an improved animal, a glorified baboon. A clerical and medical dropout, Darwin's real interest was natural history—bugs and barnacles. Early in his career, he had come to question the existence of a soul; now he called consciousness (in his notebooks) "a secretion of brain." Thought, then, must be the flux of cerebrum, as bile is of the liver. Unlike Wallace, Darwin dismissed any ethereal component behind mind and the founding of our species.

And so, from Darwin's time forth, evolutionists have busied themselves with the distinguishing *physical* features of humans—most notably bipedalism and our almighty brain. Man's almighty soul doesn't even get honorable mention.

> I came to the earth to develop the soul of man chiefly.
>
> OAHSPE, THE LORDS' FIRST BOOK 1:23

CHILD OF THE STARS

Evolutionists do not believe we possess a soul, or if we do, it is irrelevant to our origin. The theory of evolution is designed and tailored by and for the agnostic or frankly atheistic position. Yet, we call ourselves *Homo sapiens,* and *sapiens* means "wise." *Homo sapiens sapiens:* man the double wise. And so, it *is* mind that principally distinguishes us from animals, and while brain is certainly the *vehicle* (the apparatus, the housing, the machine), it is not brain that moves us to think. Man, unlike all the animals, is something more than a zoological phenomenon. A meta-explanation is in order.

> *Humans appeared on the scene suddenly and without evolutionary ancestors.*
>
> MARVIN L. LUBENOW, *BONES OF CONTENTION:*
> *A CREATIONIST ASSESSMENT OF HUMAN FOSSILS*

There is a reason why our species had AMH genes from the beginning, yet no one seems to know where those genes came from! Significantly, Louis Leakey thought the only proper ancestor for humanity would be an ancient "true man"—quite an un-Darwinian idea. Mind and AMH were concurrent, thought George Frederick Wright, and came about without evolution; the intelligence of man, he said, implies "creative interference . . . a direct ingrafting of divinely related qualities."[11]

I spent a lot of time in my last book (*The Lost History of the Little People*) explaining the phenomenon called euhemerism, which is the deification of the ancestors. But why were they—these AMH ancestors—deified? Because they were a holy people. The Ihins, now a lost race, were invested with soul force (and even immortality) from the beginning. A revered race, they were "capable of prophecies and miracles to such an extent that all other people called them the sacred people."[12]

What is man and his destiny?
Born of the earth: earthly. Freed from the earth:
   his inner part, the soul, ascends
and dwells in the soul of things.
What, shall a dead man live?
—Yes, and rejoice that he so lives.

SUN DEGREE CEREMONY,
OAHSPE, BOOK OF SAPHAH

Called the immortals, the sacred little people simply continued on after death. Immortality did not come packaged with the first race, Asu man; he did not have the spark of eternal life. Some (writing about the Atlantean age) identify this creature (with scarcely any forehead, no frontal development of brain, and just a slope from eyes to pate) as a being without light, called *ong* in the Panic language.

Early religion can usually be traced to traditions of the Great Ancestor. Tellingly, the oral history of the world's people sharply contradicts Darwinian anthropogenesis, for the true ancestor was thought

People of the Light

Ong signifies spiritual light in both Algonquian and Hebrew. Ong varies with ang as in *ang*el or the Batek T'angoi. These little Malaysian Batek people named themselves after their angelic helpers the T'angoi, the name intriguingly similar to the Algonquian word for medicine man: ang'hoi. The earliest meaning of ong/ang is, simply, "Light from above." By extension it came to represent intelligence or people of light, specifically denoting that segment of humanity in touch with celestial beings, their ancestors, their teachers—who are then called Or-ang Hidap, meaning "the immortals." North America, Malaysia, Asia, Africa, and Oceania are chock full of ang/ong-named places and tribes.* Orongo is the ceremonial center on Easter Island. In the Orient are Hongshan (Chinese culture), Hong Kong, Mongol (M'ongol), and the Onge Andamanese, and in America are Monongahela (Pennsylvania) and Ongwee: "Behold, the Ongwee had the gift of prophecy; and of seeing visions and of hearing the voice of the angels."[†]

In the Mohawk language, *ongwe* means "person," indicating a being of light. In the time of Apollo, in America about 18 kya, the Ongwees came suddenly into the world, possessing all the symmetry of the Ihin and the savageness of the Ihuan: "for the Lord caused his chosen [Ihin] to display the mold of their thighs and their short shapely arms, and together they brought forth heirs of more shapeliness. . . . And they were called Ongwee-ghan, signifying, good-shaped men."[‡] "The Ongwees came into the world by the thousands and thousands . . . in the north, south, east and west. They had long hair, black and coarse, their skin was brown, copper-colored, and their arms were short, like the Ihins."[§]

The Honga or Hoanga priesthood of ancient America also took its name from *ong*, "spiritual light." For it came to pass that he who heard the Voice was made high chief prophet for the tribe and was called Hoanga. Through him, the voice was heard for many, many generations in North Guatama.

---

*Some of these in Malaysia are the Semang, Pahang, Selangor, Penang, Jeransang, Batang, Teliang, Tampang, Mekong, Langka (ancient name of Sri Lanka), Kenderong, Temangor, Trang, Mahang, Sadang, Pangan, Kangsar, Kupang, Hangat, Lenggong, Kampong, Belong, Tiong, Siong, Ruong, Chong, Bong, Gunong, and Yong.

[†]Oahspe, The Lords' Third Book 1:29.

[‡]Oahspe, The Lords' Third Book 2:3 and 2:2–5.

[§]Oahspe, The Lords' Third Book 2:2.

to be a kind of divine personality—not an animal—with an off-planet origin. World mythology, with remarkable consistency, portrays the tribes of men as children of the gods (or stars).

> *A perplexed professor . . . I look up across the moon and*
> *Venus—outward, outward into that blue-white glitter. . . .*
> *And as I look and shiver I feel the voice in every fiber of*
> *my being: Have we come from elsewhere?*
>
> LOREN EISELEY, *THE IMMENSE JOURNEY*

To the ancient Hebrews,* it was the "daughter voice" that made God's will audible on Earth. Their sages said that the voice issues from the glory with great force. An angel receives it in his hands, flies with it to the seven heavens, voice after voice. And the last angel takes it, now faint, and transmits it to the angel messenger. This is the daughter of the voice, the voice of the voice. It is eternal and powerful, full of majesty. It is only when it arrives at the ear of the prophet that it is a subtle voice, a murmur.

> Behold, I come not as a sound to the ear; My voice hieth into the soul from all sides.
>
> OAHSPE, BOOK OF APH 1:8

As Jim Dennon, the late Oahspe scholar, summarized these teachings: Via a starship we would now call a UFO, etherean angel volunteers were brought to Earth to teach the Asuans to walk upright. Materialized with mortal bodies, in innocence they (the volunteers) mated with Asu.

This scenario resonates with the theosophical description of incorporeal beings, also volunteers, who incarnated in human bodies. Called Lords of the Flame, they brought immortality and all the high

---

*See Acts 9:4 and 7, a voice heard by Saul.

The Voice

In the olden day the spiritual standard for those humans with "blood of the gods" was the ability to hear the Voice (of the angels). In the time of Apollo, "the Ihuan was first capable of hearing the voice understandingly."* (But they lost the voice by retrobreeding.)

My voice is in all places. The light of the soul of man hears Me. I speak in the vine that creeps, and in the strong-standing oak. . . . Hear the voice of your Creator, O angels of heaven. Carry the wisdom of My utterance down to mortals. Call them to the glories of the heavens and the broad earth. . . . That which speaks to your soul, O man, teaching you wisdom and good works; reproving you for your faults, and enchanting you with the glories of all created things, is the voice of your Creator. . . . As a man among you employs a thousand men to do his bidding, so do I, your God, have billions of angels to speak in my name. . . . Do not put off my words, saying: It is only your conscience speaking. My angels speak to you in spirit, with my very voice and words.

OAHSPE, BOOK OF APH 1:6,
THE LORDS' FOURTH BOOK 2:4,
GOD'S BOOK OF ESKRA 4:17–18

*Oahspe, The Lords' Third Book 3:11 and The Lords' Second Book 2:2.

points of culture to the ancient Lemurians. In fact tribal myths all over the world describe contact with a foreign entity who came from somewhere else in the universe to fecundate Earth. In Ottawa Indian tradition, for example, it is said that the inhabitants of Earth were first on other planets; all human beings descend from people of distant worlds.

Genes are to be regarded as cosmic.

FRED HOYLE AND N. C. WICKRAMASINGHE,
EVOLUTION FROM SPACE

However, they were not, as recent theories have enthusiastically proposed, ancient astronauts, helmeted and strapped up in their space suits. As no one seems to know the actual source of sapient man (the suddenness of Cro-Magnon and the Great Leap Forward), it's open season for extraterrestrial proponents and ancient cosmonauts, various writers suggesting a deliberate genetic intervention by aliens. One favorite spaceman is the (cuneiform) Anunnaki—"those who from heaven to earth came," or the Watchers described in the Book of Enoch. These visitors are interpreted as a master race from another world, who came to mix their divine genes with the indigenous ape-like people of the planet.

Maybe it's a minor point, but angels are different from ETs; writers have portrayed angels as spacemen, god as the captain of the spaceship, and ethereal fireships as spacecraft. In this connection, Paul Von Ward's book *We've Never Been Alone* has chapter 5 titled: "Are ETs Angels by Another Name?" The answer is yes. The human race, as one Canadian mystic believed, may have been approached by beings from another region of creation who wanted to interbreed, perhaps to upgrade humanity.

In *The Genesis Mystery,* Jeffrey Goodman proposes that "instead of evolution through natural selection, some sort of outside intervention was responsible for modern man's most distinctive characteristics. . . . One must ask whether man's sudden appearance was truly the result of random gene mutation or the design of some outside force, some purposeful intervention."

There is a legend from India and Afghanistan relating that an immense "spacecraft" (an angel fireship, not material, but made and powered by the finer etheric currents) came to Earth bringing perfect human beings who mixed their genes with the first earthly race of man.

In the beliefs of many other ancient peoples, the stars of other worlds are inhabited by celestials who long ago contributed to our genesis. For example, in the Hebrew Midrash, angels from an ethereal realm descended to Earth, thus accounting for anthropogenesis; these were called Elohim, those who came down from above and interbred with humans.

TIP OF THE RAY

Of earth and starry heaven, child am I.

ORPHIC RITE

To the Inuit and Athabascans of North America, man is the product of a primal creative power and never literally derived from an inferior species. There are many related accounts, describing earthly/celestial unions; it could easily fill a volume of its own. Mircea Eliade, in *Myths, Dreams and Mysteries,* found this to be "a motif of universal distribution . . . the celestial origin of the first human beings." In Japanese lore, entities from the abode of the gods came down and mated with earthlings, just as India's *Rig Veda* and *Vikramaruasi* extol the light beings who once became enamored of mortals. In the Sumerian tradition, a divine strain is said to have intermarried with "Adamic" (earthly) females, the angels led by god Ea, whose plan was to engineer a hybrid race by combining lowly savages with the seed of the gods.

> *The very beginning of evolution . . . the mystery of man*
> *and his origin . . . [occurs] when the One Eternal drops its*
> *reflection into the region of manifestation.*
>
> H. P. BLAVATSKY, *THE SECRET DOCTRINE*

Many religious traditions worthy of our attention embrace the idea of man somehow *receiving* a soul, receiving a small piece of infinity like the tip of the ray. Humanity as child of the stars is one of the most widespread and persistent legends among the tribes of man. Variations on the theme of this extraordinary amalgam, this hybrid of heaven and Earth, are found in every form.

Figure 7.3. (A) Assyrian genius with double set of wings found at Nineveh in the Palace of King Sargon, shows man as part mortal, part angel (B) Sphinx figure at Nimrud (C) Eagle-headed genii from an Assyrian bas-relief. "What is the sign of half a dog, of half a horse, and a man's head?—That man at best is two beings, a beast and a spirit."[13] As portrayed in iconography, human beings are two things: body and spirit (the latter sometimes represented with wings).

TABLE 7.1. CELESTIAL PARENT PORTRAYED
IN WORLD MYTHOLOGY

Where/People	Celestial Parent	Details
Babylonia/ Babylonians	God Marduk	Mates with mortal maiden
Egypt/ancient Egyptians	Sky god Nut	Mates with Earth goddess Geb, produces mankind
Europe	Fallen angels	Mate with mortals, producing fairy children
Greece/ancient Greeks	God Poseidon	Mates with mortals Halia, Cleito, and many others
Greece/ancient Greeks	Sky god Zeus	Mates with mortal Danae (and many others), produces Perseus
India/Hindus	Sun god	Mates with Kunti, producing mortal kings
Italy/ancient Romans	God Mars	Mates with vestal virgin, produces Romulus
Japan/Ainu	Ancestors	Fell from the sky
Java/Javanese	Moon goddess	Marries a mortal; similar stories in Bali and Moluccas
New Zealand/ Maori	Sky Father Ranginui	Mates with Earth Mother, produces semidivine children
North America/ Cherokee	Star Woman	Fell from the sky
North America/ Kickapoo	Culture hero Wisaka	Fell from the sky
North America/ Seneca	Sky Mother	Brought life
Polynesia/Moriori	Tawhaki	Came from seventh heaven
San Blas, Panama/ San Blas	Moon god	Mates with Indian mother, produces the white Indians
Scandanavia/ Norse	God Freyr	Mates with mortal women
South Sea Islands/ Native Islanders	Angel Hapai	Mates with a mortal man
Sumer/Sumerians	God Enlil	Mates with mortal maiden Meslamtaea

According to the eighteenth-century Great Chain of Being (which evolutionists love to discount as a "thoroughly discredited" school of thought), humans were a little lower than the angels. This resonates with the second race of man (Ihin), which race, according to Afghani esoteric lore, possessed ethereal bodies, though the third race was purely physical,[14] giving us a rare memory of the otherworldly component of the Ihins, the second race, as against the rank earthiness of the Druks, third race, *H. erectus.*

In the Mayan Popul Vuh, it is said that our forefathers were created and formed from the heart of the sky. The Maya hold that early humans were etheric beings, their sight encompassing the whole of Earth. This ether, to Wallace, consists of "minute vibrations of an almost infinitely attenuated form of matter . . . [a] recondite force. . . . [with] a power of motion as rapid as that of light or the electric current. . . . [T]he whole material universe exists for the purpose of developing [such] spiritual beings."[15]

Ethe (ether) is indeed the medium of *sargis* phenomena, meaning: materialization by which a being of the spirit world draws upon the ambient energy in our world to manifest itself in a humanlike form.* It is this sargis power that has so often been mistaken for ETs, gods, spacemen, and so on. Sargis is the particular power of certain angels who are themselves nothing more than the immortal translation of deceased humans: "the spirits of the newly dead shall have power to take upon themselves the semblance of corporeal bodies."[16]

"Human spirits intervened in the life history of certain hominids," wrote William Fix in *The Bone Pedlars,* for they "had the power to draw into themselves finely dispersed elements of matter in the earth's environment, somewhat the way magnets materialize a magnetic field if iron filings are spread on a paper above them. . . . Such an emergence of man could have happened far more suddenly than anything currently dreamed of in the bone rooms . . . [or] in man's evolution from the apes."

---

*In Tibetan tradition, the divinities (read angels) have the power of materialization, taking everything they need for transubstantiation from the life of trees.

ETHE the subtlest element

Figure 7.4. Ethe: man must learn to master the elements. Ethe is the
most rarified of all substances, it is solvent of all things and it fills
Etherea. Ethe or ether is sometimes called the fifth element;
it may in fact be the same as dark matter.

Figure 7.5. Sargis: God's proxy, a medium; a person who sees
or hears angels, likewise an angel who can be seen or heard; an
apparition or materialization. The sargis angel can take its bulk
from the air, electric works, vegetation, or even the cellular energy of
a medium's body. They can do tricks with it before your eyes, a kind
of biological production. Oahspe states, "And whithersoever the Ihins
went, there went the angels . . . [who] often took on sargis . . . and
man talked with them face to face. And the angels told man what
was good for him, showing him the way of righteousness."

Thus Spake Zarathustra

While Asha [the king] was thus speaking, behold, the soul of [deceased] Zarathustra came and stood before them, and he was arrayed in the semblance of his own flesh and color, and in his own clothes. And he spoke, saying: Do not fear; I am the same who was with you and was hanged and died, whose flesh was devoured by the lions; I am Zarathustra! Do not marvel that I have the semblance of a corporeal body, for its substance is held together by the power of my spirit. Nor is this a miracle, for the spirits of all the living each hold in the same way, its own corporeal body. As iron attracts iron, the [risen] spirit learns to attract from the air a corporeal body of its own like and measure.*

Figure 7.6. Zarathustra. Illustration from Oahspe.

---

*Oahspe, Book of God's Word 29:21.

The very first fruit of mortality was Asu or Adam; but after contact with the angelic, the second race was born and called Ihin. Thus does author Alan Hayward in *Creation and Evolution* propose that "Adam was personally selected for upgrading to *homo spiritualis.*" Dr. John B. Newbrough (himself a gifted medium and sargis), in *The Hidden Prophet,* put it this way: "The origin of man is . . . the result of the cohabitation of angels with Asu. Modern spiritualism has demonstrated the angel appearance, the coming out of the side of the sargis. . . . It is said Eve was made out of a rib."

Monck Materializations

In John B. Newbrough's day were the famous Monck materializations: Phantom forms issued from the rib (left side) of the physical medium Francis W. Monck in the 1870s, the heyday of these phenomenal "trans-figurations." Among Monck's sitters and supporters were none other than Alfred Russel Wallace and Hensleigh Wedgwood, Charles Darwin's brother-in-law. No concealing cabinet was used in these sessions where amazed sitters witnessed "the extrusion of spirit form into this world of matter. . . . nebulous at first, but growing more solid . . . into full-sized [human] forms,"* which showed all the attributes of a man (though the hands were stone cold). Arrested for fraud in 1876, Monck was backed by Wallace who deposed that there was absolutely no trick involved; the phenomena were genuine, teaching us of the nonmaterial world and at the same time making "natural selection inadequate as a causal agent to explain the origin of the human mind." We humans, said the spiritualists, are a duality, not only flesh bodies but also an organized incorporeal form. There was no turning away for Wallace; this knowledge of the nonphysical world "must modify my views as to the origin & nature of human faculty." Wallace, as it happens, also served as star witness in the 1876 trial of another medium, Mr. Slade, for which Charles Darwin discretely donated money to the prosecution; Darwin also tried to wreck the careers of two other prominent mediums.†

*Fodor, An Encyclopedia of Psychic Science, 245.
†Shermer, In Darwin's Shadow, 197–99.

Manifesting angels in the Holy Bible include the angelic "guests" of Lot, who ate and drank with him. Hebrews 13:2, moreover, gently advises: "Be not forgetful to entertain strangers: for thereby some have entertained angels unawares." The angels who appeared to Abraham (Genesis 18) looked like men, as did the one that appeared to Samson's parents (Judges 13).

As far back as we can go, the paintings and engravings of Western civilization are replete with heavenly beings who came to Earth and "mingled" with mortals. The Oahspe Bible recounts in great detail the consorting of angels and Asuans (adams), which produced the Ihins—the object of it all, to make man "upright." For earliest man, Asu was tree-dwelling, sometimes quadrupedal, long armed, slope headed, and inarticulate—"dumb, like other animals; without speech and without understanding, even less than any other creature."[17] Though he was a man, physically, he was not yet a sapient being. And as the oldest traditions hold, it was for this very reason that the gods set about raising up that rude creature to the light. Parallel to these arcane histories is one Slavonic tale, in which man was first created from the earth (Asu) by God who then commanded the angels to take some human couples ("ambassadors of the Supreme Being") to Earth so that they could multiply.

RAISED UP

There alighted upon the new earth millions of angels . . . [to] deliver first man (Asu) from darkness. For he shall rise in spirit to inherit the ethereans worlds [everlasting life].

OAHSPE, BOOK OF JEHOVIH 6:15

Siberian mythology speaks of their own descent from space visitors who were transformed by magic into human beings (as told by the Yucaghiri people). Protohistorian Paul Von Ward has recently published his own theory of advanced beings (ABs), his ideas evidently based in part on Judaic sources that describe the progeny of angels and humans: "*Homo sapiens*," says Von Ward, "did not just arise naturally, but were products of deliberate intervention by ABs . . . [who] somehow . . . splic[ed] hominid DNA with their own." Von Ward's favorite example is the Sumerian Anunnaki, ethereans who "took a brutish Earth being and mixed its genes with some of their own." In this scheme, Von Ward seems to have lit upon the Ihin race, for he ascribes "the Caucasoids' appearance [to] the widespread intermarriage between Anunnaki [ABs] and humans. . . . The AB bodies [possessing] smaller bones . . . and a proportionally larger brain capacity, less body hair, thinner and lighter skin."[18] All of which reminds us very much of the Ihin *bauplan,* in particular the observation that the gracile little people in the fossil record had relatively large brains. No doubt, Ward's extrapolation was informed by the accounts of Berossus, the Babylonian priest/historian, who recorded that some higher intelligence brought architecture, writing, geometry, laws, religion, and science to the ancient people of Sumeria, who, until that time, lived a humble if not abject existence.

This is the Oahspe version of how the Ihin race came into being:

> I looked over the wide heavens . . . and I saw countless millions of spirits of the dead, who had lived and died on other corporeal worlds before the earth was made. I spoke in the firmament, and My voice reached to the uttermost places. And there came in answer . . . myriads of angels from the roadway in heaven . . . I said to them . . . deliver Asu from darkness, for he shall also rise in spirit.
>
> OAHSPE, BOOK OF JEHOVIH 6:12–15

Here we can see that the angels (spirits of the dead) were sent down to teach, not to mix, this fine point faithfully recorded in Incan tradition: "Know that in ancient times the people lived like brute beasts without religion nor government nor towns nor houses. Our father the Sun, seeing the human race in its lowly condition, sent down a son and daughter to *instruct* [e.a.] them, with these commands: Do good to the whole world . . . you shall imitate this example, as my children, sent to earth *solely for the instruction* [e.a.] and benefit of these men who live like beasts."[19]

These circumstances are further explained: "I sent angels to man, to teach him who he was, and to rouse him up to his capabilities. . . . And my angels took on corporeal form and dwelt with man as helpmates, to make man understand. But those [angels] who had never learned corporeal things, comprehended not . . . and they were tempted, and partook of the fruit of the tree of life, and lo and behold there was born of the first race (Asu) a new race called man [Ihin]."[20]

These angels of heaven, as the saga continues, "took on forms like unto man, having all the organs and attributes of mortals, for it was the time of the earth for such things to be. . . . The earth was in the latter days of semu* and the angels could readily take on corporeal bodies for themselves, by force of their wills, clothing themselves with flesh and bones out of the elements of the earth . . . [and] by majesty of their own wills. And in innocence they mingled with the people Asu who were of the earth."[21]

Will man ever know he has been raised up?
Will he be believing?
Or will he, too, need to go to some new world and raise up its first fruits and toil his hundereds of years with naked mortals?

OAHSPE, BOOK OF SETHANTES 10:6

Adam Sedgwick, Cambridge geologist, friend and mentor of Darwin, saw his student's theory of evolution as "grievously mischievous," its advocates "led by the nose."[22] Sedgwick believed that the so-called transmutation of species was a "frenzied dream" and that, instead, a provident power, some "creative interference," was the cause of species, "as if matured in a different portion of the universe and cast upon the earth."[23]

---

*Angels could materialize at the end of the Semuan age (explained in chapter 10); "the conditions no longer exist today which once upon a time [enabled] them to pay us a visit," as one esotericist put it. (Kolosimo, *Not of This World*, 58.)

Now a new race descends from the celestial realms.

VIRGIL, *FOURTH ECLOGUE*

We hear more of this dramatic event in the New Testament (Jude 1:5) as well as in *The Book of Jubilees:* "the angels of the Lord descended upon earth—those who are named Watchers—that they should instruct the children of men . . . [for] uprightness upon the earth"—which is not unlike the rabbinic version where angels invested with sublunary bodies descended to Earth and mingled with the daughters of men.

SONS OF HEAVEN, DAUGHTERS OF EARTH

When you think of it, even the biblical Virgin's immaculate conception repeats the old idea of divine impregnation of an earthly woman. In the apocryphal Book of Enoch, Watchers are mentioned in the passage that recounts how the antediluvian patriarchs were reminded of their *white-haired* ancestors when Lamech's wife bore a son, Noah, whose flesh was "white as snow" and long locks "white as wool." Radiant, "like a child of the angels," the infant rose from the hand of the midwife, opened his mouth, and praised the lord of righteousness. (In the early day, the righteous ones were the Ihins, "the sons of the righteous . . . sons of God . . . called God's chosen. . . . The Lord brought the Ihins together in lodges and cities, and he said unto them: Henceforth ye shall live upon the earth as an example of righteousness."[24])

But the prodigy proved quite disturbing to Lamech, who feared the wonder of his son's strange birth. The superstitious people murmured: Was this child from the watchers? Distraught, Lamech went to his father Methuselah and pleaded for an answer. Methuselah, in turn, went to *his* father Enoch, the scribe and righteous man, who told him: "To be sure, in the days of my father Jared, heavenly beings did indeed come down to earth and seduce mortal women and begat sons like giants."[25]

Giants? I don't think this is quite right. Bear with me as I try to untangle this oft-quoted myth. Let's look at the standard genealogy for a moment: Adam begat Seth begat Enos begat Cainan begat Mahalaleel begat Jared begat Enoch begat Methuselah begat Lamech begat Noah. Seth (apparently half-brother of Cain and Abel) is the leader of the second generation of patriarchs. All right, but after the birth of Seth's son Enos (see chapter 9 and Genesis 5:10), the righteous ones strayed, and the worship of God was "wretchedly corrupted by the race of Cain," the legendary "infidel giants" who became condemners of everything divine.

Now, if these histories can be deciphered, it seems that the godly line of Seth's son Enos broke the rules and took wives from the ungodly line of Cain (read: the little Ihins married out with the large Druks and degraded Ihuans). According to some biblical scholars, the sons of God indeed denotes the lineage of Seth; but in the following generation a measure of intermarriage took place with the daughters of men, that is the line of Cain (Druks and barbarians).

Sorting out these old legends, we realize that the sons of God were actually the holy Ihins; while the daughters of men were the Druks (erectoid tribes). "The ground people came to the Ihins beseeching for food. . . . And the chosen were tempted."[26] The Druks fetched the root of babao (an intoxicant) and brought it to the little Ihins to eat and get drunk, saying "Lest the white and yellow people fall upon us, and our seed perish on the earth, make us of flesh and kind, bone and bone, blood and blood"—the seduction reminiscent of Hebrew lore (following on the Legend of Alconuz), which recalls the depraved daughters of Cain, women of a lower order, who had *charms and enchantments* with which they seduced the "sons of God."

The faithful, the sacred tribes who lived apart, were long called the children of God (sons of God); but the rest of humanity were termed children of men (daughters of men)—as in Genesis 4:26, 6:1, 2: "And it came to pass when men began to multiply on the face of the earth, and daughters were born unto them, that the Sons of God saw the

Daughters of Men . . . and took them wives of all which they chose . . . and they bare children to them . . . [who] became mighty men, men of renown."

Who were these mighty men? They are the same as the giants spoken of in the Book of Enoch, (gigantic compared to the tiny Ihins). They are the people we call Ihuans (Cro-Magnons of the Aurignacian)—the tall and well-formed progeny of those Ihin-Druk "exchanges." "Half-breeds betwixt the Druks and Ihins, the Ihuans were taller and stronger than any other people." Their size and strength might well be a matter of hybrid vigor, as discussed in chapter 5.

Why did Enoch give the time of Jared in connection with these mixed marriages? At this time the angels withdrew[27]—leaving man to his own devices (backsliding). This we find in Hebrew legend, which says that in the days of Jared (perhaps 50 kya), the depravity of men increased monstrously "by reason of the fallen angels." But let us stop right here for a reality check. I can make no sense of these stories unless we take the "rebel angels" or the "fallen angels" or "sons of God" as no more than figures of speech representing, really, the holy Ihins who strayed from their high mounds (the heights) and cohabited with the more carnal races of men. When the Lamech Scrolls (among the Dead Sea Scrolls) say "ye left the lofty heavens, slept with women. . . . and begat sons like giants"—the lofty heavens are hyperbole, which we can easily translate as the heights, the reclusive mounds upon which the chosen people lived.

As for giants or mighty men, they can only be the Ihuans aka Cro-Magnons (who resulted from the forbidden Ihin-Druk mergers). They are the same, in fact, as the biblical Nephilim (giants), men of renown: "The Ihuan shall inhabit the whole earth in time to come; and he shall have dominion over everything on earth. And it came to pass that the Ihuans were a very prolific people, four times more prolific than the Ihins . . . and they spread rapidly over the earth . . . prospered, and they became *mighty* [e.a.] in many countries."[28]

Finally, the holy watchers are yet another hyperbole, standing for the

Ihins themselves, whose name is actually equivalent to the kabbalistic angels called Irin. In one linguistically interesting passage in the Book of Enoch, the holy watchers, called Irin, are a class of angels inhabiting the sixth heaven. (Well, the deification of the Ihins made them angels, if not gods!) Irin appears to be nothing more than a Mediterranean variant of Ihin, corresponding also to *ir* in Sumerian, which means "guardian" (watcher). In the Aegean, Ayia Irini is the ancient name of a Bronze Age settlement in the Cyclades; this place was the renowned home of poets, philosophers, and physicians, indicating the elect, a seat of ancient intelligence, if not enlightenment, as seen also in Huz-irina, the ancient Assyrian center of learning (at Sultan Tepe), with "a proper library."

Although the Ihins were certainly mortals, the hand of myth *deified* them, euhemerized them, making them sons of heaven, partly because of their peculiar angelic feature, the vail or caul. A good number of the gods of the ancient world were actually mortals or ex-mortals—deified.

But lost as a race, the Ihin little people are only vaguely remembered, one protohistorian, for example, noting that "at one time . . .

*Figure 7.7. Sign of hin'kwa (with the same root as Ihin)
designating the unseen part within corpor, that is, spirit within
body. Oahspe teaches, "they were called Ihins, because they were
begotten of both heaven and earth." The Ihins were the first on
Earth to be endowed with soul power, an etheric, inner connection
to the cosmos, the thing within a thing—that spark that gave the
human race everlasting life in the hereafter.*

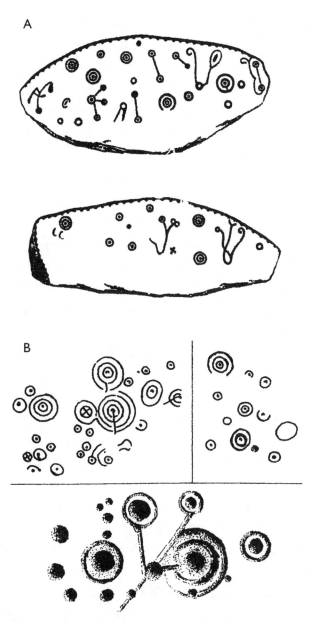

Figure 7.8. The diagram of Hinqua is frequently seen in undeciphered petroglyphs: (A) Hinquas from Forsyth County, Georgia. (B) Cup marks similar to these have been found in England, France, Italy, Spain, Algeria, Israel, China, India, and the Gobi Desert, epigraphers sometimes inferring an otherworldly or shamanistic meaning.

there was actually a group of men on earth who, for reasons unimaginable to humankind, undertook to guide and teach them . . . these people came . . . out of the sky."[29] The Greek Hesiod spoke of a golden race of humans created in the time of Kronos, who were later hidden away to become beneficent spirits. These, we know, were the immortals of the Ihin race, their spiritual influence unfettered by the thing called death, hence known as sky people—with some truth.

When the Holy Bible speaks of Enoch walking with God and not seeing death (Genesis 5:23, 24, Hebrews 11:5), this also is a metaphor of immortality, conferred by the caste of patriarchs, all with Ihin blood, not unlike Indra, the Hindu "god" who bestowed immortality. The little sacred people came to be worshipped among the barbarians, to whom they appeared as very gods. Thus do we find *dwarfs* sacred to the sun god in Peru, just as the Mexicans and also the Egyptians had their dwarf gods (Ptah and Bes, in Egypt).

Among the Cherokee, the little people who were called Nunnehi are remembered as "the immortals." In Europe as well, it was the pint-sized Tuatha De Danaan (people of light, heroes of the old Irish) who had all the powers of the immortals, for they and their forebears were the first men on Earth to attain everlasting life. This is the reason that "fairy" land (abode of the Ihins) was remembered in myth as pleasurable and without death and the reason that the fairies themselves were called the Ageless Ones, the elves who "lived a thousand years."

We have seen that the chosen (Ihin) were commanded to marry among themselves only and to withdraw from other peoples: "The Ihins shall not mingle with any other people on the face of the earth."[30] This commandment goes all the way back to the beginning: in the best known story of Adam and Eve, the pair was driven from the Garden of Eden for disobedience. What was that disobedience?—giving in to forbidden marriage, crossbreeding. "Death" through Adam's sin actually stands for loss of immortality through back breeding—a spiritual death. The same marriage rule (endogamy), in a latterday context, appears in Genesis 28:1: "You shall not take a wife from the daughters of Canaan" (that is, descendants of Cain). And in Ezra 9:1–2: "The people of Israel," complained the Old Testament, "have not separated themselves from the people of the lands, with respect to abominations of the Canaanites . . . so that the holy seed is mixed with the people of those lands."

Concerning "separate people": circumcision was introduced with the sole purpose of keeping the races distinct*; thus would endogamy assure offspring of everlasting life. "And the Lord commanded the male Ihins, old and young, to be circumcised, that woman might not be deceived by the Druks . . . the mark of circumcision being as a limit to the line of my chosen."[31] Circumcision was later given to the Ihuans as well, to separate them from the Druks. But this practice outlived its usefulness many thousands of years ago; for, as we have seen, the races did thoroughly intermingle, ultimately erasing those early differences in spiritual potential.

ETERNAL LIFE

To know eternity is enlightenment. . . . The decay of the body is not to be feared.

LAO-TSE, *TAO THE KING*

Jehovih has said: To man I gave an earthly body that he might learn earthly things; but death I gave to man that he might rise to the inheritance of My heavenly kingdoms.

OAHSPE, BOOK OF JUDGMENT 35:32

---

*"Those of the lesser light were called Cain (Druk), because their trust was more in corporeal than in spiritual things. . . . And those with the higher light were called Faithists [led by Ihins], because they perceived that Wisdom shaped all things and ruled to the ultimate glory of the All One" (Oahspe, The Lords' First Book 1.19).

The Asuans were not created to everlasting life[32]: we see this very clearly in Buriat anthropogenesis where first man, made of red clay, was *without a soul*. This was Asu man, otherwise described as "the nearest blank of all the living . . . like a tree. . . . Behold, their souls go out of being as a lamp that is burned out."[33]

A similar concept of first man is echoed in the Mayan Popul Vuh, where the first creation saw people made of mud, "but they were unable to walk, talk, or breed." Oh, breed they did, but they were devoid of sense, devoid of soul, without the spark of the divine. Significantly, one Quiche Mayan version of creation has the Maker first fashioning men of wood ("like a tree"), but not succeeding until making humans from yellow and white corn (symbolic of the Ihin race). The Kayan people of Borneo also have the first human pair born of a tree, as does Norse mythology and Algonquian legend—both of the latter forming first man of the Ash (Asu?) tree.

Asu and Ihin, as we know, were the first two races on Earth; the next, third, race (Druk) came about through the amalgamation of the first two—the spiritual (second) and the physical (first) races. Similarly, man in Genesis 1:26–7 is radically different from the Adam of Genesis 2:7; the former was created in the image of God, but the latter was formed of the dust of the ground and became a living soul only after the Lord God breathed the breath of life into him. What is indicated here is that one race was of a purely physical stamp, the other, spiritual, even godlike. And "the union of these two races produced a third . . . sharing the natures of both."[34] As Dixon said, intermediate races "have arisen from the blending of two *extreme* [e.a.] forms."[35] Just so, the third race (Druk) was a blend of the first two extremely different stocks—Asu and Ihin.

Jim Dennon, charting these "beast" versus "angel" extremes in the early races, calculated that all mortals on Earth today have more than one-third angelic blood, and are thus all endowed with the spark, capable of everlasting life. It was precisely these ratios that caused the ancient sages to designate the beast numerically as 666 (Revelation 13:18)—meaning that

ORIGIN OF HUMAN RACES DURING THE FIRST ARC OF SEFFAS, AND HOW ETERNAL LIFE DETERMINED

	Adam (Asu'ans) 100% Beast	Extinct after 8,000 years	Eve (Ethereans) 100% Angel	
The threshold between eternal life & dissolution:	**Dissolution** 66⅔% Beast		**Eternal Life** 33⅓% Angel 333	666/1000 = Threshold number of man.
666		**I'Hin** Yellow & White Vegetarians 50% Beast	**Races (Abel)** 50% Angel	Extinct after 60,000 years.
		Druk Races (Cain) 75% Beast Flesh Eaters	25% Angel	Extinct after 72,000 years.
		Yaks ("Sasquatch" or "Bigfoot") 87½% Beast	12% Angel	Long extinct.
		I'huan (Copper-colored) 62½% Beast Flesh Eaters	**Race** 37½% Angel	Largely exterminated; nearly extinct.

Genetic Result Today (Kosmon):

Of all colors: Black – Brown – Red – Yellow – White

GHAN Average: 60% Beast Most flesh eaters, some vegetarians	**RACES** 40% Angel	100% of Humans capable of Everlasting Life today

Figure 7.9. Chart displaying Dennon's calculations of beast vs. angel percentages.

any person with *less* than one-third angelic blood or with *more* than 66 percent "beast " genes (the lower hominids) would be incapable of immortality. This refers to spiritual growth both here and hereafter. The Asuan Adam was, in this scheme, 100 percent beast (animal-man—without the spark), but when they commingled with the angels, they begat the Ihin race (who were 50 percent beast and 50 percent angel). Soon after, when Asu and Ihin, in turn, cohabited, they begat the third race, Druk or *H. erectus* (Cain), who was then 75 percent beast and 25 percent angel—hence incapable of eternal life (being more than 66 percent beast). But when Ihin and Druk mixed bloodlines, they begat the fourth magnificent race, the Ihuan (62.5 percent beast and 37.5 percent angel)—who, with more than 33 percent angel ratio, was then capable of everlasting life.

If not for all the backsliding (retrobreeding), we might be able to construct neat tables based on these percentages. But the Ihuans (with more than 33 percent angelic blood) retrobred again and again—losing the spark, the hope of immortality. Even the Ihins lost the light when they back-bred with Ihuans, 12 kya—their offspring inheriting neither "the silken hair nor the musical voices of the Ihins, nor the light of the upper heavens."[36]

In regard to this loss of holiness, Stephen Oppenheimer found a distinctive feature in Austronesian mythology: Man, it was said, already had immortality, but lost it. This really is the same as the biblical "fall": when humanity held the prospect of everlasting life—but lost it by back breeding. Indeed, the Ihuan race, before the flood, was capable of everlasting life. But they mixed with the Druks "until the seed of the spirit of eternal life became exhausted, and they brought forth heirs incapable of self-sustenance in heaven. . . . Behold, the Ihuans can [no longer] hear the voice."[37] After the flood, however, a new race of Ihuans came forth, and they were at first capable of all light. But they also did not keep the commandments and mixed with the Druks (the ground people), and they descended rapidly.

Earlier on, the physical contrast between *H. habilis* and *H. sapiens* "was immense"[38]; earlier still, within *Au. africanus* (South African australopiths) there was an "amazing amount of variability . . . astonishing in its diversity." And this is a fact I wish to stress: "the variation in early humans. . . . exceeded what variation there is today." If the early heterogeneity within the human family had been properly understood, "many of the mistakes in anthropology could have been avoided"[39]—meaning, this great diversity led theorists (mistakenly) to take mere varieties as *separate species,* falsely multiplying the number of human taxa. For there was only one, albeit highly varied, human species. *And with incessant crossbreeding, all the important differences have now been neutralized.*

> *We have become homogenized.*
> DONALD JOHANSON AND BLAKE EDGAR,
> *FROM LUCY TO LANGUAGE*

In today's world, the races are, genetically, much less different from each other than they once were. Dixon, Weidenreich, and Hooton all stressed this point—that races were at one time much more distinct and well defined, Dixon underscoring "the existence in extremely early times of sharply differing varieties of man."[40] Today, the amount of genetic code that humanity shares in common is impressive, characterized by Stephen J. Gould as "profound equality." Indeed, Alfred Russel Wallace saw evolution eventually producing "a single nearly homogeneous race"[41]—something like Pierre Teilhard de Chardin's omega man, entailing a harmonized collectivity of consciousness, a single unit. Is this not the destiny of the human race? Unity through amalgamation?

Although all that early mixing was prohibited, back of it all was divine purpose—to immortalize the human race as a whole. Man, as Dixon held, has "passed from an early condition of relative heterogeneity . . . to become more homogeneous through the long amalgamation of the originally discrete types. . . . The peoples of the world to-day . . . are the result . . . due to the fact that the elements have been blended so variously."[42] And this is not evolution.

"I WILL HAVE NO CASTE"

Man's spirit crystallized into separateness . . . [and he] rose up and ascended into the firmament.

OAHSPE, BOOK OF JEHOVIH 4:4

It is interesting that in the Babylonian scheme of things, the hero Utnapishtim was granted immortality *after the flood;* in many other myth cycles besides, the flood hero is made immortal. How can this be interpreted? Quantifying the "blood of the gods," Jim Dennon determined that *before the flood,* as much as half of humanity remained in the lower state, without the spark.[43]

But after the diaspora, things changed dramatically, thanks to the dispersal of Ihin genes, raising the ratio on every hand. These percentages remind me of Milford Wolpoff's idea of "sapienization" of all races—by gene flow.

In the |hin race | laid the foundation for the redemption of the whole earth.

OAHSPE, THE LORDS' FIFTH BOOK 4:25

Beginning in flood time, 24,000 years ago, the percentages in the table below indicate something of a spiritual quotient of the human race, the percentage of humans capable of immortal life. (See table 7.2.)

God has made of one blood all nations of the earth.

APOSTLE PAUL, ACTS 17:26

In French esotericism, no race is deemed superior or inferior, for we are all of mixed blood; whether yellow, white, black, or red, we possess "that fragment of the divine Spirit that makes us humans and not animals," according to Glenn Kreisberg, in *Mysteries of the Ancient Past.* All the races of man are now one and the same, not only in all their organs and capabilities, but also in spiritual potential.[44]

TABLE 7.2. PERCENTAGE OF
SOUL-POWERED HUMANS

Years Ago	Percentage (%)
24,000	54
24,000	54
21,000	64
18,000	72
15,000	85
12,000	87
9,000	90
6,000	95
3,500	97
Present	100

The sum total of comparative study also indicates that "no genetic differences in mental ability among peoples has been found, despite much effort."[45]

One loose end: What became of those earthbound angels after the genesis of the Ihins? "And the lord commanded the angels to give up their forms, and to be no more seen as mortal. And it was done."[46] Thus does Hebrew sacred history record that the Shekinah were induced to leave Earth and ascend to heaven, amid the blare and flourish of angelic trumpets. A like scenario obtains in Sumerian tradition, which has the god Enlil ordering all the "pure gods" living on Earth to abandon their human families and leave the planet at once. In Brahman tradition an Aspara would sometimes wing down to Earth and bear a child with a mortal only to return to the skies.

This is not too different than the Navaho version, which tells of beings who came from the sky and stayed a long time on Earth but finally returned to their world. Likewise is it told in New Zealand by the Tawhaki people that a goddess named Hapai descended from the seventh heaven to spend nights with a handsome mortal; but after giving birth to a daughter, she returned to the cosmos. All of this is so much like the European tales of fairy brides who, by and by, return to their heavenly home, after conceiving mortal children on Earth.

The seed of truth and light on Earth, the Ihins were the foundation for raising up prophets and seers unto other peoples.[47] From the beginning, the great plan was to redeem the tribes of darkness, for "the Druk had not the light of the Father in them . . . they [can] not dwell forever . . . except where they cohabit with the Ihins, whose seed is born to everlasting life."[48] The Ihins, teachers or initiates, taught the Druks about spirit and these ideas "quickened their souls within them, so that they brought forth heirs unto everlasting life."[49]

Vernon Wobschall put it this way: "The Ihins were vital to the development of humankind. . . . After the Faithists of Pan [the Ihins] landed in . . . Africa, India and China, the [resulting racial mixtures] were used to replenish those who had become extinct [and exterminated]; the Ihins were yet needed for the maturing of man."

This view, then, is almost the *reverse* of the pessimistic Platonic one, which held that: "[g]radually the blood of the gods which flowed in their [ancestors'] veins was diluted with the mortal admixture, and they became degenerate." While Plato stressed the *loss* of holiness incurred by these mixings, these faithist histories, to the contrary, remind us that all this interbreeding served a higher purpose; it was overall a *gain,* not a loss: "Jehovih said, I condemn thee not because ye have become joint procreators . . . for ye have done [a] service unto Me . . . having caused the earth to become peopled with such as are capable of immortality."[50] Indeed, in the approach to historical time, the law of inbreeding (endogamy) was almost reversed: "I will have no caste, for the races are becoming impoverished in blood; marry here! Marry there!"[51]

Are we special? Or just another species, glorified baboons? "Man in his arrogance thinks himself a great work," said Darwin.[52] Angel genesis, today's Darwinists argue, is a most presumptuous theory, not a very "humble" approach to the origin of mankind; we are inordinately prideful to align ourselves with the divine, they say. Jared Diamond calls it "our smug self-image" that makes us believe we "were specially created by God."[53] And this they call anthropocentrism: nothing but vanity, they say, prevents us from realizing that human beings are just another animal. Our hubris "is absurd," for we are nothing but "creatures of chance."[54]

This chapter was written to give the reader an alternative perspective. Your choice: ape or angel.

People often say we will never know about these things. But this time in which we live is the age of truth: we will attain to know *all* things. It is not so far from reach.

> *There must be understanding. . . . why else for the*
> *universe to utter us into existence?*
> JACK KERLEY, *THE HUNDREDTH MAN*

Ashley Montagu, on the preface page of his book, wrote: "There are some anthropologists who declare that it is not the business of anthropology to tell anyone what he was born for. . . . I do not agree. I think one can and should study man in order to learn what he is."

Sure, man may fit physically into the order of primates; but from his birth he also belongs to the order of angels. We come from a divine root, and we are gods in the making, gods and goddesses in training. It is unlikely that the question of origins will be solved until the science of soul is also understood.

> *Within his house of clay, there is an everlasting life.*
> JAMES CHURCHWARD, *THE LOST CONTINENT OF MU*

"thou art both a flesh-man
and an es-man"

Figure 7.10. Es: the unseen worlds, where angels live; all that is beyond. Openings in the circle represent beautiful shafts of light coming from the emancipated heavens.

No, the soul of man cannot be left out of the definition of *Homo sapiens*. You cannot neuter man of his transcendent nature. Strip man of soul force, and we are back to Asu man—a blank, like a tree. The missing link, as ever, is in the unseen. So let us turn now to that eternal debate, to which chapter 8 is devoted.

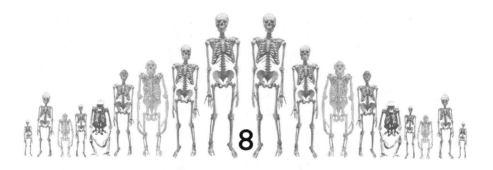

"THAT MYSTERY OF MYSTERIES"*

Science vs. Intelligent Design

Think not O man that I did not foresee the time when men should question, and say there is no Great Spirit.... For I foresaw these things and provided ... in advance to show ... that the cause of evolution came from the Great Spirit and was directed unto righteousness.

OAHSPE, THE LORDS' FIFTH BOOK 4:21

Here is another hypothetical debate between POE (professor of evolution) and A (adversary).

POE: It is generally agreed that we can rule out extraordinary processes in the ascent of man—touched by nothing more than tangible,

---

*Quote from Darwin on the origin of life.

observable nature. Some of my colleagues and friends find it appalling that any serious publication would dignify the creationists' agenda, presenting their views on an equal footing with evolution. And when those uninformed folks say evolution is "just a theory," this is amazing to scientists. Why, it is a phenomenon, a fact.

THE GOD PARTICLE

A: Each man decides for himself. Some people can *picture* descent with modification. Others can *picture* man brought forth by the hand of Creator. Some, like my colleague, John W. White, try to reconcile the two: "It doesn't matter whether the various human species were natural mutations or special creations. . . . God is the motive force of all history, including evolutionary history."[1]

POE: Sorry, but I have to agree with Professor Jerry Coyne and many others besides who see creationism as folly. Evolution is science, not philosophy or belief; it is a child of the scientific method.

Figure 8.1. Intelligent Design vs. Evolution.
Cartoon by Marvin E. Herring.

A: Scientific method? It reads more like rhetoric. The books by Richard Dawkins, although filled with zoological cases (regaling us with long-winded examples of so-called evolution in the animal kingdom—insects, birds, fish, bats, rats, moles, gazelles) still read as turgid, *rhetorical* tracts, brimming with hypothetical scenarios, impossibly lengthy analogies (highly recommended for insomniacs), mathematical models, arguments from probabilities, analogies, and the composition of "luck"! Not to mention other philosophical intangibles and sci-fi scenarios. Dawkins's "explanations," as one biochemist saw it, could be "filed alongside the story of the cow jumping over the moon."[2] In one debate over the God particle, he responded "Why would a theologian have anything to contribute to any worthwhile discussion, on any subject whatever?"[3] And in reviewing Richard Milton's brilliant book, *Shattering the Myths of Darwinism,* he calls the author a "harmless fruitcake [who] needs psychiatric help."[4]

POE: Dismissal, condescension, insult, or ridicule certainly should not be the primary weapon. Nevertheless, I do agree with Dawkins that evolution should be taught as science, and intelligent design should be taught as philosophy.

A: But in reality *both* are philosophy—notwithstanding all the scientific trappings—for they both deal in *causality,* first causes. With actual scientific evidence admittedly too far removed in the past for direct testing, evolution itself is just as hypothetical as creation-by-design.

POE: We can nonetheless *approach* the facts by indirect means, and ultimately prove evolution by comparative analysis and extrapolation.

A: Yet evolution, of all scientific theories, is the least provable or proven. You cannot prove, or even demonstrate, the gradual transformation of species, any sooner than opponents can prove creation. It is a matter of ideology. Not science, really.

WHAT ARE THE ODDS?

A: Even Dr. Einstein said, famously, life isn't a crapshoot. Kurt Gödel, mathematical genius, devoted to rationality, and good friend of Einstein, established that mathematical systems are essentially incomplete, simply because—*not everything that is true can be proven to be so*. It seems that some facts of natural numbers are true but unprovable. Gödel struggled, probably unsuccessfully, with an ontological argument for the existence of God, and also looked into the world of the paranormal; psychic factors and intuition, he thought, might be as important as elementary physics. Brains are a machine, but minds aren't. Gödel: "I don't think the brain came in the Darwinian manner . . .the laws [of life] . . . are not mechanical"[5]; the mind, he thought, is immaterial (incorporeal), not measurable in any orthodox sense.

OK, intelligent design may not be science; it could, though, be the truth. Science can't explain everything. Paleontology or genetics is only one discipline concerned with the origin of man. The way you are saying "science" or "scientific," it sounds like you are out to exclude all evidence from other lines of inquiry: oral tradition, ancient tablets, epigraphy, psychogenesis, protohistory, cosmogony, philosophy, and so on. The origin of man is not something anyone can monopolize. And something else: As I see it, we have been laboring under the delusion that science and *reality* are the same thing— that evolution is objective while creationism is subjective. But either side is a matter of conviction. The materialist's disbelief, in short, is the *belief* that no higher power is behind the order of the universe; and its corollary of creation by *accident* is just as much ideology as creation by design.

POE: Perhaps so, but we have tried to maintain religious neutrality in the study of evolution.

A: Why not let impartial thinking include the possibility of things being created?

POE: God simply does not belong in the sphere of rational discussion. Our parameters must be real—not imagined.

A: People talk about the real world. I wonder if they even know what the real world is. To say that a creator was not involved in any way is, itself, a judgment. There's no "neutral" position, as you suggest. You know, Darwin turned to materialism, partly in reaction to the fundamentalism of the overbearing Captain Fitzroy, whose conversation and evangelical dogmatism poor Darwin had to endure, in close quarters, for five years on the HMS *Beagle*. There is also the matter of Darwin's father Robert, an overbearing man who was a closet atheist, though maintaining an orthodox appearance. The son would come out of the closet. Indeed, Darwin's pal and watchdog Thomas Henry Huxley invented the word *agnostic,* and Darwin cheerfully adopted it. But I'd like to ask: Can science alone clear up all the mysteries of the universe? And how qualified is the agnostic professor—unaware of developments in parascience—to judge questions of cosmogony or supernature or the unseen dimensions?

POE: Ah, but we are concerned with no higher authority than nature herself in this scheme called evolution. Science deals with the material content of the universe and must not overlap with matters spiritual or irrational.

A: I say it is no more presumptuous or irrational to assign the order of the living to a great Designer than to the blind forces of nature, which we are then asked to believe could have accomplished (accidentally) something as complex as the human eye or human brain.

> *Man is the product of causes which had no prevision of*
> *the end they were achieving.* . . . *His origin, growth, hopes*
> *and fears, his loves and beliefs are but the outcome of*
> *accidental collocations of atoms.*
> BERTRAND RUSSELL, *WHY I AM NOT A CHRISTIAN*

Species are evolutionary accidents.

JERRY COYNE, *WHY EVOLUTION IS TRUE*

Speciation is an accident.

NILES ELDREDGE, *DARWIN: DISCOVERING*
THE TREE OF LIFE

I know of no reputable science that pins all on accident and randomity. Could mere chance produce elaborate—almost perfect—design? The coordination of parts into a functioning whole is characteristic of intelligence—not blind chance.

POE: You must grasp the way natural selection operates—toward the improbable. Is that not marvelous? Nothing in nature is unlikely!

A: You lost me there.

POE: Let me put it this way: The evolutionist trades design for nature's power of selection. Indeed, there are scientists who believe in a supreme being but draw the line with creation, giving nature alone that power, for it is utterly unacceptable to bring an unknown deity into the scientific equation.

A: If species traits can be correlated to climate and environment, wouldn't an intelligent designer be able to do the very same?

The cases of the South American four-eyed frog and the snake-tailed caterpillar raise the possibility that these species exist as we know them today because of an artful plan. A caring higher being could have easily equipped these animals with such features to improve their chances of survival.

BALAZS HORNYANSZKY AND ISTVAN TASI,
NATURE'S I.Q.

POE: We must hold supernatural forces in abeyance in order to be scientific, to observe the laws of nature impartially.

A: Nature's law? Or man's assumptions? Like the law of selection; the law of succession; the principle of common descent (really, a chain of inferences); the biogenetic law that ontogeny recapitulates phylogeny, which, disproven, is now no longer a law; or the law of collateral evolution, which *assumes* that an alleged common ancestor held the latent possibility for its descendants to manifest a trait it didn't have!

> Thou hast blockaded the way against Me on every side. Thou hast put Me away, and said: Natural law! Moral law! Divine law! Instinct!. . . I say unto thee: I have no laws; I do by virtue of Mine own Presence. I am not far away; behold, I am with thee.
>
> OAHSPE, BOOK OF INSPIRATION 10:16–20

POE: Let me tell you: No mind planned it.

A: How do you know that? Can it be by sheer chance that human beings ended up on this planet? How could a mindless process produce mind? Impossible. Purposeless order is really a contradiction in terms: organization *means* a plan. Some of Richard Dawkins's circuitous arguments could actually be taken to prove the *opposite* of his belief. The complexity of the eye, for example. It's really stupid: both sides are using the *same* examples! ID (Intelligent Design) says life-forms are too organized and complex to have arisen by purely natural circumstances. Evolution says life-forms are too organized and complex to have been created suddenly. To me, the remarkably complex design of organisms coming along by pure chance is actually more irrational than a purposeful, planned genesis.

> *There's a divinity that shapes our ends.*
>
> SHAKESPEARE, *HAMLET*

LUNATIC FRINGE

POE: ID is a superstition that has long since been discredited by thinking people. Most intelligent folks frankly think the creationists are gulled, humbugged, duped—in short, religion and its dogmas are not our turf and certainly have no place in science.

A: Let me clarify something, I am using the term *believer* not to mean church or fundamentalism or Christianity or any other particular creed, but to encompass all—regardless of faith—who trust in a supreme power. That is not the same as superstition. It is faith.

POE: And in contrast, evolution comes from a long tradition of rationalism. Reason veers people to science and empiricism and away from religion and doctrine.

> *Nothing that is destitute itself of life and reason can*
> *generate a being possessed of life and reason.*
>
> ZENO

A: Yet reason, deep reason, may also be a path to God. Francis Bacon said of atheism: "A little philosophy makes men atheists; a great deal reconciles them to religion." It is not the irrational mind, but the rational one, that infers a creator from the grand design of the universe. I am ready to argue that belief is actually part of our makeup. All I'm saying is unbelievers do not have a monopoly on reason! It is no more outlandish or "uncritical" to infer a divine, unknowable hand in all things than it is to strike all things sacred from the record of life. To me, the spontaneous transmutation of species is more a piece of magical thinking than the creation of life by an intelligent designer!

POE: Yet for us scientists, it is standard operating procedure to place ID in the category of mythological and supernatural trappings.

A: Treating it like the tooth fairy or snake oil. This is insulting, as if creationists were kooks or cranks. The pop science mags, especially *New Scientist* in the UK, are full of demeaning, condescending, and hostile remarks on ID as if it were a nasty piece of work by the lunatic fringe.

POE: Admittedly, the problem for us evolutionists is we cannot prove there is no creator. It cannot be falsified (Popper's rule). It is beyond the realm of science, in the realm of the invisible, intangible.

A: Well, yes, perhaps the answer does lie beyond the domain of science, and since science divorces itself from religion, it thereby relinquishes the right to declare one way or the other on the existence or nonexistence of a creator. The Cartesian way would be to recognize "a fundamental core of unknowability," the idea that corporeal beings may be incapable of using reason to discover the sublime. Sometimes hidden things are the most potent—intangible but real.

No man can measure my mysteries.
OAHSPE, BOOK OF SAPHAH EMP'AGATU:1

POE: Science does not traffic in intangibles, unknowables. But what we really dislike is the religious intolerance displayed by fundamentals, especially in the U.S., where ID advocates see the hand of a higher being throughout nature, making scientific investigation pointless. Settling for a given such as an unseen intelligence would be, well, just giving up.

> *Organisms are not less mechanistic for being manifestations of the Creative Power.*
> MARTIN RUDWICK, *THE MEANING OF FOSSILS: EPISODES IN THE HISTORY OF PALAEONTOLOGY*

A: Nonsense. Even if Creator made this world, it is still subject to scientific scrutiny! Natural law is not suspended by an original act of creation. You know, Schopenhauer called Darwin's *Origin* "soapsuds," missing the

Figure 8.2. Intelligent design. Cartoon by Marvin E. Herring.

"hidden force." Adam Sedgwick, a beloved teacher of Darwin's, regarded *Origin* as "a string of air bubbles." Sedgwick, "perhaps the best teacher in England,"[6] concluded that man was utterly unaccounted for by the "laws of nature." My point: there is nothing unscientific about God.

> *The more I study nature, the more I stand amazed at the work of the Creator. . . . A bit of science distances one from God, but much science nears one to Him.*
>
> LOUIS PASTEUR, QUOTED IN *THE LITERARY DIGEST*

There is no conflict between transcendent knowledge and science, only the conflict we ourselves have created in our darkness.

POE: You must realize that science is mechanistic and religion isn't; science is neutral of belief.

> *[T]he stream of knowledge is heading towards a non-mechanical reality; the Universe begins to look more like a great thought than like a great machine.*
>
> SIR JAMES H. JEANS, *THE MYSTERIOUS UNIVERSE*

A: I'll grant you this: Evolution is the greatest triumph of atheism in the world today.

POE: Would you deny that man can be known from his parts: femur, dental arch, cranial bones, genes?

A: This gives us materialism per se, reductionism, which will never produce a satisfactory account of the origin of mind. Never. As long as science and religion maintain their apartheid, the inexplicable will remain just that. The failed marriage of science and religion has a lot to do with intransigence *on both sides*. It is emphatically not due to any intrinsic conflict.

OF ALL THE ANIMALS

We are not unique. We are animals.

COLIN PATTERSON, *EVOLUTION*

A: Isn't man worthy of a bit more than mere biological, zoological explanation? Should all scientific research be limited to the material, visible realm?

POE: Yes. Nonphysical ideas should not be admitted to science.

A: Like mesons and dark matter? Or subatomic particles? Or neutrinos—particles with no mass, that move about with the speed of light?

POE: To Darwin, humans were quite simply domesticated animals; their social instincts, not really unique, are deeply rooted in animal behavior. We are actually more like chimps than chimps are like lower monkeys, according to degrees of consanguinity. You may not like it, but Boule saw comparisons between the most intelligent modern apes and "savages in the natural state." Darwin himself found the rudiments of architecture, dress, and even language among the pongids.

A: Man is different from all the rest of animal creation. Darwin's contemporaries, the scientists Sir Richard Owen, Alfred Russel Wallace, Georges Cuvier, and the Duke of Argyll, held man as a *separate* phenomenon, distinct from the animal kingdom, a difference in kind, not just degree. Even Darwin's most avid American supporter, Asa Gray, believed in the special creation of man as against the natural evolution of plants and animals.

POE: Well, that was the nineteenth century; we now know that we are just another kind of animal—different in degree only. It is a matter of critical mass. By careful study, Darwin even found "a mind of some kind" in worms.

A: I wonder why evolutionists are so ready to humanize animals and worms, and at the same time, dehumanize man!

POE: Simply because the rudiments of all our mental faculties can be seen at work in the animal kingdom. Japanese monkeys carry sweet potatoes to the stream and wash them.

A: Hey, raccoons wash their food religiously. Does that mean raccoons are on their way to becoming human?

POE: But chimps can *make* tools to perform certain tasks and also use stones to crack nuts.

A: So what: Otters use rocks to open clams; beavers cut sticks to make dams.

POE: Yeah, but chimps will sharpen a branch with their teeth, wield it like a spear, and stab their prey. They also dig for bulbs.

A: Can they change a lightbulb?

POE: No, but they can thread a needle and roller skate! Besides, primates also have a form of communication. Chimps, our closest living relatives, have sign language, many gestures to communicate, and a wide range of vocalizations used in the wild. The great apes even show an incipient degree of conceptual thinking. Romanticism aside, man is, quite plainly, an improved animal.

A: A cultured ape? As I see it, this idea springs from powerful nineteenth-century racialism, which took the Andamanese, for example, as "very little above animals."[7] Evolutionists rarely quote Darwin's comparison of orangutan behavior with that of "naked wild savages," or Huxley's equally racist references, such as "the thoughtless brains of a savage."[8] Is there any valid evidence at all that human beings evolved from apes?

POE: Plenty. They are our closest cousins in the animal kingdom.

CHIMPS AND CHUMPS

A: Well, I am not aware of any fossils that link apes to the australopiths ca. 6 million years ago. Every hominid ever found was a primitive man and not an advanced ape. Are we improved chimps or just scientific chumps? Did we really evolve from a tree shrew? (See figure 8.3, page 312.)

POE: It is perfectly obvious that man's anatomy is on the same plan as other primates. The chimpanzee's short arms and structural details of the skull quite nearly approach man's.

Noses and Spines

The black African and the native Australian (the latter once considered the most archaic of living races), in their platyrrhine nose, are actually the diametrical opposite of that found in the apes. "The nose of Neanderthal Man, far from resembling that of anthropoid apes, differs from it much more than does that of living Man."*

Australopithecus, the protoman, judging from *Au. africanus* (STS-14), had six lumbar vertebrae—longer than in modern humans: "This is certainly at odds with the evolutionary pressure leading to the shortening of the lumbar spine of the apes."[†]

---

*Boule, *Fossil Men*, 205.
†Filler, *The Upright Ape*, 248.

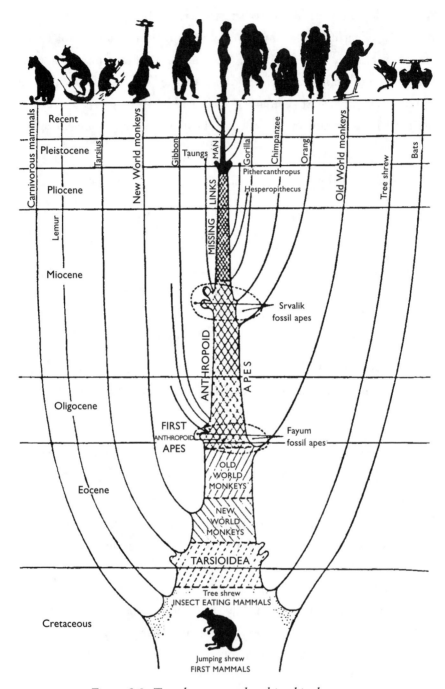

Figure 8.3. Tree shrew seen placed in this chart as a progenitor of ape and man.

POE: More than 98 percent of the human genome is shared with chimps.

A: And we share 99 percent genes with mice! Although gibbons, of all the apes, are *genetically* farthest from man, they are the most bipedal of the apes! "It might not be correct," cautioned Jeffrey Schwartz in *Sudden Origins,* "to try to sort out the evolutionary relationships of organisms on the basis of overall genetic similarity."

> *Similar genes mastermind the development of wildly different creatures.*
>
> JERRY FODOR AND MASSIMO PIATTELLI-PALMARINI,
> *WHAT DARWIN GOT WRONG*

More than 80 percent of all the proteins shared by chimps and humans *differ* in at least one amino acid. There is a gross difference between human and chimp chromosomes.[9] Even so, maybe the apes *were* the Creator's prototype for Asu man, a permutation of the same idea. "Even the Creator may use a good device more than once."[10]

POE: Only evolution and descent can throw light on the many homologs:* The chimp's lower humerus, for example, is very similar to the human's.

A: I find it curious that evolution, a science that loves to find minute anatomical *differences,* indulges wholeheartedly in gross *similarities* when it comes to comparison of man and ape. Man and ape are analogous, not homologous. Homologs, to the German school of *Naturphilosophen,* were simply archetypes/leitmotifs in the Creator's inventory; in which case, similars might just as reasonably be attributed to an efficient, parsimonious Creator as to a common ancestor. Even

---

*Homologs are similarities acquired presumably from a common ancestor (a good discussion of homologs is found in Fix, *The Bone Pedlars,* 188–91). Analogs, by contrast, are similar but not related, for example, the wing of a bird and airplane are analogous, but are not related.

Professor Mayr, after giving Darwin credit for solving the question of homology (from a common ancestor), demurred: "Homology cannot be proven; it is always inferred." Other authorities see the human pelvis as so distinct from that of the apes that it is impossible to derive them from a common ancestral stock.

POE: Yet the human and orangutan shoulder is strikingly similar. Same holds for orang brain, whose convolutions are numerous, and the frontal lobe (seat of intellect) is more prominent than in any other anthropoid ape.

A: But not because of a genetic link.[11] If a common ancestor split into *Homo* on the one hand and today's apes on the other ca 6 mya, why do apes have much more genetic variation than we do?

POE: I view the pongid as prologue to man, prefiguring him in shape and form. Do you deny the similarity?

> *Resemblance does not always imply descent.*
> Marcellin Boule, *Fossil Men*

A: These homologous organs may have very different histories. Resemblance does not guarantee an actual relationship, just as Joe Blow in Kalamazoo may be a dead ringer for John Doe in Oshkosh—without any relationship whatsoever. It is entirely gratuitous to assume that similarities between species indicate a common ancestor. At best, pongids are similar—same basic blueprint—but unrelated to australopiths, just as whales (mammals) are similar but unrelated to fish, and sharks (fish) look similar to but share no genetic heritage with porpoises (mammals).

American vultures look a lot like Old World vultures, even though the former are related to storks and the latter to hawks. Falcons may behave like other birds of prey, but are genetically unrelated to them. Conversely, some birds that look very different, such as hummingbirds and nightjars or songbirds and parrots, are more closely related.

POE: We are still looking for that common ancestor—a hominoid type like *Ramapithecus, Kenyapithecus,* Proconsul—who gave rise to the australopiths.

A: Weren't those beasties ruled out as ancestors, recognized as a false alarm back in 1979? Nothing more than anthropoid apes? *Ramapithecus,* last time I checked, was too old to be a hominid forerunner. Any comparison of Ardi or *Australopithecus* to apes is superficial, for *Homo* features are dominant even in the earliest men: cranial height, shape of occiput, poise of head on vertebral column, structure of pelvis and limb, face and teeth.

POE: Darwin believed our common ancestor was "furnished with a tail" and our coccyx is what's left. With that understanding began our attempts to link man and simian homologously, phylogenetically. As an example, the region of a macaque's brain that controls jaw movements is a direct homolog of Broca's area, which controls human speech; these findings contradict the theory that speech evolved (from novel neural structures) specific to humans.

> *There is nothing unique about human evolution.*
> CHRISTOPHER WILLS, *DARWINIAN TOURIST*

Indeed, we may well have found a missing link in *thick-skulled* archaic hominids; and no doubt the eyebrow torus, the "awning," of the anthropoid ape persisted in *H. erectus* and Neanderthal man.

A: So why is the gorilla's cranial wall *thinner* than modern man's? The Miocene apes also had a fairly thin skull and no browridge. Proconsul lacked the bony torus: his forehead was smooth. And in many monkeys we find the same shaped browridge as in modern man. In fact, if the heavy brow is supposed to be more apelike, I wonder why *H. erectus* brow is larger than that of austrolopith, his predecessor. Don't tell me it's more specialized.

Some nineteenth-century Europeans believed in direct continuity from ape to man: An 1824 visitor to India spoke of "wild tribes which the native names liken to the Orang-Utang, and my own knowledge certainly bears them out . . . the individual I saw might as well pass for an Orang-Utang as a man." A like-minded contemporary thought "the pendulous abdomen of the lower races . . . shows an approximation to the ape, as do also the want of calves, the flatness of the thighs, the pointed form of the buttocks, and the leanness of the upper arm." The first Europeans in Australia called the Aborigines "tailless chimpanzees." Even today the Semang Negritos, say the (haughty) Malaysians, are descended from *siamang* (monkey). In India, too, there is talk of dwarf tribes (Negritos) who descended from the monkey god Hanuman. In Africa, the S-shaped spine of the Akkas (who have long arms and short legs) inspired some to posit a link between man and ape. Darwin himself thought the Alacaluf Indians of Tierra del Fuego could hardly "boast of human reason."* Indeed, all these popular etymologies of the nineteenth century are just one step removed from today's misguided search for a common ancestor linking apes and man.

> [They are] looking and hunting for that which was forefather to both man and monkey. What sort of beast they expect to find I cannot imagine.
>
> JAMES CHURCHWARD, *THE CHILDREN OF MU*

---

*Goodman, The Genesis Mystery, 45.

PONGIDS AND PUNDITS

POE: Of course, we are not from apes as such, but we share a *common ancestor* with them. There is a difference. *Ramapithecus* of India—say, 12 to 30 mya—was one candidate for that common ancestor: its molars, canines, and jaw had a decidedly human cast; the curved dental arcade made him look quite human.

A: But molecular dating asserts that nothing hominid could have existed more than 7 mya. Anyway, new evidence, as you well know, overturned these *Ramapithecus* hopefuls. They weighed no more than thirty pounds; the mandible was finally judged not manlike. It turned out to be some sort of primitive orangutan.

POE: These fossil anthropoids, in Hooton's time, were called dryopithecines; they were spread over a wide zone of the Old World and probably evolved into the ancestors of today's apes on the one hand and to several varieties of early man on the other. That apes and humans share a common ancestor is now undoubted fact. Each new fossil tells us more about the continuity of the lines leading from hominoid types to the races of man.

A: Yes, yes, there are always candidates, but why no trace of their *descendants* until the time of Ardi or Asu man? Why the gap? It was the same with *Ramapithecus:* there was that yawning gap—millions of years—between that "common ancestor" and any of his supposed human descendants.

POE: Still, Africa's Lucy and Ardi are close in appearance to that common ancestor—who, I'm sure, will turn up sooner or later. Be patient. The pongids had to have split from the human lineage between ten and five million years ago. Molecular anthropology supports this very nicely: the ancestor of man, according to twenty-first-century genetics, was something rather apelike, something from which both chimp and gorilla also descended. But after the chimpanzee and human lineages diverged, both underwent substantial evolutionary change. It was not until the early 1990s that *Australopithecus*'s predecessor Ardi, a million years older than Lucy, was discovered, yielding a total of forty individuals. Don't you think a few more decades will suffice to hand us Ardi's predecessor?

A: No, I don't. I'll tell you why: Because *Ardipithecus* was true Asu man, and he *had* no predecessor. He was the first of his kind with all the earmarks—albeit roughhewn—of the *Homo* lineage.

POE: Well, there were two kinds, two different species: *Ardi ramidus* 4.4 mya (early Pliocene) and *Ardi kadabba* 5.6 mya (late Miocene). All right, Ardi had manlike canines, but its jaw was ape-like, and the female weighed 110 pounds; she was larger than Lucy.

A: Well, there you go: Lucy was younger and mixed with gracilizing Ihin genes, but Ardi (older) was pure Asu—first man ever—with no gracile features, and the female, at 110 pounds, was a far cry from 30-pound *Ramapithecus*.

POE: But so apelike, with a chimp-size brain (350 cc) and prognathism—a fruit eater.

A: Precisely the description of Asu man.

POE: But showing an *improvement* on the anthropoid: *Ar. ramidus* upper canines were less sharp than chimps. Thinner tooth enamel, too, reflecting a diet rich in easy-to-chew fruits and vegetables. It was a fantastic mosaic of ape and human traits—the foot, with its human toes, but lacking arches and with big toe splayed out; the pelvis is also a mixture of human and ape. Isn't that proof enough of descent from anthropoid ape?

A: I don't think so. Ardi's hand was humanlike. And while chimp feet are made for grasping trees and branches, Ardi's feet were actually better suited for walking.

POE: Ardi was what we call a facultative biped—climbed trees but also walked upright, though inefficiently, like modern chimps and gibbons. Ardi's lower hip was adapted to climbing, but upper hip (the ilium) was broad, which is to say, adapted for walking. Ardi's adaptations did all the hard evolutionary work for her successor Lucy. No one was surprised that Ardi showed a mix of chimpanzee-like and human traits, for Ardi was a link—way closer to an ape than to an australopithecine.

A: Not according to Tim White, its discoverer, who thought it was not particularly chimpanzee-like according to overall morphology.

POE: OK, at best, Ardi shed some light on our last common ancestor—the source of both us and chimps. Ardi tells us a bit about what that creature looked like.

A: Pure assumption. A leap of faith. The details of the alleged pongid split are still a complete mystery. There's a good reason they never found the missing link. There *is* no missing link.

Isn't it curious that some zoologists with no ax to grind for primate evolution have found as many similarities between man and ape as between man and dog.

> *I've seen cocker spaniels who looked about as human as Zinj.*
> WILLIAM FIX, *THE BONE PEDLARS*

Figure 8.4. Dog and his man. Cartoon by Marvin E. Herring.

So why not argue our descent from dogs? After all, dogs, like humans, are the only mammals who are all in one species; dogs share 80 percent of their hemoglobin sequence with humans. And dogs are more like humans emotionally than any other animal—capable of shame, remorse, curiosity, enthusiasm, grief, hopefulness, jealousy, deceit, and even some sort of conscience. The love and devotion of a dog for its master, Darwin thought, represents a kind of rudimentary religious impulse! He also thought that the beginning of reason was demonstrated by sled dogs who diverge when they approach thin ice. In 1285 Marco Polo said that the Andaman people "are no better than wild beasts, and I assure you . . . [they] all have heads like dogs, and teeth and eyes likewise; in fact, in the face they are just like big mastiff dogs."[12]

POE: You must be joking. There is hardly any morphological resemblance to dogs, whereas structural resemblance to the anthropoids virtually proves shared descent.

WE WHO WALK ON THE "HIND PAIR OF HANDS"

A: Ardi's wrist could bend *backward*. Wholly unlike any known apes (who are knuckle-walkers), Ardi could walk on her *palms*.

POE: Nevertheless, paleontologists are closing in on the split. We're in the ballpark. Ardi is the closest we've come to the first manlike thing that split off from the common ancestor. Only after the split did chimps evolve *specializations*—like knuckle walking and broad incisors. Many features of the australopiths are markedly different from both humans and apes alike; the australopiths are rather unique among hominids.

A: Excuse me for noting that *Australopithecus* was never proven to be a link to the higher apes, or to any apelike common ancestor whatsoever. His elbow and anklebone are human, not at all gorilla- or chimplike. His hand bones and jawbones are slender, teeth quite modern.

POE: But *Australopithecus* had a decidedly simian look.

A: Sure, he was the first edition of man, a creature of the wild, but he was not an improved ape, dog, or pig—which he also resembles chemically, by the way: most xeno-transplants (for humans) are from pigs. So what, if apes and human beings have similar DNA. So do humans and bacteria.

POE: Perhaps you forget that man and all other vertebrate animals have been constructed on the same model. Nor can you deny that the Pliocene Asiatic apes present dentition in some respects approximate to the Hominidae. The real question then is whether these resemblances indicate an actual phylogenetic relation. We think it does.

A: Pongid dentition is different from man's, for which reason Le Gros Clark postulated they must have diverged earlier than the Miocene. A fine ad hoc solution. South Africa's Swartkrans australopith had a pattern of dental eruption the same as *H. sapiens,* which never occurs in Pongidae. There are fundamental differences between man and anthropoid, especially in the canines and lower premolars. The tooth patterns of early man, noted Dr. Weidenreich, "even in the most primitive forms, remained basically the same as those of the later phases."[13] Don't you think the man-ape resemblance has been overplayed? Isn't the ape basically the atheist's substitute for God's own creation—man?

POE: Let me refer you to Dr. Gould's pertinent remark that the subject of evolution "doesn't intersect" that of religion.

A: Let's not beat around the bush: it comes down to atheist versus believer, not science versus religion or rational versus supernatural!

> *To be a true evolutionist, a man must be an atheist. If a man believes in God, and [that] he has a soul and a hereafter, he is not an atheist, nor is he an evolutionist. He only thinks he is. He is only professing to believe in evolution to be considered orthodox.*
>
> JAMES CHURCHWARD, *THE CHILDREN OF MU*

Figure 8.5. James Churchward was not only the modern founder of Panology but also a sage and prophet in his own right.

BEYOND THE PALE

We are born to search after the truth.

MICHEL DE MONTAIGNE

A: With the sciences so specialized and departmentalized, knowledge has become fragmented and a bit of tunnel vision has set in. We should all be generalists, don't you think? Probing the origin of man *must* be an interdisciplinary effort. But do we really know how to work together? Think together?

POE: Frankly, most of us feel that ID doesn't qualify as a competing theory because it doesn't offer natural explanations for biological phenomena that can be tested scientifically. Our methods must be empirical.

A: Empirical? This is precisely what evolution is not—based rather on assumptions, special jargon (which favor Darwinism), extrapolations, surmises, comparisons, hypotheses, tautologies, probabilities, computer models, and might-haves. Just for a moment, let us not worry so much if it is science but if it is truth. Darwinism was enshrined to legitimize the age of science and industry. But now we are entering a new time, the age of truth—Satya Yuga. This cut-and-dried *Homo sapiens* of yours is not the same person as my human being—with a soul and a mind and a heart.

POE: Even though the doctrine of ID thankfully passed away in the nineteenth century with the coming of Darwinism, here it is back again—impeding the advancement of true science, fundamentalists demanding a literal reading of the book of Genesis: the six days, the young Earth, the Garden of Eden, Adam and Eve, the rib, all stars made on the fourth day (Genesis 1:16), as well as the idea that Earth and species will last forever, utterly rejecting the fact of extinction.

A: Let's not make the mistake of equating all religion with the Holy

Bible. . . . Those who believe that Christian dogma is the *only* alternative or rival to evolutionary origins or only complainant in the court of Darwin have not looked very far—to other religions and world traditions. Creationism is not about Christianity. Forget Christianity for a moment. This is about all faiths, philosophies, and cosmogonies that admit a higher power. To understand life and death, you need to go outside the straitjacket of science. And evolution is about life and death.

GODLIKE PRECISION

POE: If you are comforted by the idea of deity, fine, that's your business.

A: My dear fellow, it's not a matter of comfort; it all comes down to whether we view man as an intended product or not. To me, it is perfectly inconceivable that humanity is here without reason or purpose.

POE: That is your mistake. There was nothing inevitable about it. Nature follows no purpose. Evolutionists have rightly stood together to strike anything purposive from the ascent of our species, for this would suggest planning, intention, teleology. But yes, we are here for a reason, even though that reason lies in the mechanics of engineering rather than in the volition of a deity.

> *Mechanical force makes the dog's tail wag, but something different . . . makes the dog wag his tail.*
>
> GEORGE FREDERICK WRIGHT,
> *ORIGIN AND ANTIQUITY OF MAN*

A: Max Planck, *the* Max Planck, said all matter originates from and consists of a force, and we must assume a conscious intelligent spirit behind this force. I see theories all over the place that are trying to account for nonphysical things in physical terms.

POE: Such as?

A: Oh, brain versus mind, for one.

POE: The theory of evolution holds that the design of man, including his brain, was simply a product of life on Earth, not its ultimate purpose.

A: But design and purpose go hand in glove.

POE: Well, natural selection may give the appearance of design, but that design is often imperfect, and imperfection is a sign of struggling, hit-or-miss evolution—and a major flaw in the argument for "intelligent" design.

> *Whence had man his understanding, if there was none in the world?*
>
> SOCRATES, *XENOPHON*

> *If it's unreasonable to believe that an encyclopedia could have originated without intelligence, then it's just as unreasonable to believe that life could have originated without intelligence.*
>
> JONATHAN SARFATI, *REFUTING EVOLUTION*

Imperfect design, vestigial organs—all argue against special creation: for example, appendix, hernias, or tonsils in no wise reflect "intelligent" design; such imperfections and inaccurate workmanship— the screwy wiring of the retina, the sacroiliac, the wisdom teeth—are unworthy of a designer. And parts created for no use? That doesn't make any sense. Thymus, pineal. There are too many instances in the animal world where a conscientious designer might have provided better organs, but didn't, and species are stuck with what nature's own process provided.

A: You might as well say we should not have been made bipedal, for all the fallen arches, slipped disks, and knee and back problems we suffer. But let me ask: Who says we can assume the Creator made things perfect for us?

POE: I'm only saying this: The idea that evolution has been guided by divine power can easily be squelched by this objection. Indeed, most lines of descent end in extinction. What a senseless effort on God's part to fabricate species and then let most of them die out! This does not suggest the work of an intelligent designer, still less of an almighty, compassionate one. It doesn't seem so intelligent to design millions of species that are destined to go extinct—and then replace them with other species, which will also vanish.

A: I dare say, it is not the creationists who need to explain extinction, but the evolutionists—and they have not yet done so. Besides, it seems pretty naive to think any species or any planet is here forever, or should be.

POE: If this designer is a loving, caring God, why has he given us stinging wasps, poisonous plants, slimy worms, and creeping parasites to make us miserable? Did your God, so interested in perfection, make the tapeworm? Why would a good God make the serpent or the mosquito or the germs of typhoid?

A: "The serpent bites to death. . . . This is no sin, for it fulfills its labor; it is the remnant of poison of other eras."[14] Even the lowly serpent played its part in the preparation of our world.

> I made the serpent the lowest of living creatures. . . . When the earth was encircled with poisonous gases . . . I drove the poison of the air down into vegetation . . . and I created the serpents . . . and they were poison . . . *thus I purified the air of heaven . . . [e.a].* Then I overcast the earth with falling nebulae, and covered up the poisons growing upon the earth, and they were turned to oil and coal.
>
> OAHSPE, BOOK OF INSPIRATION 6:8–15

Figure 8.6. Thomas Edison—"I know this world is ruled by infinite intelligence. Everything that surrounds us—everything that exists—proves that there are infinite laws behind it. There can be no denying this fact. It is mathematical in its precision." Thomas Edison's parents were spiritualists; interested in the matter, Edison himself conducted remarkable experiments in clairvoyance with the Polish-American medium Bert Reese.

Science pioneers who were devoutly religious include Kepler, Newton, Boyle, Galileo, Linnaeus, Cuvier, Comte de Buffon, Pasteur, Lord Kelvin, Joseph Lister, Blaise Pascal, Michael Faraday, Gregor Mendel, Leonardo da Vinci, Lord Francis Bacon, James C. Maxwell, and Sir Humphry Davy. Richard Owen (who identified the fossils Darwin brought home on the *Beagle*) thought Divine Mind planned the archetypes of species and even their modifications. Sir Richard said: "It is He that hath made us; not we ourselves."* In other words, things did not create themselves or organize themselves or work out their own design under nature's umbrella. "If developing the precision instruments of an airplane requires many plans and a highly developed intelligence, how could a substantially more complex apparatus [like the avionics of birds] have developed by itself?"[†] Louis Agassiz (*Methods of Study in Natural History*) expressed his disbelief in Darwinism this way: "The resources of the Deity cannot be so meager that in order to create a human being endowed with reason, he must change a monkey into a man."

Figure 8.7. Known for his famous Sterkfontein find, Robert Broom thought you could just as well call nature's adaptations the wonderful designs of a supreme intelligence.

"There is intelligence somewhere," said Robert Broom in *The Coming of Man*. Though an evolutionist, he believed that life on Earth was the work and concern of a divine creative force.

More recently, John White in *Enlightenment 101: A Guide to God-Realization and Higher Human Culture* has stated, "Evolution is a divinely driven process by which God as Spirit expresses itself . . . not blind forces and random events . . . or mere chance. They happen because God wills it intelligently, creatively and lawfully. Science has recognized some of the laws of the cosmos, but has not yet recognized the lawmaker. . . . God is the creator-artist behind the entire panorama of the cosmos."

---

*Quoted in Switek, *Written in Stone*, 230.
[†]Hornyanszky and Tasi, *Nature's I.Q.*, 87.

Natural science and theology, in an earlier day, were one and the same pursuit—though on an unsure footing. I believe the twain shall meet again, but this time on solid ground and in all truth.

Thoughts on Evolution from an Oahspean Viewpoint
by Carl Vostatek
8/3/2011

As students of the book Oahspe, we learn that evolution is not a concept that fits in with the story of mankind's beginnings on Earth and subsequent development and growth over time. Per Oahspe, each and every creature and plant was created perfect in its own time and place and did not "evolve" or change into a new form as time went on . . .

I offer this perspective because, as an architect, I am naturally inclined to look at things with an eye to design: composition; color; balance; symmetry, form and function. As a designer I am absolutely in awe and overwhelmed by the incredible infinity of shapes and colors and forms of all things. I can only imagine what a "design team" there must be up in the heavens somewhere! What an effort to come up with all the variety, all the necessary moving parts and features to deal with their respective environments, all the beauty, all the visual enjoyment. From a practical standpoint, there is simply no way all these facets could have "evolved by natural selection of the fittest" on their own. It is just too big a design project. . . . There has to be some conscious mind and conscious direction behind it all.

It is then I remember a quote from Oahspe which goes like this:

"*. . . for until thou hast created a firmament, and created suns and stars to fill it, thou hast not half fulfilled thy destiny.*"

As I ponder the enormity of this statement, I come to realize that one day I can be one of the "designers"; that is, I can be one of those who actually plays a part in creating the infinity of forms and shapes and colors that is found throughout the universe. For I am part and

Figure 8.8. Carl Vostatek. Courtesy of Carl Vostatek.

parcel of the creative force, the Creator! The Creator isn't outside of us, He is in and of us and we are in and of Him. We and the Creator are one.

This is the teaching Oahspe offers mankind; it further says that after due education, training, development, and spiritual growth in realms beyond the Earth plane, this is what I and all of humanity can look forward to. We can each play a part in the design and creation of new worlds and all that occupies them, from the beings to plants and animals and their multitudinous environments.

It is almost too immense to think about, too majestically, staggeringly, beyond the grasp of comprehension. But to us Oahspeans, it is real, it is reality. And in this light, evolution simply evaporates into non-existence. The alternative is a lot more inviting.

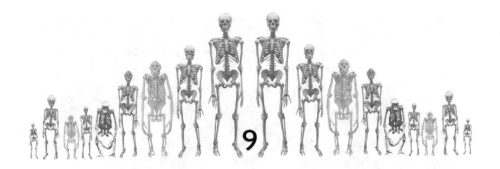

MUTANTS, MONSTERS, AND MORPHOGENESIS

Back to the Facts

*If I lived twenty more years, and was able to work, I
should have to modify the Origin, and how much the view
on all points would have to be modified!*

CHARLES DARWIN, ORIGIN, POSTMORTEM EDITION

THE WORM WHO WOULD BE MAN

That creatures respond, adapt, and adjust to their environment in
certain ways is, I would think, a given—nothing new or profound
there—and nothing so trenchant as to demystify our origins, indeed
nothing "more than a statement of the obvious," as far as author Francis
Hitching is concerned.[1] Empedocles of the fifth century BCE, as well
as other classical Greeks, wrote about it. Darwin did not invent it or
discover it.

In fact once he realized the weakness of natural selection, he actually moved away from it, preferring to pitch sexual selection; between the time of *Origin* (1859) and *Descent* (1871), Darwin switched from natural to sexual selection, the latter entailing competition among males for females. The human acquisition of a beard, for example, was presumably by sexual selection, an "ornament to charm the opposite sex." Darwin's concept of sexual selection was used even to define the races: selection, guided by tribal standards of beauty, would set the pace for morphological change. The sexually attractive feature would confer a higher reproductive rate on its owner, and hence eventually become incorporated in the race as a whole.

But this was overplayed, if not completely misguided. We know, for instance, that even though male bustards can impress females with their flashing feathers, this extravagance has its cost: scientists have found the most flamboyant master cocks, in fact, produce a greater amount of abnormal sperm. It has also been found that among the red deer of Scotland, the most prolific males sired daughters who had fewer offspring.

As for early humans, there was probably little competition for females among Au (*Au. afarensis*) males.[2] In fact, sexual selection seems to work for only a few animals and none of the plants. Biologists have asked: If it really did alter the species, wouldn't *everyone* be bright colored or fancy feathered in a few generations? Sometimes, too, the darn hens are *not even watching* the male display; even vividly colored male fish are *not seen* by the females whose eggs he fertilizes.

It has also been argued that fighting between males does not confer any special advantage on the winner; the "loser" simply goes elsewhere to mate. Often enough the female will mate, willy nilly, with the loser. The vanquished male, moreover, may have as many offspring as the "victor."

Some theorists have objected to this business of sexual selection as little more than a set of out-dated male assumptions, Victorian values (masculine bias) masquerading as science. In "deep thought, reason, and imagination," thought Darwin, the male of our species attains a "higher eminence," while women's faculties conform more with "the lower races."[3]

Earnest Hooton, for his part, refuted Darwin's argument that sexual selection led to hairlessness; in most species, Darwin had surmised, the less hairy female, with greater exposure of naked skin, constituted a special sexual attraction. But Hooton concluded "there are few indications that preferential mating could have brought about such profound modifications in the amounts of body hair."[4] Darwin had gone so far as to explain human *language*, beginning with the cries and gestures of animals, as developing out of the emotional stress of courtship—the sweeter voices of females having been acquired to attract the males! (Later in this chapter we'll take a more realistic look at the origin of language.)

Figure 9.1. Harvard physical anthropologist Earnest Hooton was also a cartoonist and author of science fiction under a pseudonym.

Do not suffer thy judgment to mislead thee as to a law of Selection. There is no law of Selection.

OAHSPE, BOOK OF APOLLO 3:6

The ultimate question is: Could the process of natural or sexual selection or any other imagined mechanism really change things enough to produce an entirely new species? Or even a new genus? That animals can change and turn into other animals seems a bit more like magical thinking than any part of science.

Each species develops according to its own kind.

LUCRETIUS, *ON THE NATURE OF THE UNIVERSE*

Isn't it interesting that natural selection, held as the key to the evolution of species *in the wild,* was so influenced by *artificial* selection as directed by human agents—animal and plant breeders, who *hybridized* preferred types by, of course, mixing! How ironic, since Darwin himself emphatically *rejected* the mixing (crossbreeding) model as an explanation of the varieties of men and animals. Here I might add, his own expertise was in the diversity of living animals and organisms—not man, and not early man. Darwin was a world expert on barnacles (he devoted eight full years to documenting their minute anatomical variations) and quite the master of beetles, pigeons, and earthworms. The worm who would one day be a man.

All right, natural selection is valid enough for minor changes, say, in the case of disease-thwarting genes against malaria, or smallpox, or when insects develop resistance to pesticides. But is that evolution or just modification? Sure, people of the Andes and Tibet have adapted to thin air with larger hearts and lungs and a greater volume of blood. Is that evolution? Whatever it is, it's not speciation.

Also in the Andes, the llama once had five toes (10 kya); now it has two. Is that evolution? The llama has not changed to a different species! So what, if natural selection is the mechanism that changes the coat color of mice: they're still mice. Enthusiasts find small differences in the wing shape of birds and call it "evolution in action." No, it is modification in action. Horizontal changes only. They're still birds.

We think of natural selection as tuning the piano, not as composing the melodies.

JERRY FODOR AND MASSIMO PIATTELLI-PALMARINI,
WHAT DARWIN GOT WRONG

All such changes are horizontal, which is to say: DNA is encoded for changes only within the range of that species (there are limits, as discussed elsewhere). The genome is a conservative thing, not innovative. Indeed, DNA is structured to *prevent* vertical variation. In light of these well-known facts, it is not unusual nowadays to hear the top people confess "that early evolution was driven by forces very different to those we usually associate with natural selection."[5]

But what are these forces? Here's where the fudging begins: We are told, for example, that it is a recombination of factors that exist in the gene pool, or repatterning of the genotype, or genetic drift that accounts for random changes in the proportion of genes. Grasping at straws, these formulas are makeshift at best.

In the 1980s, British-Australian biochemist Michael Denton mounted an incisive case against the "implausibility of selectionist explanations." This brilliant molecular biologist makes mincemeat of Darwinism's imagined phylogeny, which he suggests is on a par with "medieval astrology."[6]

No, I don't think the orangutan ever held the "potentiality . . . [of] becoming man,"[7] nor does the worm "strive to be man," as Ralph Waldo Emerson mused in a moment of poetic madness. The idea that we ultimately descend from a wood louse or insectivore sounds, to one critic, like "a Kafkaesque joke."[8]

Some critics say the supposed evolution of complex structures—from flagellum of bacteria to the human eye—is mathematically impossible. The eye, to England's William Paley, was a designed instrument, like a telescope. Darwin himself, who admired Paley's writing, owned that "to suppose that the eye with all its inimitable contrivances . . . could have been formed by natural selection, seems I freely confess, absurd in the highest degree."[9] Even his great American admirer the Harvard botanist Asa Gray thought that Darwin was unable to explain the eye by natural selection: How could "the eye, though it came to see, not be *designed* [e.a.] for seeing?"[10] One mathematician argued that there was insufficient time for the number of mutations apparently needed to make an eye. Another mathematician has said that a simple adaptive change involving only six mutations could not occur by chance in less than a billion years. In fact "the probability of evolution by mutation and natural selection is inconceivable." The odds of the eye evolving by chance are ten billion to one.[11] When testing this possibility on computer programs, "it just jams . . . [indicating] zero probability."[12]

ENVIRONMENTAL DETERMINISM: PULLING THE CLIMATE CARD

Nature does not seem to have achieved any significant modifications of existing hominid species.

JEFFREY GOODMAN, *THE GENESIS MYSTERY*

With nothing else—other than Divine Power—to explain the wonderful differences between species, environment then became anointed as the god of change. Pulling the climate card, theory now tells us that human evolution, occurring in Africa, was caused by major aridification: men evolved in adaptive response to the drying out of forest land and the subsequent appearance of savanna land. The whole argument is based on the assumption that man evolved in response to a drying Africa (see chapter 11). But since such different groups as Neanderthal and early *Homo sapiens overlap* in time and space, it became harder to explain their differences according to climate—considering their shared environment.

It has also been pointed out that, rather than *adapt* to drastic environmental changes, species tend to move away in search of a habitat like the one they were in. The fact that animals usually choose their habitats and *flee* from inhospitable environments (those all-powerful selective pressures) just about "pulls the rug from under the whole Darwinian hypothesis of natural selection."[13] Creatures like to investigate their surroundings, changes arising not from the selective pressure of the environment, but from the initiative of the living organism.[14]

Despite all the damning evidence trouncing natural selection, boilerplate explanations still abound—improvisation at its best. Neanderthaloid prognathism, for example, evolved because it "kept cold air away from the brain."[15] Only a few problems here: (1) cold climate cannot account for Neanderthal's sloped forehead, chinlessness, large browridge, retromolar gap, and round eyes; (2) The German Neanderthals living in the Valley of the Ilm "enjoyed a warmer climate than now."[16] This adaptation business is "pure speculation . . . the famous Neanderthals of Krapina, in Croatia, actually enjoyed a pleasant climate."[17] As did the Neanderthals in the Near East and North Africa.

> *The environment plays only a very limited role in the actual creation of new genetic variants.*
>
> AARON G. FILLER, *THE UPRIGHT APE*

What about the assumption that people like the Inuit developed bulky bodies and short limbs to help conserve heat, thus properly adapting to a cold environment? I would argue, though, that length of limbs, like stature, is quite simply an inherited trait. Short legs are the unmistakable legacy from early man. And since the Inuit were probably the *last* arrivals to Native America, they came on the scene (less than 12 kya) looking rather as they do today.

If, as the evolutionary argument runs, short and stocky Neanderthal got that way due to a cold climate, why then are northern Europeans (in the coldest clime) long legged, tall, and slim? And if taller is better suited to heat, why are the northern Mongoloids taller than the southern ones? The first *H. sapiens* in Europe, it is argued (to smooth away the contradiction), were slimmer than Neanderthal because they had previously adapted to the warmer climates of the Mideast and Africa. The alternative explanation, of course, is inheritance: gracile AMH genes made them so.

The climate card also cranks out a case for "cold-adapted" Mongolian features (like the broad noses), even though many Asian races live and lived in hot and moist tropics. In Europe, as one moves north, noses actually narrow (rather than broaden, as theory predicts). Scandinavians have small, narrow noses. So do the Inuit. I guess they weren't around when evolution was happening.

The races of man, in short, give no evidence that differences in nose form have anything whatever to do with cold air. Why do the little white monkeys, adapted to China's *coldest* region, have such cute tiny, turned-up, snub noses? It is the opinion of some, moreover, that a *smaller* nose is actually more adaptive to cold—to reduce the risk of frostbite! At all events, the wide Neanderthal nose would not really heat incoming air, it would dissipate it. And with all this evidence piling up against Neanderthal's "specialized" snoot, anthropologists are saying, yeah, we were wrong about the "radiator" nose, it was probably just a trait inherited from their ancestors. Back to square one.

Along the same lines, William Howells suggested that large-bodied, big-headed people in the western Pacific must be "the result of a little natural selection . . . [given] the constant cooling breezes that blew from the open Pacific over the islands of Polynesia . . . [whereby] loss of body heat could be a serious matter. Large bodies have a relatively smaller surface-to-bulk ratio, so that heat preservation . . . is better."[*18]

The same school of extemporizing submits that exertion under the tropical sun led to reduction of body hair; yet man's "peculiar larval nakedness is difficult to explain on survival principles,"[19] and the loss of a hairy covering is actually disadvantageous, even in the tropics, since it exposes the body to the scorching sun. "It is improbable that this denudation could have come about through natural selection . . . there is no appreciable relationship between climate and the amount of body hair."[20]

As flattering as it may be to think of ourselves as adaptable, flexible creatures, we are not so plastic as argued by these overplayed "explanations,"

---

*The Polynesian is not large bodied due to any adaptation, rather they are strongly Ihuan-blooded.

which ignore man's inherited or intrinsic attributes. Magazines and journals are lousy with adaptive explanations for everything from stubby legs to long noses. But may I ask: Why do women have a wider pelvis than men? Will you explain that by natural selection? No, it is simply inherent design. Even Lucy's modernized pelvis cannot be chalked up to evolutionary pressure: Why is the human pelvic opening larger than that of the apes? The ready answer is that it was an adaptive change allowing for the larger-headed human infant to pass through. Yet the head size of Lucy's people (australopiths) was no larger than an ape's.

"Adapative" is so broadly used, it loses all meaning: "Intelligence is obviously adaptive,"[21] but consider this—apes are smarter than monkeys but are *going extinct* while monkeys abound. Go figure.

ON THEIR HIND LEGS

There is no end to these contrived explanations: the teeth of first man Ardi (Asu) were short and blunt (humanlike) *because* the males "no longer needed to bare sharp fangs to scare off competing males" (and get the gals). Instead, they now won the gals by going far off "on their hind legs,"[22] bringing back enticing gifts of food! This fishy line of reasoning is part of the vaunted but improbable savanna hypothesis, which argues that human bipedalism evolved simply from the *need* to walk increased distances across open territory (the dry savanna of Africa). Part of evolution's presumptive logic and circular reasoning is the habit of explaining things by a *need* for them.

Climate change, it is asserted, having transformed Ardi's Africa into open grassland (savanna), made standing upright a great advantage. Ardi thus became bipedal due to walking and food gathering, or perhaps by looking over tall grasses (constant surveillance against predators favored a standing posture), even though: (1) George Gaylord Simpson himself, the acclaimed comparative zoologist and champion of neo-Darwinism, says bipedality is an adaptation to *desert*, (2) "bipedalism is not really the best way of getting around in a hostile world,"[23] (3) there are *disadvantages* to bipedalism: a monkey can outrun a human being, (4) baboons get on quite well on four legs in the savanna, indeed making faster escape from the toothed predators of that open environment, and (5) finally, upright bipedal posture may go deeper into primate history: the "common ancestor" of both chimps and humans could have been already bipedal, say some theorists. Nevertheless, it is held that humans got bipedal because of the *need* to carry things in their hands, or perhaps the *need* to expose less body surface to the sun. The final pitch is high drama: If our ancestors had remained in the forest and not gone on to savanna life, "we would not be here."[24]

But who can believe such fables? That these conjectural scenarios led to major structural change is as whimsical as the Lamarckian principle of acquired characteristics being inherited.

Down from the trees, but not out of the woods: Another problem with all this is that Ardi (Asu) probably lived in *woodlands,* not open plains, at least according to paleobiologist Tim White. For example, Laetoli (Au) was *already* bipedal while living in Tanzania's woodlands, not in savanna. Earliest man was a creature of the woodlands, a closed habitat (only later was his mixed descendant, Au, associated with open country). Soil research says the African savanna itself is not even 3 myr, while Ardi has been dated to 4.4 myr.

The philosophy of evolution stands or falls on the platform of natural selection, which postulates that the changes that define species come about through adaptations to the rigors of life. Hence, survival of the fittest, meaning: those best equipped to *adjust* to a given environment have an edge in survival and a better chance of reproducing after their own kind. Which is fine, theoretically; but evolution then boils down to a *negative* meaning, really, which is to say: The important changes in the gene pool will be the *elimination* of characteristics too feeble or ill suited to meet the demands of life. Natural selection, in other words, selects or filters *out* the weaker strains. But how does it account for *new and better* genes? Or for brand-new species? Where does the *improved stuff* come from?

Being a conservative force, natural selection merely *prevents* the survival of extremes, removing defective organisms in order to keep the status quo. Its work is to weed out deleterious genetic information, not innovative or creative at all, but pruning. It "chooses" from among a pool of variations—nothing actually new is introduced.

So where do novel genes come from (like *H. sapiens'* prominent chin, high forehead, bigger brain)? They presumably turn up by mutations, which of course are incidental, haphazard events. Despite this randomity, genetic mutations are royally crowned as the grand force behind the development of humankind—the human *brain!* Could chance events have fashioned the cerebral cortex in all its complexity, with more than ten billion cells all carefully coordinated? Darwinists say yes; let's have a look at their reasoning.

THE TOOL-FOOD-BRAIN CONNECTION

Recent workers, such as Meave Leakey (along with Richard himself and anthropologist Robert Martin), tell us that the brain requires a great deal of high-energy food; therefore, early man's increasingly creative use of tools (for hunting, hence for meat getting) must have been the impetus for brain growth. The energy provided now by hunted meat supposedly fueled the development of a bigger brain: "Meat and bone marrow gave them the extra energy to grow larger brains."[25] Thus did technological advances and hunting allegedly give us not only more gracile bodies but also better brains.

But have we got the cart before the horse here? This is the same sort of spurious causality argued by Louis Leakey in 1960—that tool use sped up the evolution of the Zinj hominids (he later abandoned this fruitless argument). Indeed, Ernst Mayr pointed out that freeing the arms to use tools could not have been the main reason for the increase of brain size, for even apes use tools. Nevertheless, a minute later, Mayr said Au's *need* for "ingenuity" (against predators) is the deciding factor that "created a powerful selection pressure for an increase in brain size"![26]

No hominid species shows any trend to increased brain size during its existence.

STEPHEN JAY GOULD, "EVOLUTION:
EXPLOSION NOT ASCENT," *NEW YORK TIMES*

But let me stop and ask: Could tools really have refined our anatomy? Or is it a fact that those people who used tools *in the first place* were already AMH, the earliest Ihin-blooded races? I agree with George Frederick Wright who contended that the more modern behaviors are the *results* of greater mental capacity "rather than their cause. It is not the use of tools that has produced his mental capacity. It is his mental capacity which has invented tools."[27]

It is hard to believe such patent chimeras (tools made the man), given that Hooton, so long ago, cautioned that no "gorilla will ever invent a knife or fork; and if he did invent them and use them, I doubt if his jaws would shrink. . . . Handling things does not necessarily produce thought, nor do tools make the brain grow." Why do today's workers ignore his warning against "facile mechanistic interpretations of . . . evolutionary changes . . . evolving an entirely new species out of a new habit."[28] Carl Sagan, for example, had man's knowledge of tools "eventually propelling such feeble and almost defenseless primates into domination of the planet Earth."[29] Tool use, as we saw, is also claimed as a cause of bipedalism (according to C. Loring Brace and many others), even though bipedal *Au. afarensis* (Lucy) *had no tools*; tool use actually came "long *after* bipedalism."[30]

HOT FOOD AND BIG BRAINS

What stimulated human evolution from our apelike ancestors? Cooking, says Harvard University biological anthropologist Richard Wrangham, a former student of Jane Goodall. That way, more nutrition reached our energy-hungry cerebrums, allowing them to "evolve" to their present capacity. This is how *H. erectus* got such dramatically larger brains than Au (a jump of 400 cc): their use of fire for cooking.[31]

In Darwin's own time, the Archbishop of Dublin saw Darwin's theory as "Lamarck's cooked up afresh . . . [especially] the conversion of the unaided savage into the civilized man."[32] Even though modern theorists have thoroughly discredited the old Lamarckian doctrine of habitual behavior influencing genes, here it is again, driving the theory that Neanderthal's ruggedness (large incisors, forward position of jaws, long face, supraorbital torus, and low cranium) developed out of his habitual use of teeth as tool; and also that fire and cooking eventually reduced the massive jaws of early Pleistocene man. I think we made more sense *one century ago,* following Arthur Keith's "full examination of all the facts [which] has compelled me to reject such Lamarckian explanations."[33] Mayr, for one, doesn't much care for Wrangham's assumptions: "Almost everything in this scenario is controversial . . . [especially] the date when fire was tamed."[34]

Figure 9.2. Sir Arthur Keith.

Nonetheless Wrangham seems to have good company: Milford Wolpoff found an "evolutionary" trend of cheek teeth getting smaller as AMHs began "evolving," the progressive change supposedly reflecting improvements in food preparation. According to Brace, excellent tools reduce the size of molars and the bulky shape of face and jaw.[35] Brace then used these "changes" to explain why Australia's northern Aborigines, with greater "technological elaborations," like seed grinders and nets, have smaller teeth than southern Aborigines, who lack this technology. I'll stick with Hooton, who, concerning this supposed dental reduction in African groups, inferred that "this size diminution may

have been caused by *hybridization* [e.a.] of the large-brained type with
. . . pygmies."[36]

All these primordial cooks are a misdirection: "There is no positive
evidence . . . the Pithecanthropines even knew fire."[37] At the famous
Peking site, the hearths (the ash deposits in these caves) could have
just as well belonged to a more advanced type (like the Ihuan hordes)
who hunted *H. erectus*! After all, Weidenreich showed that Peking man
coexisted with AMHs. Or Peking Man's remains "may have been introduced by carnivores. . . . The ash layers are not hearths and may not even
be ash. . . . Animal bones in the deposit are not evidence of hominid
diet."[38] Lumps of burned clay could simply represent tree stumps con-
sumed by brush fires; "baked" clays and ashes could be natural objects
of volcanic origin, prairie fires, or lightning ("geofacts" versus artifacts).

Druks did not cook their food, though they ate all manner of flesh
and fish and creeping things; probably their main meat was carrion.
According to a recent study, they snacked more extensively on termites
than on meat. Even late in the game (around 6 kya), the ground people
(in Egypt) were still eating fish, worms, bugs, and roots. Druks may
have been carnivorous (*H. erectus* sites suggest baboon kills), but not
necessarily hunters; heck, even Au ate baboon brains.

Actually, current theory *needs H. erectus* to possess fire, allowing him
to have made the great migration out of Africa; fire would have been
essential for their widespread dispersal into colder regions. If *H. erectus*
colonized Europe and Asia, how else could he have survived during the
glacial winters there? Yet we cannot use logic alone to reconstruct history, especially presumptive logic. Although it has become fashionable to
announce unambiguous evidence for controlled fires by *H. erectus,* it is by
no means certain that they could actually *start* fires. Even if Peking Man
did have fire, he may not have known how to *start* one: the depth of the
hearth layers indicates he did not permit his fires to go out, which is the
custom of even historical tribes who cannot start a fire. Today, African
pygmies on the Epilu River do not know how to make fire; nor did the
Andamanese and Tasmanians of the nineteenth century.

Regular fire use began in Neanderthal times[39]; only in later Neanderthal deposits are there signs that they ate cooked grains (bits found in their teeth). Neither were the Ihuans full-fledged carnivores until around 22 kya, while actual fire making in Africa is probably no older than 20 kyr.[40] And in America, it was not until 18 kya, that their angel mentors taught the Ongwees how to cook flesh and fish to make them more palatable. And this was the first cooked food since the days of the flood.[41] Similarly did the Greeks credit an unseen host, the god Phos, with their ability to control and use fire, just as the "advanced beings" (ABs) gave fire to man. It was also in this manner that the South American Tukano Indians learned the art of fire from their god Abe Mango, while the Kuikuru of Brazil say they had fire brought to them by the god Kanassa; likewise did the Aztecs have a fire god—Xiuhtecuhtli.

MAN THE HUNTER

Just as fire and cooking may in fact be more recent than current theory credits, man the hunter is not old enough to figure in the evolution of our species. Was *H. habilis* a "competent hunter," thus setting humanness in motion? Anthropologists claim that *H. habilis* brought back meat to be butchered and shared. However, follow-up digs at Africa's Olduvai Gorge uncovered additional *H. habilis* skeletons, which revealed that this "mighty hunter" was tiny (under four feet tall). Here, on the open savanna bristling with lions and saber-toothed tigers, *H. habilis* was likely spending a lot more time cowering in the trees. Cut marks superimposed over tooth marks indicate they were probably *scavenging* scraps of carcasses. Nor were the Neanderthals particularly organized hunters, but rather opportunistic ones, killing prey on an encounter basis—"rotten hunters" according to University of Chicago's Richard Klein. Even the "great mammoth hunters," as critics point out, may not have been hunters at all but marginal scavengers, at least until the late Mousterian. The much later people of Klasies River, and all other South African hominids for that matter, were not efficient hunters until 50 kya,[42] long past the age of *H. habilis*.

Sherwood Washburn, who led the dubious school of causality claiming that culture shapes physiognomy (rather than vice versa), argued that the "success of this adaptation" (hunting) dominated the course of evolution, resulting in "our intellect . . . and social life."[43] This way of thinking includes, of course, the notion that our bigger brains were developed on savanna land, as the forests, with its many fruits, retreated. With the switch to hunting and meat eating, hominids required "even greater cooperation and longer hunts." Or so they say. A great deal is made of this cooperation. Did early man hunt cooperatively? So what if he did; wolves, lions, hyenas, orcas, hawks, goatfish, and apes hunt cooperatively. Even one-celled animals form collectives; such grouping "appears very low down in the scale of development indeed."[44]

In three short sentences, the clincher word *evolutionary* is used five times by an anthropologist pitching the importance of hunting and eating meat: "The incorporation of meat-eating in the diet seems to me to have been an evolutionary change of enormous importance which opened up a vast new evolutionary field. [It] ranks in evolutionary importance with the origin of mammals . . . it introduced a new dimension and a new evolutionary mechanism into the evolutionary picture."[45]

But just the opposite, really: it was the *vegetarian* Ihins who were the real thrust of humanity; the sacred little people ate no flesh food, teaching that all life was sacred: "Neither killed they man nor beast nor birds nor creeping thing that breathed the breath of life."[46] Among their descendants, for example, the Atlantes people of North Africa, as mentioned by Herodotus, "ate no living creature."

TALKATIVE CHIMPS

Analysts would have us believe that humans improved their communication skills by hunting together; that we got our enhanced gray matter because we "had to obtain foods [large animals] that were even harder to find," and therefore we cooperated and communicated more fully. Thus is it argued that "diet accounts for the different paths taken by humans and the apes." This is how "humans became talkative chimps."[47] Believable? Even clear thinkers like Ashley Montagu and Ernst Mayr have turned causality toward agenda: "Speech . . . exerted an enormous selection pressure on an enlargement of the brain."[48] To get around the circularity of these arguments, they call it a "feedback loop": brain develops language and language develops brain!

We are told that because greater resourcefulness was needed (in drying Africa), bigger brains were selected to deal with the problems, the challenges. This "need" thing blunders into the supposition that human speech came along as a response to "adaptive pressures" and "demands on communication"[49] (in the movement away from the free-and-easy forest). Speech now arose because it was *needed* in the hunt; perhaps the threat of animal predators on Au led to such useful adaptations as "the evolution of increased vocalizations."[50] The need to share vital information on food sources is what "stimulated the development of fluent speech."[51] And these are the flimsy fictitious factoids taught today in institutions of higher learning.

Today's interpreters, then, are saying that the Great Leap Forward was *caused* by our capacity for speech: "development in language . . . propelled early modern humans to cultural dominance."[52] This inverted causality is an affront to both reason and scholarship. Recent researchers have gone so far as to suggest the bizarre idea that "early human language evolved as a way to effectively *groom* several people at the same time. . . . Having a larger vocal repertoire allows you to have a more complex social set-up. . . . Language arose as a form of group grooming."[53] (The idea, of course, is based on reciprocal grooming as observed among *apes*). Once the group got larger, say these theorists, vocal exchanges replaced the bonding created by one-on-one grooming. And this moonshine passes muster in today's prestigious journals. Seeing that population levels increased at sites of early mods, Erik Trinkaus, for example, suggests that "larger groups inevitably *demand* [e.a., there's that "need"] more social interactions, which goads the brain into greater activity . . . creat[ing] pressure to increase the sophistication of language."[54]

Darwin believed that the "social feelings" were innate: "Most of our qualities are innate."* Language, I would argue, is innate, too, hard-wired into our brains, like thought. "The premise that cognition *evolves* [e.a.] . . . is not only unproven—there is actually no evidence for it."† According to Noam Chomsky, the great MIT linguist and Nobel Prize winner, language was not acquired through natural selection, nor did it evolve. To Chomsky, "deep grammar" is built into the human brain only—not the ape brain. Indeed, Stephen J. Gould endorsed this "universal grammar." Language is simply one *aspect* of big brain, not a *cause* of it, as argued by evolutionary scientist Christopher Wills: "the force that accelerated our brain's growth . . . [includes] language, signs, collective memories."‡ But cognition cannot be "explained" by evolution; language, thought, and consciousness are *givens*—part of the intrinsic inheritance of the human being; we come thus equipped, hence, the hyperbolic myths of Krishna and Noah, who spoke as soon as they were born. The givens, irreducible and fundamental, do not need to be explained. They are self-evident, like the Buddha's "suchness of reality": "Ours not to reason why."

*Darwin, *The Autobiography of Charles Darwin,* 43.
†Dingir, "Hubris," 14.
‡Quoted in Leakey, *Origin,* 25.

If you've had enough of the "experts'" wizardry, let's see how speech really fits in the human picture.

THE MECHANICS OF SPEECH

When man first became widely diffused, he was not a speaking animal.

CHARLES DARWIN, *THE DESCENT OF MAN*

Darwin was right: Only *fully* upright man is a speaking man: the vertical column must be erect, not only to support his large brain, but to permit the mechanism of speech. It is for this reason that the talking birds (starling, raven, parrot) can utter words, for their larynx (voice-box) is vertical. Asu and Au, however, had the larynx rather high in the throat, limiting the range of sounds; you need a larynx low in the throat for humanlike articulation.

> And man [Asu] was dumb, like other animals; without speech and without understanding.
>
> OAHSPE, BOOK OF INSPIRATION 7:2–3

Lucy, an Au, was not much of a speaker, judging from the air sacs attached to her hyoid bone. Neither is articulate speech possible without a fairly flexible neck. This flexion at the cranial base is missing in early man; the bottom of their skulls, the basicarnia, is flat, limiting production of vowels and consonants. The Au palate and pharynx, in addition, are unsuited for articulation. Early hominids in general had shorter pharynges; one needs a longer pharynx to modulate the vibrations produced in the larynx.

But the modern vocal tract is not *why* humans can speak, it's *how*. The spoken word did not *make* us human, for speech is part of humanity, inseparable from brain power, an *aspect* of brain power.

Let us pause and ask if language, as claimed, was the *cause* of improved culture and complex social organization. Can we agree with today's experts who reckon that man's command of language triggered the cultural explosion of the Upper Paleolithic? We cannot, judging from the Kebara find in Palestine: a Neanderthal that came *after* AMHs in the Near East and that had an anatomically modern larynx*—yet Neanderthals never did make that Great Leap Forward.

The Max Planck people made big news with their "astonishing find" of Neanderthaloid El Sidron (Spain) specimens who share with mods a

---

*I would think Kebara was a Cro-Magnon/Neanderthal hybrid.

True to the record, in Hopi myth, the First People could not talk. The Maya Quiches also say the antediluvian race was without intelligence or language, for when Gucumatz and Tepeu first attempted to fashion human beings, the result was a dumb creature who could not speak or worship their creator. Neither did H. P. Blavatsky's third root race have spoken language: they lived, like the ground people, in pits dug into the earth. Even after the flood, the ground people had little speech (ca 21 kya).* As late as the nineteenth century, the Sri Lankan Nittevo, now extinct, had speech reportedly like the twittering of birds, while other eighteenth-century travelers mentioned Madagascan "manimals" who apparently did not speak any language. Even in our own day we hear reports of throwbacks, such as the cryptid Jacko, caught in British Columbia in 1884 (see chapter 12), who spoke in a half bark, half growl.

*Oahspe, The Lords' Second Book 1:15–6 and Book of Sethantes 9:10. Like the Yaks without speech? See First Book of the First Lords 2:7.

version of the gene called FOXP2, which contributes to language ability: "they possessed some of the same vocalizing hardware as modern humans."[55] Neanderthals also had a modern hyoid bone, enlarged thoracic spinal cord, and a Broca's area (even *H. habilis* had Broca's area, a feature of the AMH brain).

But have our paleoanthropological experts got it backward? Thinking that speech ability *led* to larger brains and modern behavior?: "Articulate speech . . . opened up whole new vistas . . . [and stimulated] brain development"[56]—the idea probably inspired by Darwin's strange logic: that speech, in time, helped perfect the mind.

Having found mod features in the vocal tract of hominids who possessed neither the culture nor the intellect* to go with it, theorists

*Neanderthal brain was large but not well developed in the cognitive regions.

had to save the day, *somehow*. Thus it is that Tattersal and Schwartz posit the vocal tract as a feature *prepared in advance*—its downward flexion traceable all the way back to some members of *H. heidelbergensis*, 500 kya. However, no symbolic or linguistic or cognitive behavior is evident before 50 kya. This 450-kyr gap is then explained by something called exaptation (or preadaptation)—a desperate gambit, inventing something that lay "dormant" in the genome until *needed* for some new purpose.

Such features, as the argument goes, were present in the ancestor, but they were doing something else initially, say breathing, and later were put to a different purpose, say eating or speaking. Thus may a population be "adapted in advance . . . a good example of a human exaptation is speech," the necessary structures having been in place "well before" language actually arose.[57] The modern vocal tract "must have been acquired in another context, possibly . . . a respiratory one. . . . Our ancestors possessed an essentially modern vocal tract a very long time before they used it for linguistic purposes."[58] Nonsense—it

---

This pseudo-explanation called "exaptation" now crops up in zoological evolution: In the case of saurians "evolving" into birds for example, the morphology of proto-wings is a "crucial flight feature [that] evolved long before birds took wing. It's an example of . . . exaptation: borrowing an old body part for a new job. . . . Bird flight was made possible by a whole string of such exaptations stretching across millions of years, long before flight itself arose."* Rubbish. Feathers, allegedly, had evolved long before they were used for flight—which only came later as a by-product of dinosaur arm-flapping and gliding to balance themselves as they made fast turns. Eventually the flapping evolved into the repetitive strokes of wings. *Yeah, right;* just like the giraffe got a long neck from repetitive stretching for the golden apple. This exaptation is craft, not science.

---

*Carl Zimmer, "What Is a Species?" 40.

appears only because of genes inherited from early AMHs. Exaptation has become a convenient catchall solution, or as cosmologist Sir Fred Hoyle ventured, a "rubbish bag for all evolution's awkward odds and ends. . . . Talking about pre-adaptation is simply a way of avoiding the issue."[59]

MUTATION: DUMB LUCK

Structural and chemical complexity reduces the chance of evolution by mutation to near zero. . . . The chance mutation of genes causing a series of concerted, appropriate behaviors would be more than a miracle.

BALAZS HORNYANSZKY AND ISTVAN TASI, *NATURE'S I.Q.*

The fortuitous character of evolution by mutation is certainly a coincidence of the highest order. Mutations, after all, are *errors* that occur during cell division. They are copy mistakes made by DNA, sometimes producing striking deformities. To get "good" evolution, you need a thousand *favorable* copying "errors" all in the proper order and in the right direction. How likely is that?

The mutation argument for evolution was famously debunked by the great cosmologist Fred Hoyle, who quipped that a living organism emerging by a series of *chance* events is about as likely as a tornado in a junkyard assembling a Boeing 747 from the materials therein. What a singular stroke of luck that out of the chaos of blind forces emerged this prince of beings—man!

By what trick of mesmerism have we fallen under the spell of such intellectual sorcery—making haphazard mutations the cause of order and design? Most new mutations actually become extinct because they only happen once; besides, some of these "mutations" may actually be recessive traits coming out!

The laws of probability rule out advantageous random changes on any but the smallest, most insignificant scale. A finely graded *series,*

Figure 9.3. Ray Palmer—"Planets hold to orbits, day follows night; the seasons progress in fixed order. Everything that lives and grows does so by a process that is consistent and not haphazard." Ray Palmer was editor of Amazing Stories, Fate, Mystic, *and* Search *magazines, as well as publisher of Amherst Press and the 1960 edition of Oahspe.*

all stages heading in the same direction, is not only implausible in the highest degree but also absent in the fossil record. Evolution teaches the improbable; that the great diversity of life is due to atypical, aimless deviations from a norm—the vanishingly small chance of accidental design features (itself an oxymoron, for design means plan)—that still, somehow, add up to superb system and order! This blind random process, "a giant lottery," in Michael Denton's words "is one of the most daring claims in the history of science."[60] Fred Hoyle, for his part, could not believe that chance (or even any earthbound theory) could produce genes capable of writing the plays of Shakespeare.

The belief in randomity, the *faith* in it, as against system and purpose, is very much a piece of philosophy, not really science. It is my understanding that the deeper a society (like our own) falls into *anomie* (a form of rootless alienation), the quicker it embraces unhinged notions of randomity and purposelessness, utterly losing sight of the universe with all its grand order and design. Even the evidence is against it, as geneticist Ronald A. Fisher put it: For mutations to afford evolution, one must postulate mutation rates immensely greater than those that are known to occur. Or as Gould saw it: "mutations have a small chance of being incorporated into a population."[61]

> *Genes are a powerful stabilizing mechanism whose main*
> *function is to prevent new forms evolving.*
> FRED HOYLE AND N. C. WICKRAMASINGHE,
> *EVOLUTION FROM SPACE*

When lab manipulations aim for eyelessness in fruit flies (*Drosophila*), it does indeed produce a strain of sightless flies. But given enough generations, some force seems to step in and once again the offspring have eyes. Called reversion, this response betrays genetics as the machinery of stability—not change, not innovation, not evolution. And "to observe a mutation" in the laboratory, thought French biologist and Nobel Prize winner Jacques L. Monod, "is a very far cry from observing actual evolution."[62] Yet so many evolutionary claims depend on experimental breeding programs with insects and little animals. Can we extrapolate from fruit flies to human beings? Do you really think that bombarding fruit flies with X-rays is a good way to find out about the history of man's evolution? Even mutated fruit flies are still fruit flies—eyes or no eyes—and have not speciated to some other critter.

Neither are large mutations—the stuff of "rapid evolution"—likely to be an improvement; chances are, a big jump, genetically, will end in death. Most macromutations, as they are called, are lethal, for any mutation big enough to cause an important change would, by the same token, be fatally disruptive. "Mutation is a pathological process which has little or nothing to do with evolution."[63] Change a gene too much, "and it will be unable to continue its existing functions."[64] Mutants are usually short lived and leave no progeny.

Only tiny changes (micromutations) have a chance. Darwin called them "infinitesimally small" modifications. Indeed, Dixon thought whatever the modifications in man's development, they were "far more likely to have affected the superficial rather than the actual structural portions of the body."[65]

With evolution pinned on random mutations, it is most damaging to find that mutations are overwhelmingly negative and admittedly rare;

natural selection actually operates against them. Are human beings the exception to the rule? "Many geneticists express doubt that genes have evolved. . . . Genetic variations are more likely to be harmful than helpful."[66] Not a few biologists will tell you that mutations are pathological and can have nothing to do with evolution.

Perhaps most telling is the observation concerning novel traits in flowers: "What looked . . . like mutants were actually *hybrids.*"[67]

INDIVISIBLE COMPLEXITY

Francis Crick, Nobel prize winner, proved that DNA was far too complex to have evolved by random chance. The accidental synthesis of DNA molecules and its associated enzymes would be a coincidence beyond belief. Ask any biochemist. The unfathomable interworking of DNA demands an explanation of order and pattern, not chaos and happenstance.

The unsolvable problem is this: How could so many complex parts evolve piecemeal yet work together toward a coherent whole? The theory of mutations is empirically and even *logically* unacceptable. We're talking about thousands of small changes—all happening coincidentally and randomly, yet evolving together in *tandem?* Please. This is a step in reasoning no thinking person should be asked to take. Minor changes over millions of years, all in exactly the right direction, producing *mutually* beneficial behaviors? Tell that to the judge.

Partial, incomplete, changes (so-called transitional forms) would be of questionable value to an organism, indeed a hindrance. "The piecemeal evolution of birds' lungs from reptiles' lungs," for example, "seems virtually impossible. The survival of [such] intermediates . . . is totally inconceivable."[68] Biochemist Michael Behe finds no living thing that can be "put together piecemeal" and provides many examples of how the complex machinery of life could not have "come into existence . . . in step-by-step fashion."[69]

Correlation of parts (as the matter is classically phrased) implies an all-or-nothing situation, and it is this *indivisible* web of inter-relationship that defies evolutionism and its cozy, safe-sounding "step-by-step" changes. Something more systematic *must* be involved.

Correlation of parts in the house mouse, for example, entails coat-color genes that have some effect on body size; they are interrelated. In fruit flies, induced eye color mutations actually changed the shape of the sex organs. "Almost every gene in higher organisms has been found to effect more than one organ system, a multiple effect,"[70] and these effects are *species specific*—which means, of course, it doesn't "evolve" and turn into any other kind of animal, any other species.

It is a question of orchestration, of interdependence.

> *Each change, taken in isolation, would be harmful, and work against survival. You cannot have mutation A occurring alone, preserve it by natural selection, and then wait a few thousand or million years until mutation B joins it, and so on, to C and D. Each mutation occurring alone would be wiped out before it could be combined with the others. They are all interdependent. The doctrine that their coming together was due to a series of blind coincidences is an affront not only to commonsense, but to the basic principles of scientific explanation.*
>
> ARTHUR KOESTLER, *THE GHOST IN THE MACHINE*

Georges Cuvier, an opponent of evolution, regarded organs as so intimately coordinated within the matrix of life that no one part can be fitted to perform a function without affecting other parts. In comparative anatomy, each major group of animals is seen to have its own peculiar correlation of parts—every structure functionally related to others and too well coordinated to survive major change through evolution. *All the elements of change must be simultaneously present,* and how incredible that would have been, a coincidence beyond belief, coordinating, in synch, all the accidental factors involved.

Figure 9.4. Georges Cuvier. Zoologist, comparative anatomist, paleontologist, once known as the Aristotle of biology, the French scientist was also a statesman and public figure. Cuvier was opposed to the concept of evolution, even before Darwin came on the scene.

Irreducible complexity, said Behe, involves a "meshwork of interacting components . . . matched parts that block Darwinian-style evolution." Change one link in the chain, and the system is seriously endangered. Gradual evolution, Behe goes on to argue, cannot by any stretch of the imagination account for the origin of: the immune system, metabolism, blood clotting, antigens, photosynthesis, DNA replication, vision, cellular transport, protein synthesis, metabolic pathways—all of which participate in an integrated, irreducible matrix. Darwin's tiny steps, analyzed biochemically, are a sham, "wildly unlikely."[71] Nor could the symbiotic relationships we find throughout nature have developed gradually, by "incipient stages" or trial and error.

AN INTRICATE WEB

As book review editor of a research journal, I was once asked to review Balazs Hornyanszky and Istvan Tasi's *Nature's I.Q.: Extraordinary Animal Behavior That Defies Evolution.* Outside the English-speaking world, French, German, Swedish, Russian, and East European biologists and ethologists are not so sold on Darwinism. This colorful book by two Hungarians, Hornyanszky and Tasi, advances the view that species do not and never did evolve from one another, citing "countless instances of complex . . . mechanisms whose origin the theory of evolution is helpless to explain. It seems much more reasonable to conclude that a being possessing higher intelligence equipped all species with the organs, knowledge and abilities they need."[72]

Nature's I.Q. was translated from the original Hungarian to English in the "Year of Darwin" (the 2009 bicentennial of Darwin's birth), a time when audiences were focusing attention on the undying controversy of Evolution versus Intelligent Design. From the notorious 1925 Scopes "Monkey Trial" in Tennessee (where the two views locked horns) to ongoing curriculum wars in Texas and other states, this great debate just won't quit, and with the publication of Hornyanszky and Tasi's delightful animal study, it becomes crystal clear *why* the great debate has continued without stint for 150-plus years. Gradual evolution of traits and behaviors does not hold up to scrutiny; rather, they appear to be quite simply innate, givens.

Never letting the facts get in the way, the Darwinists ask us to believe that a series of chance events (mutations) have resulted in the highly intricate mechanisms enjoyed by members of the animal kingdom. Thus does science continue to ignore the *inherent* order in things. How is it that the house of evolution, built upon the shaky, unscientific principle of blind chance, luck of the draw, has endured? How can the honest man say that one anomaly or mutation—freak, really—after the other has created the incomparable architecture of each and every species?

The matchless efficiency in nature that these Hungarian scientists regale us with, its interactive attunement, harmony, and wonder—in weaverbirds, night moths, scorpion fish, horned frog, mallee fowl, praying mantis, crested grebes, humpback whales, pelicans, starlings, sea turtles, cicadas, and dozens more—is all consigned (by godless science) to random transmutations within each species; the grand theory of evolution thus teaching us and our children that nature's precision and majesty are nothing more than a collection of fortunate accidents, a series of chance events, a chain of randomity bordering on a miracle— no, make that "more than a miracle" (according to Hornyanszky and Tasi).

Blind chance, to these authors, is perfectly laughable in the face of the keenly interconnected faculties found everywhere in the animal kingdom. It is "out of the question" that step-by-step change could result in the carefully arranged and coordinated equipment possessed by living creatures, simply because the behavior is "adaptive" only in its final form; there'd be no survival advantage at all in "incipient stages." Rather than natural selection or survival of the fittest, it is an intelligent designer, say these authors, who supplied each family of creatures with their life-enhancing characteristics.

Take the South American four-eyed frog whose fake eyes on its rump help to scare off aggressors. To these scientists, it is almost a joke that the posterior eyes appeared suddenly from a mutation, in exactly the right position and with the right markings; not to mention that the frog *knows* what mask it has on its derriere and behaves accordingly— turns its back on the attacker and lifts its rump-mask menacingly. These ethologists say point blank that the theory of evolution "is on the level of fairy tales." Indeed, "the sudden appearance of such complex biological systems, as if by magic, is completely impossible."

Flower and fertilizing insect fit each other like hand in glove. Has this sort of symbiosis "evolved" over and over again in thousands of species by blind chance? No, according to paleontologist Robert Broom and geneticist Richard Goldschmidt, the latter showing that complex mammalian structures could not have been produced by the selection and accumulation of small mutations, which can only account for minor changes *within* the species boundary. Large-scale mutations would produce "hopelessly maladapted monsters."

> *If you make Chance your creator, you are likely to get*
> *nothing but monstrosities as your creatures; you cannot*
> *make an alarm clock by whirling bits of scrap iron in a*
> *closed box.*
>
> JACQUES BARZUN, *DARWIN, MARX, WAGNER*

SPEAKING OF MONSTERS

What science calls mutations (or Loren Eiseley calls "genetically strange variants") may be nothing more than unfortunate hybrids, the result of unwise mating, in which case, we ask: Have humanoid monsters arisen from mutations or simply from mixing? Some esotericists have interpreted the Sons of God/Daughters of Men myth of biblical fame as a parable of mismatched genetic codes—the merger of the holy watchers with Earth men, their offspring "not quite human." The antediluvian patriarch Enoch, reasons one writer, knew about this mating between angels and earthlings, which resulted supposedly in the birth of mutants, "hybrid beings of all kinds, contaminated, depraved, a terrifying mixture of intelligence and beast."[73] Another extraterrestrial interpreter supposes that similar unnatural couplings produced hybrid beings, half human and half animal, that "nearly brought about the disappearance of the human race."[74]

But leaving extraterrestrials out of our search for monsters, let us consider Carl Sagan's suggestion that "occasional viable crosses between humans and chimpanzees may be possible."[75] Perhaps this is where our monsters lie. The school of theosophy, accounting in its own way for legendary monsters, posited that the fifth subrace of Lemurians, being stupid things, interbred with beasts. Some of the acclaimed (and controversial) Ica stones of Peru depict such bestiality. Indeed, today's paleontologists are not above suggesting that the ancestors of chimps and humans may have interbred, about 6 mya.[76] Even in today's world, we occasionally hear of cases such as Siberian village children born from sexual matings between humans and savage humanoids like Yeti.[77]

In 2001, Michael Brunet and his team made a strange fossil discovery in the Sahara Desert of Chad: six individuals of the race of Toumai. Officially named *Sahelanthropus tchadensis,* they were dated 6 to 7 myr (the time of the supposed split of anthropoid apes and hominids). This Toumai, put forward as the "common ancestor" of man and ape, possessed a mosaic of human and nonhuman traits. He had an apelike cranium (only 370 cc), round eye sockets, very prominent brow, and a short muzzle; yet he had a flat, vertical, humanlike face* and small teeth. Thought to be an upright, walking creature, Toumai was only three feet three inches tall. Prehuman? They found that the angle at which his spinal cord connects to the brain was like humans and quite different than the chimpanzee's much more acute angle. *Sahelanthropus* also had a pattern of tooth wear like humans.

Yet, rather than accepting Toumai as a common ancestor, might we ask if he was a hybrid of Asu and ape? Stephen Jones, editor of *Cambridge Encyclopedia of Human Evolution,* on the very first page (xi), mentions the possibility of early crosses between man and ape, for we are "extremely close cousins" (sharing 98 percent DNA). There are also "rumors of test-tube crosses between men and chimps."[78]

The earliest traditions of Western civilization remembered a time when Earth brought forth monsters, be it legends of the north, with monsters in the region of the Gobi Desert; or India's memory of a world "polluted with monsters" (at the time of Krishna's birth); or the Babylonian creation epic, describing Earth as peopled with gods, men,

---

*A manlike face, though, is readily seen in the marmoset and American spider monkey, both of which are monkeys. Hmmm, we're supposed to be descended from apes, not monkeys.

and monsters; or the Chaldean goddess who sired a crop of monsters; or the classical Greek legend of the grotesque Cyclops.*

Turning to Hebrew tradition, we learn of a time when the purity of the race was imperiled by degrading sexual customs: "There was a time . . . when both men and women had sexual intercourse with animals, from which resulted the birth of monsters."[79] The Bible (Lev. XVIII) warns: "Thou shalt not lie with any beast to defile thyself therewith. . . . It is confusion." Hebrew legend, moreover, recalls that in the antediluvian period, the tribe of Seth was of noble blood, but then the children of his son Enos(h) became idol worshippers, those who "strayed," their sins changing the "countenances of men. . . . No longer in the likeness and image of god, they resembled centaurs and apes."[80] Is this a remembrance of the four-footed Yaks (see below) or Pan the satyr (of the early Greeks) who was the most debased child of these miscegenations? Myths depicting men transformed to animals are fairy tales, of course, but their ineluctable kernel of truth lies in the retrobreeding of Adam's sons. "I showed him that every living creature brought forth its own kind; but he understood not and he dwelt with beasts, falling lower than all the rest."[81] (The time frame for this Oahspe quote matches the advent of the patriarch Enosh and "apeish" men, ca 65 kya.)

Cain, Adam's son, was not above retrobreeding with the lower races: "The Druks (Cain) went away in the wilderness, and dwelt with the Asuans. . . . And because the Druks had not obeyed the Lord, but dwelt with the Asuans, there was a half-breed race born on the earth, called Yaks . . . and they burrowed in the ground like beasts of the forest. And the Yaks did not walk wholly upright, but also went on all fours [giving us eventually the goatlike satyrs of the Greeks]. And the arms of the Yaks were long, and their backs were stooped and curved."[82]

Prehistoric drawings on pottery found in southern Peru depict a queer manimal with bent back and long thin arms, rather like images

---

*For the ancients, these one-eyed giants, the Cyclops, may have been a metaphor of the huge Druks, a race with "one eye only"—the eye for earthly things, without the light of heaven.

*Figuer 9.5. Greek Pan,
a metaphor, really, of
degenerated man.*

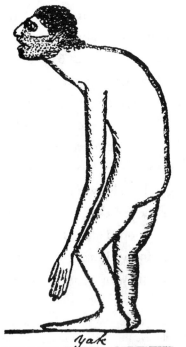

*Figure 9.6. Yak. It is
possible that creatures
like Zinj, Denisova, and
Dmanisi were Yaks, a race
produced both before and
after the flood, thanks to
retrobreeding. Illustration
from Oahspe.*

in the strange catacombs hidden in the cliffs of Easter Island, these frescoes representing a humanoid with a catlike head; the curved form has a rounded back and long skinny arms. Who was this prehistoric monster, this unknown race? Were they Yaks?

The first wave of Yaks, before the flood, resulted from the admixture of Asu (animal-man, with zero percentage of Ihin blood) and Druk/*H. erectus* (hybrid man, with 50 percent Ihin blood). The Yak therefore had only 25 percent fully human genes. I believe the name *Yak* has been retained in the languages of the world. It is curious that the long-haired, hump-shouldered, wild ox of Central Asia, domesticated as beasts of burden, are called yaks—just as the subhuman Yaks of old were made into servants,[83] to build, sow, and reap for the higher races. In fact a Japanese word for servants is *yakko*.

The name *Yak* persists in India, where the yakshas are spirits of the wilderness, and Hiran-yak-ashas are legendary giants. In Malaysia, Yak Jalang is the name of one ancestor. In Japan, the Gilyaks, like the Neanderthals, had a cult of bear sacrifices; these people were somewhat hirsute like the Ainu. In Northern Europe are the Votyak people. In Borneo are the headhunting Dyaks.

In Siberia are the dolichocephalic Ostyaks (proto-Australoid), the Koryaks of Kamchatka, and the Yakuts hunters. The Siberian Yucaghiri is phonetically Yakaghiri. It is, after all, from Siberia that we hear reports of humanoids like Yeti (almost a Yak type). In America are the Guayaki tribe of Paraguay, the Yakina Indians, and the Yaqui (Yaki) Indians of New Mexico. Among some California tribes, Sin-yak-sau was the first woman. In addition, the Yakama forest in the Cascades abounds in legends of animalistic Stick Indians; while the Yak name also seems to be recalled in places like Yakima (Washington), Yakutat Bay (Alaska), Tuk-to-yak-tuk (Canada), and the Eyak languages of the Pacific Northwest.

Oahspe makes mention of a great number of monstrosities betwixt man and beast before the time of Apollo. No, man did not *come* from the beast so much as he *became* one by these unruly mixings: thus it is said by theologians that "the sons of Adam erred," filling

Figure 9.7. The Yakuts are a Turko-Tatar tribe near the Bering Strait. Yakuzia is the name of a region in northeast Siberia.

the world with cannibals and manimals, apes, and centaurs. Such offspring were without judgment and of little sense, hardly knowing their own species. And they mingled together, relatives as well as others.

And idiocy and disease were their general fate, though they were large and immensely strong. Because of their marvelous power and brawn, we find them glorified in legend and icon, which we may interpret as a remembrance of the extraordinary strength in the bloodline of the ground people, the race of Cain, the burrowers (Druks). They were so strong, their grip could break a horse's leg, like the incredibly strong arms of Nittevo, Jacko, didi, and Orang Pendek (see chapter 12). Even the Ihuans, being the offspring of the Ihins and the ground people, inherited this corporeal greatness; they fought in popular arena games, unarmed, catching lions and tigers and "with their naked hands, choking them to death."[84]

But back breeding brought disease and deformity. According to tradition, Mexico's Cholula pyramid was built by gigantic men of deformed stature, the legend matched perhaps by the Old World cyclops and titans, also great builders, but bestial in both act and appearance, having "union with beasts." Huge, misshapen, and violent also were the Fomorians of Celtic myth.

The book of Genesis says "all flesh had corrupted his way upon the earth," much as Oahspe recounts: "They bring forth in deformity on the earth,"[85] for mortals descended rapidly in breed and blood when the Ihins mingled with the large Druks,[86] their offspring afflicted with catarrh, ulcers, lung, and *joint* diseases (i.e., deformed). Earth before Apollo was overrun with monstrosities, men and women looking like a dog that is tired, with deformed spines.*

Lesions have been found on Neanderthal skeletons as well as African specimens, which show abcesses, rheumatism, and joint disease. Paleontologists say the average life span of Au in Swartkrans, South Africa, was seventeen years! Even the lifespan of the Neanderthals was no more than forty years, while half of them died by age twenty. Numerous specimens had crippling ailments, many with rickets and syphilis. Smithsonian workers found many deformed bones and syphilitic lesions on ancient skulls in Patagonia.

In the tenth century, when the Scandinavian Vikings reached northeastern America, near Rhode Island, they found there a race "totally distinct from the Red Man . . . whom they designated Skrellingr [*sic*], or 'chips,' so small and misshapen were they."[87] It was "the same race as in Labrador . . . [whom] they contemptuously call Skralinger . . . and describe as numerous and short of stature."[88] In Mexico as well, the deformed little Puuc were hunchbacked like the Skralinger. Such crooked men in the Americas have been confirmed archaeologically: Owen Lovejoy excavated skeletons with "strange physical deformities . . . [and] many unusual bones."[89] And in Africa as late as the nineteenth century, de Quatrefages reported on deformed pygmies south of Abyssinia.

Asia also has its misshapen figures. In Tibetan legend, the Han-Dropa tribe, of unknown breed, were supposedly so weird that their

---

*Under Apollo came the beautification of mankind and the last of Earth's monstrosities. Hence are "the Cro-Magnons the *Apollos* [e.a.] of the prehistoric world." (Montagu, *Man: His First Two Million Years,* 71.)

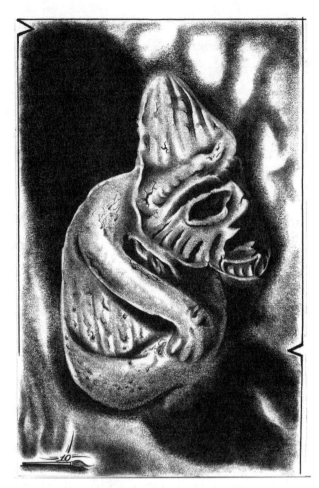

Figure 9.8. Puuc depicted in Mexico's Loltun Caverns.
Illustration by Jose Bouvier.

Figure 9.9. "Misshapen gods"
are represented in Olmec art
by "jaguar-men" and other
monstrosities depicting all types
and stages between man and
beast.

Chinese neighbors tried to exterminate them. Puny and fragile, four feet tall, they possessed disproportionately large heads. Their eyes were large with pale blue irises. A similar whispery race was known to live in North America, south of Cherokee country, a tribe of little people called Tsundige'wi, with very queer-shaped bodies, living in nests scooped in the sand. The little fellows were so weak, said the Cherokee, that they could not fight at all, and were in constant terror from the wild geese and other birds that came to make war upon them.

Figure 9.10. A large-headed African figurine.

Yet not one of these "hopelessly maladapted" races was the product of natural selection. The vagaries of hybridization—not mutation—once peopled Earth with surprising spawn, such as are recalled in our own "enlightened" age only in the fables and fairy tales of children.

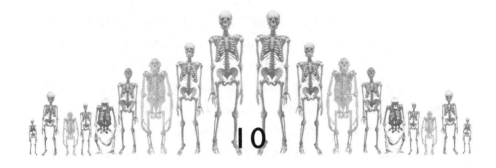

10

WHEN THE WORLD
WAS YOUNG

The Question of How Life Began

Where were you when I laid the foundation of the earth?
Tell me, if you have understanding.

BOOK OF JOB 38:4

SEMUAN BEGINNINGS

The very ancient word *semu* (the root of words like seminal, inseminate, semen, semolina, etc.) can be understood in terms of raw beginnings, for it applies to the inchoate world and is *no longer potent* once a planet has attained maturity. Think of semu as a gestative period: in the youth of our world, each species was "brought into existence . . . using special processes which are *not operative today* [e.a.]"[1] Creationists call this "a period of special constructive processes"[2] (in stark contrast to evolution's ploddingly slow and continual process of change and transmutation).

A formative age, quite different from our own, has been recognized by scientists like Immanuel Velikovsky; even the late S. J. Gould posited "developmental locking" after the Cambrian, thus ending the age of flexibility in which body plans were established. "No new phyla have emerged since the Cambrian age, 500 million years ago."[3] The Paleozoic Era (which began with the Cambrian Period) was a time of great pliancy, an inscrutable plastic power, an arcane force, which "cannot be seen . . . at the present time."[4]

The earth ceased to bear, like a woman worn out by age.
LUCRETIUS, *ON THE NATURE OF THE UNIVERSE*

Darwin, for his part, could only imagine that life itself began "in a warm little pond"[5]; indeed, the best analogy for semu today is the green scum that accumulates on ponds and pools of water.

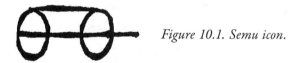

Figure 10.1. Semu icon.

Let a sign be given to man so that he may comprehend semu. And so the jellyfish and the green scum of water* . . . [are] permanently coming forth in all ages, so that man could understand the age of semu, when the earth . . . [was] covered over with commingled atmosphere and corporeal substance called semu . . . in this day the earth does not produce semu abundantly.

OAHSPE, BOOK OF JEHOVIH 5:4

Figure 10.2 shows a pool of water with a negative current escaping. What is that current? Something on the order of "fluctuating electrical fields" are thought to account for the origin of life on Earth,[6] living things having developed through the agency of some bioenergy field. Roughly

*Single-celled algae and bacteria were among the first life-forms. Algae appear in Earth's oldest rocks.

Figure 10.2. Pool of water showing negative current escaping.

equivalent to electromagnetic energy, vortexian currents escape through Earth overnight: these (atmospherean) currents are *to* Earth in the daylight, and *from* Earth in the night. A pool of water is charged during the day with the positive current, but during the night the negative current escapes upward from the water. The resulting decomposition is called semu, a mucilaginous substance that floats on the surface of the water. "In a few days this semu, by motion, assumes certain defined shapes, crystalline, fibrous and otherwise, like the strange configurations of frost on a windowpane. . . . They float against the ground . . . and take root and grow. . . . No seed was there. This new property is called Life."[7]

Indeed, Anaximander of Miletas (550 BCE) thought that life first started in sea slime; electrolytes of seawater are, after all, similar to the internal fluids of living organisms. Thus has the root of life come to be conceived of as a sort of primitive cell floating around in the "primordial soup." A soft gelatinous matter ("abysmal slime") taken from the ocean bed during dredging operations was once named bathybius, apparently an amorphous protein compound, capable of assimilating food, a diffused formless protoplasm.

Scientists like to point out that (1) comets hold organic molecules, possible building blocks of life, and (2) interstellar space is filled with clouds of dust containing microorganisms, cellulose, and other organic

Figure 10.3. Plate 39 from Oahspe, Earth in semu.

matter. Consisting of dust, air, water, heat, and colloids, a kind of proto-plasmic substance fell to Earth from nebulous regions in the firmament; as it is said in the Popul Vuh: "From fogs, clouds, and dust, creation came," just as other scriptures say that out of earth and atmosphere conjoined, "He created the trees of the earth and the flesh of animals."[8] Meteorites contain up to 30 percent organic compounds such as hydrocarbons and chlorophyll. And with precursors of DNA found in the Murchison mete-orite (which crashed in Australia in 1969), the ashes of dead stars, comets, and dust grains are now thought to contain the ingredients for life.

The new science that I am studying also looks at coalification as a nonterrestrial dynamic. "There is a time for semu; and a time for falling nebulae to bury deep the forests and semuan beds, to provide coal and manure for a time afterward. . . . If luts, a time of destruction, followed soon after a semuan period, when portions of the earth were covered with rank vegetation, it charred them, penetrating and covering them up. Most of the coal-beds and oil-beds in the earth were made this way. . . . Luts belongs more to an early age of a planet, when the nebulous clouds in its outer belt are subject to condensation, so as to rain down on the earth these corporeal showers."[9] Thus did condensed nebula (dust and stones and water combined) cover up the forests; it fell like hot molten iron, and the trees were beaten down, covered up, and burnt to blackness.

The formation of coal seams, then, belongs to formative times, in the Paleozoic's Carboniferous Period, when the first amphibians appeared. The profusion of plant life buried in the Carboniferous suggests a high concentration of carbon dioxide in the coal period; thus with CO2 locked up in coal deposits, the atmosphere would then be cleared, now suitable for terrestrial vertebrates—air-breathing reptiles.

Such deep vertical beds (some one hundred feet) are not formed today, nor is oil any longer being formed.[10] Forty-foot high standing trees buried in coal seams bely the Darwinian (and uniformitarian) view of long and slow compression by overlying rocks; for it was pressure, rather than time, that caused coalification—and quickly done! If the average rate of deposition of sediments is something like 0.2 mm per year, "such a slow rate would be quite incapable of burying and fossilizing entire forests, dinosaurs, or even a medium-sized tadpole."[11] Exposed, they would simply rot away.

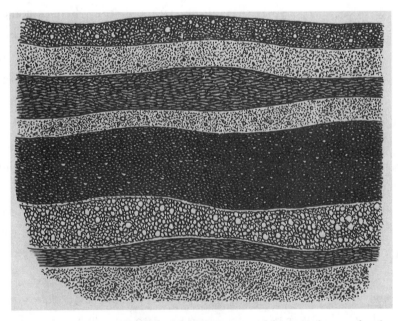

Figure 10.4. Depiction of coal found compressed between layers of rock.

SOMETHING CAN'T COME FROM NOTHING

Ex nihilo nihil fit (nothing comes out of nothing). It was during our planet's travel through the semuan firmament that life began on the face of Earth. It was not "random mutations" that led to cellular life.

> *Just as the brain of Shakespeare was necessary to produce the*
> *famous plays, so prior information was necessary to produce*
> *a living cell. . . . Life cannot have had a random beginning.*
> FRED HOYLE AND N. C. WICKRAMASINGHE,
> *EVOLUTION FROM SPACE*

Could RNA molecules have been formed by a chemical *accident* in the primordial slime? Are men "raised like Vegetables out of some fat and slimy soil?" quoth Richard Bentley (1692).[12] This idea was called, contemptuously, the "gospel of dirt" by Thomas Carlyle. Today, exponents of the "Godless particle" (Higgs Boson) say: "creating 'stuff' from 'no stuff' seems to be no problem at all—everything we see could have emerged as a purposeless quantum burp in space."[13]

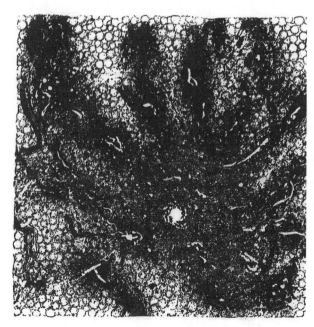

Figure 10.5. Plate 6 from Oahspe, picturing Earth in the Semuan firmament.

Could life have been burped from nonliving matter, as Ernst Haeckel thought of his *Urschleim* (primordial slime)? Or is life actually a *property* of matter, as eighteenth-century enlightenment materialism suggested? Or is there instead an animating principle, an unseen power, a force? James Churchward, so ahead of his time, stressed the difference between elements (matter) and force (energy). Isn't matter, of itself, dead, inert? Ironically, science, having thoroughly debunked the old theory of spontaneous generation of matter, invokes it once again, with molecules—godless particles—somehow self-assembling and jump-starting all life.

Yet, as biologist Michael Behe sees it, "staggering difficulties . . . face an origin of life by natural chemical process. In private many scientists admit that . . . [we] have no explanation for the beginning of life. . . . The involvement of some intelligence is unavoidable."[14]

There is no evidence in the oldest rocks on Earth of any *prebiotic* matter. Could viruslike particles have "evolved" into cells? Are we descended from viruses? If so, can anyone explain how a protein molecule evolved? "The odds of the chance evolution of a protein," said Francis Hitching "is about one in 10^{600}."[15] Did mineral surfaces magically spring to life, organizing key molecules? Did volcanic sources "synthesize" amino acids? What catalyzed the synthesis of compounds?

Can experimental design apparatus really simulate conditions of baby Earth? Yes, say those whose claim comes out of laboratories: starting in the 1950s, radiation or other energy sources were applied to simple gases, obtaining a by-product of amino acids. Critics, though, think such claims are misleading; it would be "an outrageous fluke for amino acids . . . in a polypeptide chain . . . [to mutate properly] which is well known to geneticists and yet nobody seems prepared to blow the whistle."[16]

Nurs'd by warm sun-beams in primeval caves, Organic life began beneath the waves. . . . Hence without parent by spontaneous birth, Rise the first specks of animated earth.

ERASMUS DARWIN, "THE TEMPLE OF NATURE"

Do you mean to tell me, then, that inert dust found a way to become a living thing? Evolution overall is the idea that things made themselves (with a little push from nature—*warm sunbeams* perhaps). Today it is said that nonliving matter somehow gave rise to life; somehow a few particles of matter managed to arrange themselves spontaneously into living structures.

Geologist George Frederick Wright, however, found it hard to believe "that plant life has had the power of taking upon itself the forms and prerogatives of animal life. . . . It is impossible rationally to believe that a principle of life is the product of chemical forces [which are] but the machinery of a mill. . . . If anyone wishes to believe that the marvelous adaptive capacities of plant life sprang from the dead forces of nature, he is at liberty to do so, but at the risk of his reputation for sanity."[17]

Figure 10.6. George Frederick Wright, turn-of-the-century American geologist.

Notwithstanding scientific feint, inert matter shows no potential to organize itself. How could complex information codes be generated without intelligence? I am one who doubts it. Just as consciousness cannot be put to matter, so too is life itself independent of matter, something more than "atoms in a bag."

Particles had to have a form of awareness that let them know how to behave.

PETER J. BOWLER, *EVOLUTION: THE HISTORY OF AN IDEA*

THE QUICKENING...
AND THE COOLING

The molecules needed for life are widely distributed in the universe.

ERNST MAYR, *WHAT EVOLUTION IS*

Perhaps, offers the materialist, life was engendered by the sun's ultraviolet radiation, or in a volcanic setting of "fire and mire." Perhaps it was "bolts of lightning" that turned simple chemicals to organic molecules, like the mythical Zeus hurling thunderbolts at Earth. The problem here is that macromolecules tend to fragment, not form, under such high-energy barrages. Asking chemical reactions to produce life from nonlife is about the same as asking favorable mutations to make a man out of a monkey. Organic molecules are a collection of parts. What would make them gel?

I rained down semu on the earth; and by virtue of My presence I quickened into life all the living. Without seed I created the life that is in them. . . . Thus made I him out of the dissolved elements of every living thing that had preceded him.

OAHSPE, BOOK OF JEHOVIH 5:14 AND
BOOK OF INSPIRATION 6:19

The ray of light that goeth out of Me taketh root in mortality, and thou art the product. Thou wert nothing; though all things that constitute thee, were before. These I drove together, and quickened. Thus I made thee.

OAHSPE, BOOK OF INSPIRATION 1:7–10

Elements (matter) are one thing; forces (energy) are the other. The formula of life entails not only ingredients (elements, raw materials) but also a force, an unseen dynamic, the power behind things, the punch, the spark, the quickening. Some ineffable cause or power, thought Alfred Russel Wallace, "must necessarily come into action when protoplasm appeared."[18] Indeed it was in ancient Indian cosmogony just such a quickening power, wielded by the High Lord of the Upper Heavens, when he sent whirlwinds abroad, which gathered in the substance of the world egg, and rained it down upon Earth. "Thus did Eolin touch the earth with His quickening hand, and instantly all the living were created."[19]

This quickening is also expressed in Genesis's "breath of life," as it is in AmerInd creation (Algonquin and Blackfeet) wherein Manito first molded men from Earth's clay, then *breathed life* into them. Among the Chinese, it was the deity Nu Kua who made the first humans out of clay from the banks of the Yellow River, then quickened them with the breath of life. The breath of life is also a part of anthropogenesis in Hawaii, Australia, and the Kei Islands of Indonesia.

Here is the Oahspe version of creation: "In the Semuan Age, the time of ripeness, all the living were quickened into life, for the earth had moved into its season for the bringing forth of living creatures. . . . [With] the triumphant entry of oxygen to the earth's surface. . . . [came] the gestative age for the animal kingdom, a time of propagation. . . . Those that were quickened into life, and not attached to the earth by fibers or roots, were called animals . . . created not in pairs only, but in hundreds of pairs and in thousands and millions of pairs. . . . And, in time, semu covered the earth abroad with Asu, till hotu came and the creation of new living things ceased."[20] And this was approximately 75,000 years ago. Man was the *last* creation in semu, and his advent was a thing apart from the anthropoid apes or any other of our "fellow primates." With the advent of Asu man at 80 kya, there would surely be no time for Darwin's evolution, which entails the accumulation of small changes over vast eons, measured by current orthodoxy in the *millions* of years. (See figure 10.7 on page 379.)

Some say the physical evolution of man ended 72 kya. Curiously, this is the date given for the advent of Ihin man and also for the following:

- the first AMHs in Altai, Russia
- the first modern Chinese[21]
- Europe's Fontéchevade Man, who was essentially like ourselves, allowing us to "date our species as in existence perhaps seventy thousand years ago"[22]
- the deepest divergence in an evolutionary tree, developed by genomic studies[23]
- the Welsh genome, where the biggest difference (eight mutations) is among individuals who shared a common ancestor ca 80 kya[24]
- the common ancestor of European and Australian Aborigine (dated to 70 kya); while New Guinea mtDNA traces back to 77 kya[25]
- the cognitive explosion, which some say occurred around 73 or 70 kya,[26] matching India's Ramayana, which traces the beginning of writing and culture to an ancient white race some 70 kya

Also suggesting the advent of man between 80 and 70 kya is the work of a Native Cree archaeologist who has created a database of the oldest sites in the New World—with five hundred entries ranging back as far as 80 kya, just as Native American mtDNA sequences from a Northwest tribe trace back to 78 kyr. While the earliest engraved object in America dates to 70 kya,[27] American grinding tools and pollen of cultivated corn have been dated between 70 and 80 kya.

HOW OLD IS THE EARTH, REALLY?

Just as the age of man may be a great deal younger than we are taught, the age of Earth itself has, I believe, been greatly overestimated.

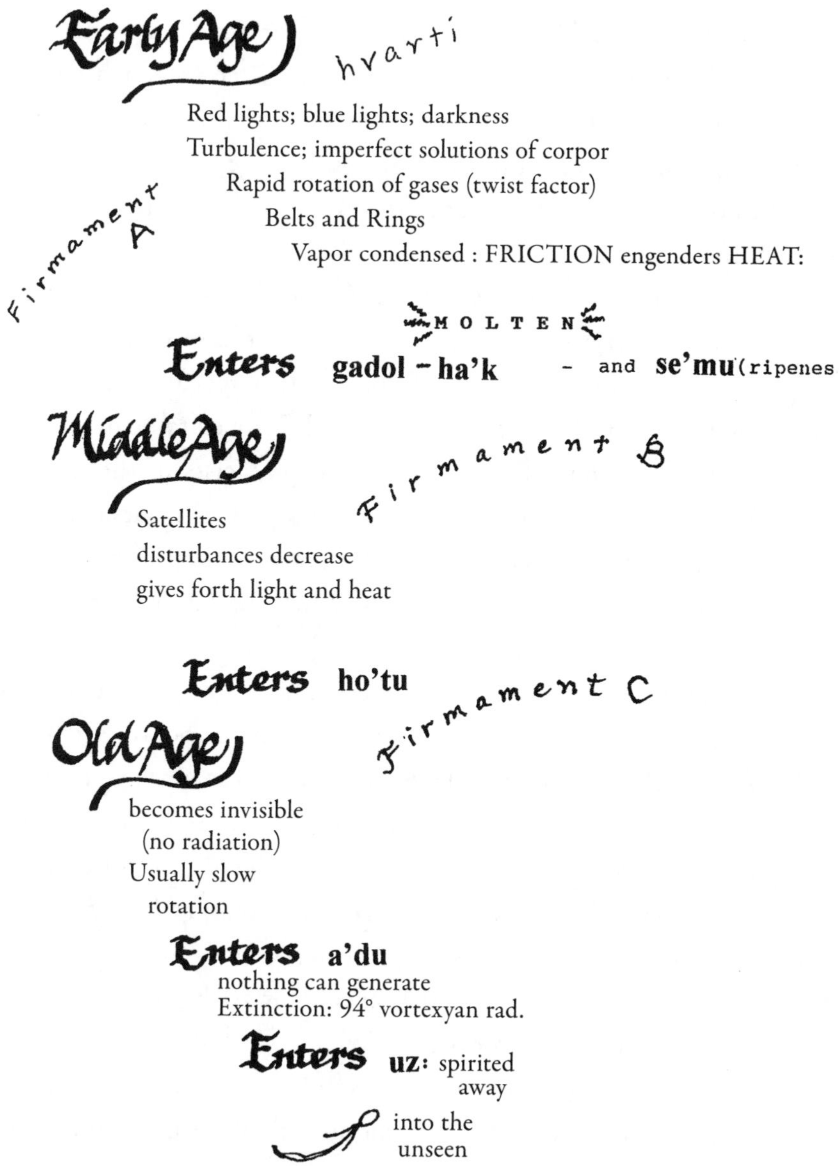

Early Age *hvarti*

Red lights; blue lights; darkness
Turbulence; imperfect solutions of corpor
Rapid rotation of gases (twist factor)
Belts and Rings
Vapor condensed : FRICTION engenders HEAT:

Firmament A

MOLTEN

Enters gadol - ha'k   - and **se'mu**(ripenes

Middle Age

Firmament B

Satellites
disturbances decrease
gives forth light and heat

Enters ho'tu

Firmament C

Old Age

becomes invisible
(no radiation)
Usually slow
rotation

Enters a'du
nothing can generate
Extinction: 94° vortexyan rad.

Enters uz: spirited
away
into the
unseen

Figure 10.7. Ages of the planet as told in Oahspe. By the ancients, the four seasons of manifestation were termed semu, hotu, adu, and uz, corresponding respectively to: birth/time of creating, growing and maturing, harvest and senility, and, finally, time of destruction and rest. In semu a planet is ripe for the bringing forth of living creatures. Then it enters hotu, for it is past the age of begetting, even as the living who are advanced in years. Next it enters adu, and nothing can generate upon it. Finally comes uz, and it is spirited away into unseen realms.

Modern science speaks of *billions* of years for Earth's age. Darwin assumed there were virtually no limits to the time in which natural selection operated, positing, in effect, a perpetual motion machine. For example, he estimated more than 300 myr for the time in which the Wealdon deposits of England had eroded, "a mere trifle of geological time."

Yet Darwin's contemporaries (and the following generation) found evidence for a younger Earth; hence, in the second edition of *Origin,* Darwin confessed his rashness; and in yet later editions he quietly withdrew his bloated estimate, which "proved to be immensely above that of the reputable geologists and physicists of the time."[28] Darwin's 300 myr was then reduced to 70 myr at the outside, while his son, Sir George Darwin, would calculate a geological history for Earth no older than 100 myr, perhaps as little as 50 myr (as against today's claim of 4.6 *billion*).

In fact, some scientists are saying that erosion rates indicate that the continents are no older than 20 myr[29] (versus the orthodox two and a half *billion*); other geologists believe that figure is somewhere between 25 and 70 myr.[30] Lord Kelvin (who discovered the second law of thermodynamics as well as absolute temperature, which is measured today in Kelvin degrees), calculated Earth's age* based on how long it took to cool down to its present crustal temperature. Kelvin's thermodynamics was itself based on the principle that energy always becomes less available; therefore hot bodies always cool down.

Kelvin counted 98 myr since the solidification of Earth's crust. His new science was a blow to the evolutionist's uniformitarianism, for Earth's age now appeared to be but a fraction of what Darwin's theory required. This, of course, was disputed vigorously by evolutionists, who required many more millions even billions of years for their theories to hold. We might also wonder why 95 percent of animal phyla are said to be no older than 40 myr.

---

*There may also be some major errors in Geologic Time (geologic column). See, for example, Eiseley's argument in *The Immense Journey,* 80–83.

The geologic column, argued one critic, is but "a public relations tool for the general theory of evolution."[31] It all comes back to the requirements of Darwinian gradualism. The great maverick scientist, Immanuel Velikovsky, an inspiration to Einstein, thought, like Kelvin, Earth much younger than generally believed. If the Darwinian establishment today says Earth cooled down enough to make oceans and rocks 3.8 *billion* years ago, there are still some intrepid scientists who contend that the maximum age of the oceans is only 62 *million* years. Darwin's "partner" in evolution, Alfred Russel Wallace, dated the oldest rocks to no more than 28 myr.

Geologists with no ax to grind for deep time or Darwinism have fixed the limits of geologic time (age of Earth) at about 70 myr. In 1862, Darwin himself became worried whether there was in fact enough time for the accumulation of such minute changes (in animal species) to amount to such large effects, when Kelvin, that year, said Earth was less than a 100 million years old—the time needed for it to have cooled down to its present temperature. Comte de Buffon was an Enlightenment naturalist who pointed out that living forms could only have been produced when Earth cooled down to a temperature suitable for life. Based on the geothermal gradient observed in mines, he calculated our planet cooled sufficiently to support life 70 kya. Although his work was criticized, and "hot origin" almost dropped out of science, his idea of gradient was later validated. Buffon also calculated that in another 70,000 years the planet would be so chilled as to sustain no life on its surface.* The gradual cooling of Earth, he thought, had eliminated the warm-loving fauna of an earlier day, while many of today's existing species will, in time, perish from the same cause.

---

*See Anderson, *Seven Years That Change the World,* 155: 144,000 years equals man's existence on Earth; 72,000 years from now, Earth will be so cold, it will no longer be habitable.

EXTINCTIONS:
A TIME TO LIVE AND A TIME TO DIE

From the time of semu to uz, creatures are born and die. Why do most species (99 percent) go extinct? (Most vertebrates have gone extinct after a tenure of about a million years on Earth.) Failure to adapt, of course, is the stock explanation. Or a species vanishes because it has successfully evolved to something else; the parent species disappears or is replaced after it changes into the daughter species. Or so they say. Evolutionists also contend that certain species went extinct because they had become too "specialized" to a particular niche, which then changed, making new (impossible) demands. Such species presumably failed to come up with "favorable mutations," as selection required. All this, in my view, is pure speculation. Species do not really overspecialize or wear out (another theory). Certain animals have simply failed to propagate, including men.

I have come to think of most extinctions, quite simply (and parsimoniously), as a function of heat loss; in this view, only those plants and animals better equipped to withstand cooler-drier conditions persist. (Indeed, Afrocentrists, as we will see in the next chapter, use *global cooling and drying* to explain the critical change from one fossil type of hominid to the next.) In the animal kingdom, the mass extinction events that eliminated dinosaurs or trilobites were, in my understanding, due to chill kill.

The story begins with hot Earth. "As a testimony to man, behold the earth was once a globe of liquid fire!"[32] In the beginning, a planet comes forth as a molten mass of hot, undisciplined gases. Once stabilized and cooled, it matures. And after the span of ages, it goes out like a candle in the firmament—cold and dry, slackening in its dotage, in velocity, moisture, heat, and light—until extinct. This should be axiomatic. Buffon, Lyell, Leibniz, Kelvin, Descartes, and many other scientists besides recognized the cooling state of Earth.

A slowing Earth is a cooling Earth: Earth is aging now, with slower rotation and weaker electromagnetic field. And as Earth's rotation slows, the solar day lengthens. Because of steady slowing, an extra second (a "leap second") is added to atomic clocks from time to time to make up for the lag. Speed also correlates with *heat,* as demonstrated in October 2006 when astronomers announced the fastest known planet, named SWEEPS-10, with a "year" just ten hours long and a surface temperature of about 3,000 degrees Farenheit.*

In the new-old science, the temperature of a planet depends on its axial velocity,† which is rapid in the youth of a world and slowing with age. Although the textbooks might say otherwise, I believe temperature loss (through slowing) is an intrinsic feature of planetary life and the probable cause of most animal extinctions. Indeed, I propose using this model to date the advent of man to ca 80,000 BP, at which time the average temperature on the planet fell to 98 degrees. Frankly, this cooling, 80 kya, is the opposite of orthodox science, which has things warming up during an interglacial period at around 75 kya (and lasting supposedly until 10 kya).

At the time of the quickening of animal life, Earth, according to the structure of coral fossils, made its daily rotation in what would now be twenty-one hours and forty minutes. This would give a difference, in animal heat, of 2.5 degrees, which is to say that large animals, now extinct, had an average body temperature 2.5 degrees higher than at present. In other words, three hours and twenty minutes' loss in axial motion produced a loss of 2.5 degrees of vortexian (ambient) heat.

As the decline in axial speed indicates, a fraction of a second is lost every century, and *as that motion is diminished, so is heat;* just so, can the aging of any planet be observed in the diminishing of its atmospheric warmth and breadth. Today, only a few miles above ground

---

*Earth is slowing at a rate of 2 milliseconds per century, according to American physicist William Corliss. Some astronomers say 13 seconds per century, others 0.00073 seconds per century. Our planet loses axial speed very slowly indeed but quite steadily, resulting in a total of 100 minutes lost since the quickening of life at 99 degrees.

†According to the new science, which is being developed principally out of the Book of Cosmogony and Prophecy, Oahspe; see our website www.earthvortex.com.

level, hardly 122,000 feet up, the temperature above tropical regions is as cold as above the polar regions. In very small increments, and over a great span of time, the atmosphere of our world has lost both amplitude and heat. Picture a time-lapse viewing of Earth in formation: a rapidly rotating fiery furnace begins to decelerate. The flashing heat subsides, in degrees, along with decrease in axial motion. Now the atmospheric canopy begins to cool. The surface world becomes, at last, habitable.

And over the eons, when the temperature finally falls to 98 degrees, Earth is ripe for the bringing forth of human creatures. According to this scheme, we can place the quickening of man at eighty thousand years ago, at the very end of the Semuan age. To find the time of first man, we place his temperature at 98 degrees Fahrenheit for optimum health. Four below normal will be the end of the period of man's inhabitation of Earth. After the vortexian radiation reaches this period (94 degrees F), man will cease to propagate, and so, become extinct on Earth.*

"The earth upon which we live is but a cooling planetary mass," stated George Frederick Wright on page one of *Origin and Antiquity of Man*. In his view, Earth "became fit for the habitation of man only during recent geological ages. A few million years ago the heat upon the surface of the earth was so great that it would have been impossible for man to have endured."

Twentieth-century scientist Louis Alvarez popularized the idea that an asteroid struck Earth at the K-T boundary (Cretaceous-Tertiary), causing major extinctions through a disruption of the food chain. Can we blame mass extinctions on catastrophes? In the Alvarez model, immense clouds of asteroid dust would have blocked sunlight. (The reader might be interested in finding out more about dark periods in

---

*With the loss of 1 degree vortexya every 18,000 years, we'll lose 4 degrees in another 72,000 years. This gives man 8 degrees of vortexya as the sum of his existence (144 kyr). By reversing these measurements, find the axial decrease of Earth in 80,000 years, which will be about one hundred minutes (i.e., Earth's decline in speed). For which reason the first race of man on Earth began about 80,000 years BP. See Oahspe, Book of Cosmogony 5.17–21.

chapter 4 of my book *Time of the Quickening.* For example, in the age of Thor, ca 15 kya, "the lands of earth were covered in darkness (nebulous period) [due to] a veil over the face of the sun."[33]) But some mass extinctions did not involve an impact. If mass extinctions are caused by catastrophes, why do some species survive and others perish? We recall that the Devonian extinction effected mostly *shallow* water invertebrates; of course, it was *coldest* in the shallows.

At the critical point of heat loss, a species ceases to propagate. The time of begetting has a beginning and an end for all the living. With cooling, "bison genetic diversity plummeted. . . . Cold and dry conditions dominated." Horses also went extinct (in Alaska) at that time.[34] Every species has its own temperature threshold (cold tolerance)— critical not only to its own metabolism but also to the food chain it feeds off. There is a time of genesis and a time of old age for all individuals, all species, even all worlds.

Scientists Comte de Buffon, Richard Owen, Thomas Henry Huxley, Isaac Asimov, and others agree on this much: cooling was the cause of the dinosaur extinction—not a giant asteroid (à la Alvarez) or solar flares charging Earth with lethal levels of ultraviolet radiation or even because they had become "overspecialized." Only this: With a loss of 2 or 3 degrees in body temperature, the empire of the saurians came to an end; other large animals also went out of existence. The apparently abrupt ending of the giant reptiles, as Gerald Heard saw it in *The Sources of Civilization,* was "probably connected with the close of a vast period of equable warm conditions and the onset of a new age in which the winters were bitterer . . . in the cold, the reptile falls into anesthetic coma."

The hulking saurians were just too big to burrow and hibernate (like the smaller animals, which did survive), and the temperature probably dropped too low to hatch their eggs. Good-bye giant reptiles, hello little mammals. Loss of heat is evident by the kinds of animals that succeeded them: the fur- and feather-bearing creatures, sheathed in fatty layers and adapted to a cooler clime.

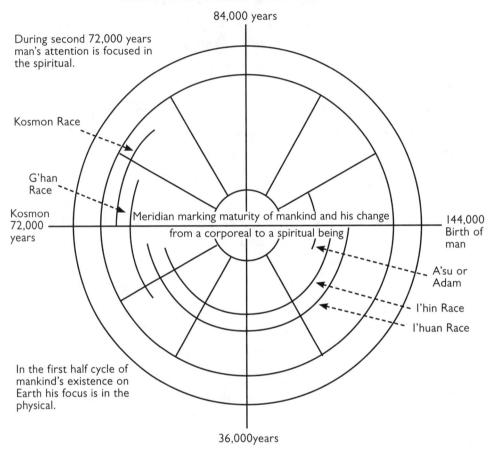

144,000-year cycle of human life on Earth illustrating
time of birth and extinction of each of the root races.

84,000 years

During second 72,000 years
man's attention is focused in
the spiritual.

Kosmon Race

G'han
Race

Kosmon
72,000
years

Meridian marking maturity of mankind and his change
from a corporeal to a spiritual being

144,000
Birth of
man

A'su or
Adam

I'hin Race

I'huan Race

In the first half cycle of
mankind's existence on
Earth his focus is in the
physical.

36,000years

*Figure 10.8. Wing Anderson's chart indicating 144,000 years as total life
span of humanity. With mankind now 72 kyr, we are halfway through the
Age of Man; in other words, our procreation period is to last another 72 kyr.
And whereas the first half of the age of man was focused on corporeal things,
the second half is to be spiritual, ushering in the world community
of nations and cosmic understanding.**

*In *Time of the Quickening* (pp. 234–37), the interested reader can find a discussion of
these periods, the current age marking the halfway point of race unfoldment.

Even as certain species of animals have ceased to propagate, and have become extinct, so shall it be with man . . . [when] the earth will have fulfilled its labor.

OAHSPE, SHA'MAEL, PLATE 40,
BOOK OF COSMOGONY AND PROPHECY

Eight degrees of vortexya, in this reckoning, is the sum of man's existence on Earth, which is a span of 144,000 years, called the grand cosmic day or human season, within which mankind is born (semu), matures (hotu), and passes into old age (adu) and death (uz). (See figure 10.8.)

HOMINID EXTINCTIONS

What has caused hominid extinctions in the past? Amalgamation? Catastrophe? Genocide? The Ihins of Shem and Ham became extinct by amalgamation (EBA) after 12 and 21 kyr respectively (more on EBAs in chapter 12). There have also been many natural disasters that have decimated the tribes of man. Asuans, for example, were destroyed by a kind of brimstone falling on the five divisions of Earth, approximately 72 kya.[35] In very recent time, several Andamanese groups were wiped out in the 2004 Boxer Day tsunami. As for genocide, all the Old World Ihins were exterminated before the flood. Eight thousand years ago all the Druks of Heleste (southeast Europe) were exterminated as well as the hoodas (Druks) of Arabinia, at the hands of the Ihuans who slaughtered them right and left; degenerate Ihuan-Druks were pursued without mercy by the more civilized pure Ihuans.[36] In addition, the extermination of Negrito groups is not hard to discover in the record: the aboriginal people in Japan, reddish-skinned pygmies, were wiped out by the Ainu. Similar histories tell of death-dealing cave fires, which eliminated the Nittevo (Sri Lanka), the Ebu Gogo (Flores), and the Taiwanese Negritos who were brutally trapped and massacred, according to their destroyers, the Saisyat. In America, the scenario is almost identical: the Paiutes once launched a war against the marauding Druks (giants cannibals): "My people gathered wood and began to fill up the mouth of their cave."[37] They set it afire, totally exterminating the tribe of red-haired cannibals.

Archaeological evidence of other massacres includes the upper layers of Mohenjo Daro, where groups of contorted skeletons were apparently massacred. The Tasmanians were exterminated by the new settlers. On Easter Island, the Short Ears exterminated the Long Ears in the thirteenth century. There have been far too many holocausts to mention—religious, political, territorial, tribal.

Have yet other races been eliminated by neutering? The Ihins made eunuchs of the ground people and Yaks; later, the Ihuans also neutered the Yaks and ground people "wherever they came upon them." Later on the Ongwee-ghan made eunuchs of their Ihuan enemies.[38]

And what about extinction through sterility? The Yaks were barren and died off—tens of millions of them. But even then, they were not extinct, so many had been produced. Starvation and disease have also killed off whole populations. An ancient Hittite text, for example, refers to such a time: "barley and wheat throve no more, oxen, sheep and humans ceased to conceive."[39] Sterility of people and barren fields is also described in the related flood myths of the Assyrians.

This rule follows on all worlds; that with the culture of the corporeal senses, man becomes vigorous, strong, and independent; and with the culture of the spiritual senses, they become weak, sensitive and dependent. In the first case, they ultimately become selfish and wicked; in the second case, they become impotent, and unadapted to corporeal life, and thus become extinct. On all worlds . . . [are] provided these two seasons: a season for the development of the corporeal senses, and a season for the development of the spiritual senses.

OAHSPE, BOOK OF OSIRIS 12:6–8

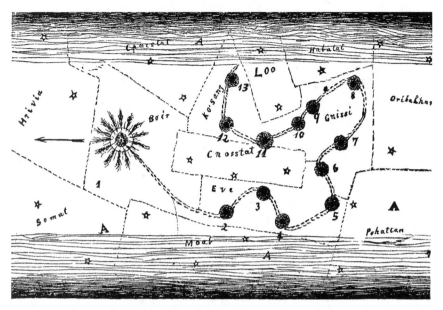

Figure 10.9. Plate 40 from the Oahspe, Sha'mael. "A time shall come when the earth shall travel in the roadway of the firmament, and so great a light will be therein that the vortex of the earth shall burst, even as a whirlwind bursteth, and lo and behold, the whole earth shall be scattered and gone, as if nothing has been. But ere the time cometh, My etherean hosts shall have redeemed man from sin. Nor shall the inhabitants of the earth marry, for the time of begetting will be at an end. . . . The earth will have fulfilled its labor, and its services will be no more under the sun."

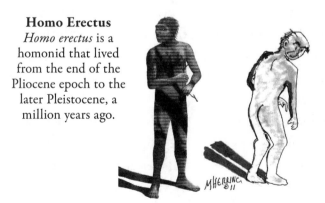

Homo Erectus
Homo erectus is a homonid that lived from the end of the Pliocene epoch to the later Pleistocene, a million years ago.

Homo Limpus
Homo limpus is a homonid that died out because of the lack of *Viagra.*

Figure 10.10. Homo limpus. Cartoon by Marvin E. Herring.

The earth gives away of its substance into atmospherea over hundreds of years; and the fields become barren and cease producing; and certain animals become barren . . . and their species go out of existence. Man is subject to the same forces; when the earth is in the giving-off period, behold, man ceases to desire of the earth.

OAHSPE, BOOK OF DIVINITY 15:8

RACIAL DIVERSITY AND SKIN COLOR

Since the ancestors of man were almost certainly tropical . . .
they may have been black-skinned.

ASHLEY MONTAGU,
MAN: HIS FIRST TWO MILLION YEARS

Man came forth in the tropics and he was naturally of rich complexion. Nahsu was the Egyptian name for the dark races: N + ahsu? Now some have said it is insensitive to represent our remote ancestors as dark skinned. The editors of *National Geographic* replied (June 1997) to such a charge by saying: "Since early humans originated in Africa, it is speculated that they displayed a dark skin adaptation to tropical climate."

Certainly in the animal kingdom, the peculiar colors of many species closely resemble the soil or foliage of their habitat, concealment giving them a better chance at survival. The skins of mortals were also "colored according to their surroundings, some light, some dark, and some red, or yellow, or copper-colored."[40]

Hooton described our earliest forebears as of brown-yellow skin tone, which accords with their tropical beginnings. Past eras on our planet, as I have argued, were warmer and wetter. With little difference in climate zones, all places on Earth were warm. Until the mid-Pleistocene, all humans lived in warm or mild climates,[41] layers below the Aurignacian exhibiting tropical fauna (rhino, elephant, hippo).

There were hippos and crocodiles in Europe in the third interglacial period; hippos and rhinos in North America as well, the giant size of many fossils indicating the tropical temperature that once prevailed even at today's temperate latitudes.

And "the skin of *afarensis* [Au] was dark."[42] The pithecanthropines, after them, also had "heavily pigmented skin,"[43] indeed the ground people were brown and black.[44] Dixon's proto-Australoid and proto-Negroid types were people of tropical origin with deeply pigmented skin. "According to their respective places and the light upon semu, so quickened I them in their color, adapted to their dwelling places."[45]

Oahspe scholar and archivist Reverend Joan Greer, who has been studying deep genealogy, says that according to DNA studies, some black people have a higher percentage of Caucasian genes than some people with white skin. Says Greer: "Skin color would seem to have little to do with our genetic makeup." The point is, for racial classification, coloring is only one of many factors. In fact, Carleton Coon discarded the term *Negroid* as useless, as the dark skin it implies is found at equatorial latitudes all around the globe, not just Africa. Caucasian he defined not by skin color but by skull measurements, nasal index, and other criteria.

Accordingly, Caucasian sometimes includes dark-complected people, as in India, Southern Arabia, Ethiopia, and Somalia. Australian Aborigines and prehistoric Americans also show various Caucasoid features. In his autobiography, *Adventures and Discoveries,* Coon pointed out that "skin color was not uniquely racial. . . . Australoids and Negroids [are both] black. Nor did it [color] go with skulls and body skeletons necessarily. Thus the Somalis could be Caucasoid in face and build and Negroid in pigmentation. . . . Some Somalis . . . trim, erect, and elegant . . . were quite European looking."

When I bring a new world into the time of semu, My presence quickens the substance into life; and *according to the locality and the surroundings* [e.a.] I bring forth the different species.

OAHSPE, BOOK OF JEHOVIH 5:13

Racial features appear to have been set in their mold from the time of quickening, which is to say, in the beginning—rather than as last-minute revisions, which evolutionists, particularly of the out-of-Africa school, maintain. Adam, in Mohammedan legend, being the first man, was created when God sent his angels to fetch handfuls of earth from different depths and of different colors; hence mankind were of different hues—from the beginning. The story is almost identical among the Hopi Indians: Spider Grandmother gathered four colors of mud—black, white, red, and yellow—with which she created the four races of man, each with its own language.

Darwin's version of racial differences was determined by sexual selection: the races slowly differentiated themselves from their neighbors, influenced by slightly different standards of beauty. Thus, by a process of sexual selection, the characteristics of tribe and race came to be fixed. Wallace, however, was inclined toward the polygenist position, whereby the various races became differentiated *before* the time of *Homo sapiens*. As noted elsewhere, it is generally believed that culture (which is virtually synonymous with *H. sapiens*) brings a halt to physical evolution. As Gertrude Himmelfarb phrased it: "The ascendancy of mind, by putting an end to the evolution of man's physical structure, fixed the different races in their different and permanent features."[46]

Now this is exactly the opposite of both (1) Brace's view that "changes in the cultural adaptive mechanism . . . [were] responsible for changes in face form and skin color," and (2) Sherwood Washburn's position that "the great antiquity of races is supported neither by the record nor by evolutionary theory."[47] Mexican Indians and Mongolians, according to this setup, share a common descent; but since crossing Beringia, "the Indians have evolved into a separate race."[48] It was also Hooton's position that the races of man were established late, in the *H. sapiens* stage (ending ca 20 kya), by means of isolation, inbreeding, and mutation. Le Gros Clark also argued that the different races came about only after *H. sapiens* was established.

Nothing of the kind happened, according to the polygenists. The French school of protohistorians, for example, attribute separate origin to the races—black, white, yellow, red—which arose on different continents. Coon and Weidenreich and today's multiregionalists also see distinctive racial traits expressed *early on* in the different *H. erectus* groups in different places: each of the four major evolutionary centers, in this view (and my own), was also a racial center. Evidence from Upper Cave Choukoutien, as Weidenreich saw it, indicated that differentiation of races had taken place long before the Upper Paleolithic. Coon, looking at the same cave specimens, saw *Sinanthropus* as the ancestor of all Mongoloids, while each of the other races evolved independently on their own turf, from an early type. Coon was able to demonstrate resemblances between pre–*H. sapiens* hominids and some of the living people of the *same region,* thus allowing racial characteristics to *predate* the appearance of modern man.

But under the strained theory of monogenism, which posits a single origin for all human beings, there should be no such resemblance. Coon, in this regard, challenged the monogenistic idea that in a short 20 or 30 or even 40 thousand years, the different races of *H. sapiens* could have taken shape. To him, the races of man were far older than *H. sapiens* and could not possibly descend (monogenetically) from a single regional stock. Neither can the highly respected, more recent, work of Vincent Sarich and Allan Wilson (on genetic distance) assign the formation of different human races to the skimpy span of 30 or 40 kyr, an implausibly short run of time for natural selection to do its work. The recency of races is also debatable on the basis of developmental locking and the end of the semuan phase 80 kya, as discussed above.

Yet out-of-Africa's monogenism (single cradle of man) requires the races to be formed later in time. But I think the polygenists are right; men were cast in their mold (Negroid, Mongoloid, and so on) from the beginning, for each was raised up in their own division of Earth. It is monogenism's spurious common ancestor that forces us to adopt "racial separation from a single root." But the races never separated; in fact, they did the opposite—they came together!

Let's look at this supposed ancestral tree: Do all living things really descend from a single entity that floated around in the primordial soup some 3 or 4 bya? The "wonderful unity in all living things" does not necessarily imply a common ancestor, but it is interpreted nonetheless as if it did, thus endorsing Darwinism "with a thundering voice for the validity of evolution theory."[49]

Scientists, finding that our cerebral cortex is similar to the neurons inside the head of the ragworm (a lowly marine creature), then jump to the conclusion that "they were too similar to be of independent origin . . . [and] we must share a common ancestor."[50]

Yes, DNA does confirm the basic biochemical identity of organisms, from bacteria to man. Fine. But these happen to be the same chemical elements throughout the cosmos, according to Harold Urey, Nobel Prize winner. Molecular unity does not prove we all come from one common ancestor, simply that this is how life is built. The fact that all life-forms have similar DNA patterns does not mean or even suggest we are all derived from the same batch. All we can conclude with certainty is that the blueprint is a mutual one.

On page 73 of his book, *The Neck of the Giraffe,* Francis Hitching makes a discovery. Based on the inviolacy of DNA's germ plasm, "biology is forced into a precarious assumption: the first living creature must have had within itself the entire genetic potential to grow into—*to create*—every one of the trillions of plants and creatures that have lived since." A bit of a stretch, don't you think?

So now it is time to reevaluate the myth of the common ancestor. Ultimately this doctrine is the only way to banish the Creator's hand from the web of life—even though we are reduced to the absurdity of postulating that all life evolved from one *single common ancestor,* say, bacteria, fungi, or the single-cell eukaryotic forms. Fred Hoyle rebuts this daring claim: "No evolutionary connection has ever existed . . . between so-called prokaryotic and eukaryotic cells, between bacteria and yeast cells. . . . Evolution from a common stock? . . . We doubt that a terrestrial evolutionary connection ever existed between the plant and animal kingdom. . . . There is no reason why some of the categories [taxa, i.e., different species] . . . should not always have been separate."[51]

Whenever it has been *impossible* to demonstrate a phyletic link (biological sequence in time) from one hominid to the next, the problem (which is really a death blow to Darwinism) vanishes simply by invoking a common ancestor! In the context of race origins, the imagined common ancestor (a dark African of the modern type, as discussed in chapter 11) simply branched out to whites on the one hand, yellows on the other, and so on. And the same falsehood posits an unknown common ancestor, even farther back in time, a creature that led to apes on one branch and to humans on another. The mythical common ancestor has become the understudy for the still missing link, but it remains promiscuous guesswork and intellectual jugglery.

> *Did Glasgow grow from a seed yielded by Edinburgh?*
> JOSEPH PRIESTLEY,
> *LETTERS TO A PHILOSOPHICAL UNBELIEVER*

Despite certain perfunctory merits of a "tree" (or bush) representing the ascent of man, such diagrams take for granted the evolution of all living things from one all-purpose progenitor, a single taproot. With man, however, we need always to look for (at least) two ancestors, for we are all hybrids!

The concept of a common ancestor is actually a relic of eighteenth century thinking, entertained by Darwin's own grandfather Erasmus and his, Erasmus's, contemporaries Diderot, Buffon, and Lamarck. Not only does this simple-minded model have all hominids branching off an ultimate *anonymous* ancestor, but the same setup is proposed for the Order of Primates: somewhere in the deep dark past, man and ape shared a common ancestor, then went their separate ways. As we've

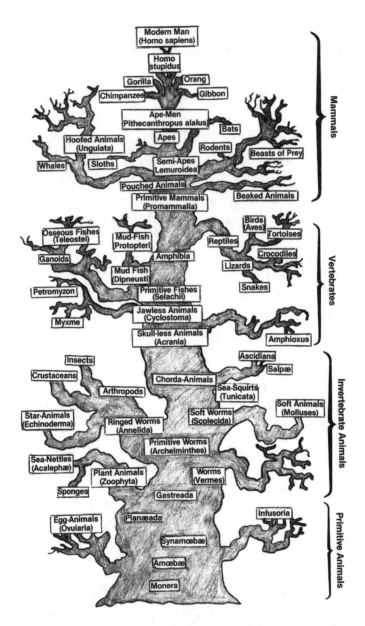

Figure 10.11. Ancestral tree devised by Ernst Haeckel in 1899. In the 1970s Norman Macbeth wrote about these supposed ancestors of ours, which are however, as Norman Macbeth points out in Darwin Retried, *"still present, although their former siblings are said to have worked up to human status. Thus one and the same ancient stock split into one group with astonishing plasticity, and another group with almost total rigidity. This is very hard to swallow."*

noted, that ancestor has never been found, despite an ambitious search. Nor have the required intermediates turned up.

This much-pampered school of anthropogenesis, Darwin's community of descent, has students of the problem forever searching for (but not finding): a common ancestor for Au and *H. erectus*; a common ancestor for Neanderthal and *H. sapiens*; a common ancestor shared by Neanderthals, mods, and Denisova, and so on down the line.

According to this scheme, time and evolution inexorably carry us further away from the body type of our shared ancestor. Darwin the monogenist believed in a common ancestry of the black and white races; Keith, along the same lines, thought "the extent of the differences between black and white indicates that the racial separation of the modern type of man must be placed far back in time." He argued that the farther back we trace European and Negro, "the more the white and black ancestral forms should come to resemble each other. . . . There is good reason to think [the different races] have descended from a common ancestor. Yet this ancestor may show no sign or trace of the new feature. [Isn't that odd?] We have in such cases to suppose that an evolutionary bias may be latent in . . . the ancestral stock. . . . We cannot explain the facts unless we accept [this] principle."[52]

Principle? What kind of sorcery is this, planting a "bias or tendency" in our nonexistent, phantom progenitor? And isn't this bias or tendency so like the Platonic essentialism, vitalism, or teleology that evolutionists laugh to scorn?

There is no coherent way to demonstrate that all the racial diversity among human beings came from a single ersatz ancestor who changed color like a chameleon. The races cannot honestly be traced back to, and did not differentiate from, a common stock; each race has its very own history. Negritos and Australians, for instance, "do not share a common ancestry but merely interbred."[53] And this is indeed what Dixon taught: "The existing varieties of man are to be explained not as . . . from a single ancestral form, but as developed by amalgamation . . . of several quite discrete types."[54] As for Coon, one researcher has written: "The only scholar who tried to offer us a coherent alternative [to Darwinian monogenism and single origin] is Professor Carleton Coon. . . . [in whose] monumental *Origin of Races* he tells us that the human race does not descend from a single ancestor but represents various types of *Homo erectus*. . . . which evolved independently of one another . . . in different parts of the world."[55]

The bogus common ancestor Eve—the subject of the next chapter—is allegedly everyone's mother, at least according to DNA spokespersons. Microbes, however, tend to "swap sections of DNA. . . . Cells have picked up chunks of DNA from completely different species to form new, hybrid genomes. Thus the idealized view of an unbroken branching history of DNA evolution from parent to daughter is invalid. There is no single common ancestor."[56]

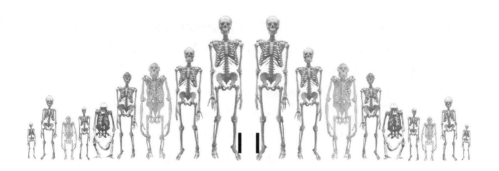

Not Out of Africa
The Many Gardens of Eden

The fads have risen and fallen around me.

Carleton Coon, *Adventures and Discoveries*

THE LONGEST HIKE

We are all Africans.

Douglas Palmer, *Origins*

Hotfooting to Africa since Louis Leakey's day, the bone people have evolved a new urban legend: we are all ultimately Africans. This out-of-Africa genesis, OOA for short, conceived for all the wrong reasons and built on a pyramid of supposition, is, as we speak, circling the drain.

Let us begin by finding out what this improbable but fashionable and politically motivated theory (already elevated to fact, if only by force of repetition) tells us about our forebears and their wanderlust: It is in two parts, actually. Part one: Early man, *Homo erectus,* departed Africa, oh, about 1.6 mya. But wait, the race of *H. erectus* was previously thought to be no more than 500 kyr,* Arthur Keith, in 1929, having shelved *Pithecanthropus* no earlier than 220 kya! Now he is eight times older! How did that happen?

Whenever it was, or imagined to be, the early hominids known as *H. erectus* (Druks) allegedly emigrated en masse from Africa, all the way to China and Indonesia—where they were eventually replaced or died out, becoming an irrelevant dead end as far as evolution is concerned (with the possible exception of the European ones, who may have evolved into Neanderthals).

The problems begin at once. Put on your thinking cap; this is tough. For starters: How did *H. erectus* manage to evolve to mods in Africa, while Asian *H. erectus* did nothing of the sort, continuing on largely unchanged? I wonder why they didn't evolve into mods in Asia as well.

The trouble—the theoretical strain of it—only deepens with Asian *H. erectus* being more primitive in some respects than African ones. Why are some Javanese *H. erectus* *older* than African ones? Richard Leakey's African Skull 3733 (at 900 cc), or OH 9 (at 1,050 cc), should have the same size or smaller brain than the (younger) *H. erectus* of Java, whose size is 850 cc and who had a much more apelike forehead than Africa's 3733. (Peking Man's forehead is also more primitive than Africa's 3733). Being the ancestor, 3733 should be the more primitive, but appears actually to be more advanced, *lacking* Asian *H. erectus*'s long and low-slung cranial structure, thick skull bones, and robust face. African *H. erectus* is the more progressive.

Explanation? Incredibly, the African version is now certified the more primitive, on the bogus plea that it is generalized, as opposed to specialized, even though the African skulls are higher domed and thinner walled than their East Asian counterparts! They are also less mas-

---

*Brace, *Stages,* 5 (in the 1970s); *Sinanthropus* dated 600 kyr in Tomas, *We Are Not the First,* 73; *H. erectus* in Le Gros Clark's time was no older than 500 kyr; and in Oxnard, *Order,* 332, he was only "a few 100 thousand years" old; Richard Leakey made that 1 myr.

sive in face and brow, which the OOA wizards make bold to call "more primitive or less specialized."[1] Here we catch Afrocentrists daringly, counterintuitively, calling *H. erectus*'s massive brow a *more advanced* feature because specialized—despite this "advance" being *away* from the modern human form! Confusion and humbug! A shameless feint of jargon, intellectual sleight of hand.

Farther down the rabbit hole, anthropologists cannot agree whether African and Asian *H. erectus* are even of the same lineage; some claim they are entirely different groups, with different skull dimensions. Tattersall and Schwartz see *H. erectus* of Africa and Java as completely different types. In fact, the morphological distinctness of Asian *H. erectus* led to doubts about the existence of *any H. erectus* from African sites! As a solution to this little problem, they *renamed* the African one *Homo ergaster*. But *H. ergaster still* has a larger brain—up to 900 cc—than his Javanese "descendant." Indeed, Kenya's Turkana Boy (though primitive in head) was tall, well-proportioned, and slender, hardly a forerunner of rugged squat bulky Java or Peking Man.

And how can you make *H. erectus* an African émigré when his Javanese "descendant" is actually *older*?[2] Let's see: East African *H. erectus* is dated 1 myr[3] while Java's *H. erectus* is up to almost 2 myr (1.9 myr). Shouldn't it be just the other way around? Chinese *H. erectus* from Sichuan Province were also dated 1.9 myr (at Longgupo Cave), while Java's Modjokerto Man, and even Georgia's Dmanisi Man, are dated 1.8 myr. Well then, the tailor-made solution is that *H. erectus* left Africa "earlier than we thought." Just like that—a major migration and new time line is created. Theory dictates.

Yet this is only the beginning of difficulties—and ad hoc solutions. This pipeline from Africa to everywhere else (monogenesis) just doesn't wash, but when conflicting evidence turns up, theory is salvaged with pseudoexplanations, like "specialization" or entirely specious waves of migration.

AFRICAN EXODUS PART TWO

For the second phase of OOA, it is taught that some time between 130 and 30 kya (a good window or "scatter" for later deniability), African *H. sapiens* went forth and supplanted *all previous hominids* in the world. According to this replacement model, outmoded species were *replaced* by new, freshly adapted ones. But *H. sapiens coexisted* in Java with *H. erectus,* who were therefore not replaced by their "betters." Java's *H. erectus,* in theory, is supposed to have been extirpated, without contributing in any way to the modern genome.

But given a ballpark figure of 100 kya for this second exodus, how do you explain the appearance of AMH features in European specimens much earlier than that? Why do European hominids start showing *mod* traits long before the world's supposedly *only* mods left Africa? Examples include fossil men from Petralona, Vertesszollos, Atapuerca, Terra Amata, Arago, and Swanscombe—all said to be around 300 kyr, and all showing at least some AMH traits. Dated ca 400 kya, the various *Homo heidelbergensis* specimens in Europe (a pre-Neanderthal type) also have *H. sapiens* traits. But that's way too soon (for the 100 kyr part two exodus), and it also places mods *before* Neanderthals in Europe.

When faced with hominids like Heidelberg Man who look rather like their contemporaries in Africa, it again signals "a fresh wave of African immigrants"[4] who, theory claims, evolved into Neanderthals in Europe, but whose sisters and brothers evolved into *H. sapiens* in Africa! How does that grab you?

According to OOA, nothing hinting of mod got to Europe any earlier than, say, 50 kya. So what do we make of those Atapuerca bones in Spain, dated 1.2 myr? These people had fairly large braincases at 1,390 cc and jawbones with a real chin. Apparently, there was some sort of AMH in Europe much earlier than OOA theory allows. To save theory, hominids, they now said, after leaving Africa, came here to Europe and evolved into the newly christened *Homo antecessor,* much earlier than we thought.

If all this is difficult for you, it is difficult for me, too. Let's back up a moment and try to get a grip on exodus, part two. It finds us in Africa (at least one and a half million years after *H. erectus*'s departure), waving good-bye to the second set of émigrés: a host of black AMHs. The OOA theory, inaugurated in 1987, dispatches this Eve out of Africa, naming her the single mother of all! Using mtDNA to determine the molecular clock of time, the resulting genetic tree tracks *us all* back to this common ancestress: mitochondrial Eve. Comparing ages of lineages from different regions, analysts came up with a time line for the outward bound migrations of this black Eve, or Mother Superior, as some have dubbed our mutual ancestor.

While the Eve theory assumes that all changes in her mtDNA (as observed in her living offspring, today's races) were the result of steady mutations over time, these "changes" could just as reasonably be due to nothing more complicated than race mixings. Geneticist Alan Templeton maintains that precisely such mixing could have scrambled the DNA sequences, to the extent that "a date for Eve can never be settled by mtDNA."* Others say the computer program that worked out this family tree was biased by the *order* in which the data were entered. Enter it a different way, and it does not give an African origin at all. Henry Gee, from the editorial staff of *Nature,* described the results of the mtDNA study as "garbage." Even Mark Stoneking, who did the original mtDNA research in 1987, acknowledged that African Eve has been invalidated.

And there's this: Some estimates put the grand exodus OOA (part two) at 160,000 years ago; though it may have been only 133,000 years ago.[5] Others say 200,000 years ago, and yet others arrive at an age "much greater than 200,000 years," that is, three million years ago![6] Read a different book or article, and that changes to 50,000 years ago[7]—hardly

---

*Some analysts say that mtDNA may mutate much faster than has been estimated, making the clock in error as much as twentyfold! Mitochondrial Eve could be a mere six thousand years old!

enough time for all the different races of the world to take shape from the Negroid original.

One might also ask: If a troop of *H. sapiens* ventured OOA say, 100 kya, why is there no sign of their *culture* (artifacts normally associated with the modern type: art, ornament, burials, storage pits, quarries, tools, trade, long-term occupancy) for another 50,000 years? Why the gap? This is unexplained. Why are Mousterian (Neanderthal) tools found at some sites inhabited by men of the modern type? These are fossils with the modern physique, but without modern *behavior*—for tens of thousands of years. "They tell you that modern man has been in existence for 200 kyr. Holy cow, what do they think these fully structured humans were doing for the past 190 kyrs?"[8] South Africa's Boskop Man, for instance, shows no culture to go along with his mod features and huge brain (1,830 cc), only rudely worked Mousterian tools. The same holds for South Africa's Klasies site (as well as Israel's Mt. Carmel, Krapina, Crimea, Borneo, and Malaysia).

In 1970, before African genesis came into fashion, Neanderthal-type tools in Africa were dated with no difficulty to 25 kya. But that figure changed to *200* kya—to better suit the newly minted OOA theory. Well, the dates are no more consistent than the theory itself, which (part two) goes on to claim that after these hardy AMHs left Africa, and trickled out to populate the entire world, becoming (or mutating into) the South American Indians, Australians, Pacific Islanders, and everyone else. All from the African hub! Thus was the whole world settled by these great migrations. Or so it is taught.

The New Guinea population, to take one example, presumably became established in this manner some sixty thousand years ago—overlooking the geographic fact that Sundaland (insular Southeast Asia) and Sahul (New Guinea and Australia region) are separated by deep ocean channels swept by strong and swift north to south currents. How exactly did these folks (with not much culture or technology) get out of Asia and Sundaland across to Sahul? Boats? Island-hopping across Indonesia to Australia? I doubt it.

If they arrived by water, why are there no traces of such boats or rafts on the northern Australian coast? Proponents say, disingenuously, we simply do not know their ship-building abilities; but no decent watercraft in the region can be dated that long ago. Neither do today's Aborigines' boats seem likely candidates for crossing to Australia or Oceania from any part of Southeast Asia. Did Micronesia's little Palauans cross 370 miles of shark-infested deep ocean just to get to rocky Palau?

Today phylo-geographists claim they can track and identify long-distance migrations that bands of humans made in prehistoric times. Much of this science, however, is set up in the first place to prove a point—evolution—and its latest Garden of Eden: Africa. Thus does current genetic mapping "prove" African origins, the family tree beginning with the San (Bushman) and branching out.

The Bushmen are thought to have the oldest mtDNA in the world. Hence African Eve—everyone's ultimate mother. I believe, however, that the great age of Bushman DNA can be explained otherwise: They are the last surviving people in Africa who lived *before the flood,* relics of a separate race, as Coon suggested, having come from a different pre–*H. sapiens* stock, the true aborigines of Africa's grasslands. (The younger age of all other African DNA devolves on postflood hybrids, which I go into later in this chapter.)

Since the late 1980s population genetics has been mapping the earliest migrations undertaken by modern humans OOA, following "the inexorable progression of colonizations" from continent to continent. This concept of replacing the natives or colonizing the world is compatible, not with protohistory, but with the rule of expansionism and imperialism, which began only in recent time—the Holocene. "The oldest tribes are invariably peaceful. . . . The seed of conquest is not in them. Mankind's desire to make himself master of [others] is a comparatively late development in the world."[9] Though OOA ostensibly credits the black man with conquering the world, all this is really the white man's burden in disguise.

BORN TO ROAM

Proto-Negroids have never ventured beyond the tropic lands.
ROLAND DIXON, *THE RACIAL HISTORY OF MAN*

Why would they leave dear Africa, anyway? The reasons given are food shortages, "skirmishes with other groups,"[10] scarcity of drinking water— pure guesswork, bald conjecture. Why would *H. erectus* (part one) abandon his warm home? Hypotheses include drying or cooling climate, which forced them out to Western Asia. Failing that, this man was simply "born to roam."[11] How lyrical: "The spread of *Homo erectus* into the northern continents was an inevitable consequence of evolutionary momentum. There was already kindled in the human mind a spirit of adventure, a real curiosity about the world around them,"[12] this fancied wanderlust totally contradicting the reverse impulse, known as nosto-phylia, meaning, the deep-seated reluctance to abandon one's ancestral home, which the Filipinos call *bungungot,* spiritual homesickness.

A spot of common sense: Hominids at the low level of Druk or *H. erectus* were more the stay-at-home sort than globe-trotting adventurers. How can we credit these brutish types, hardly advanced beyond their ground-burrowing, worm-eating ways, with such challenging migrations? How did *H. erectus* cross the forbidding desert barriers of the Middle East? The awesome rivers of Asia? The major mountain barriers east out of the Levant? If they went north and east out of Africa, why are their typical hand axes not found in Southeast Asia or the Caucasus or China? Neither can anyone figure out how those early humans reached Spain and Italy (Ceprano) "without leaving traces in Turkey, Greece, or other points en route."[13]

Which route *did* they take out of Africa?: "Your guess is as good as mine," says expert Brian Fagan, who nonetheless declares that *H. erectus* OOA "possessed all the awesome mental abilities of modern humanity."[14] I must say, when it is useful to have *H. erectus* appear as a link to the anthropoid apes, he is drawn in books and journals with a "brutal muzzle," primitive canines, small brain, and curved back—"a large-brained ape that looked rather like a man."[15] In the nineteenth

Figure 11.1. Gabriel Max's painting of Pithecanthropus allalus, *a missing link, reproduced in Ernst Haeckel's* Naturliche Schopfungsgeschichte.

century *Pithecanthropus allalus* was regarded as sufficiently distinct from humans to warrant an altogether separate *genus*. Marcellin Boule, in 1923, didn't even think Java Man (*H. erectus*) was prehuman—just an anthropoid ape, a giant gibbon!

> *It is indeed hard to comprehend Homo erectus's means of transport . . . how a creature of his limited intelligence could have crossed the waters separating Java, Australia, and Africa.*
>
> JEFFREY GOODMAN, *THE GENESIS MYSTERY*

> The ground people traveled not.
>
> OAHSPE, THE LORDS' SECOND BOOK 3:3

Only consider Darwin's sensible thoughts on the matter, stated in *The Descent of Man:* "The wilder races of man are apt to suffer much in health when subjected to changed conditions. . . . Take them away from their . . . homes, and they are almost certain to die. . . . Man in his wild condition . . . is susceptible when removed from their [*sic*] native country."

Instead of giving up OOA in the face of overwhelming evidence to the contrary, theorists prop up hypothetical migrations at every turn:

- Since Australians and New Guineans are so different genetically, it must be a *separate migration* that got them to their respective lands.
- The DNA of Denisova or X Woman, being different from mods and Neanderthals, calls for a separate migration (500 kya), Pääbo suggesting that her people migrated out of Africa at a different time than Neanderthals or mods. If not, Denisova alone could overturn the OOA theory.
- Another study posits a different wave of early Africans who made

the exodus and interbred with the *H. erectus* who had left Africa a million years earlier (so much for parts one and two).

- With the discovery of Indonesia's hobbit, "proponents now believe . . . the first human ancestors to leave Africa may have been far more anatomically primitive [than *H. erectus*]—and may have left *far earlier*—than previously thought. If they are right, the Flores [hobbit] remains rank among the most important paleoanthropological discoveries of all time. . . . If they are wrong, it will be worse than Piltdown," referring to the 1912 hoaxers who combined modern human and orangutan fragments into one fraudulent fossil.[16]

Those little hobbits supposedly "evolved from a normal-size island-hopping *H. erectus* population that reached Flores around 840,000 years ago. . . . The ancestors of hobbits probably left . . . on foot about 2 million years ago, ultimately crossing treacherous ocean waters . . . With a brain just one-third the size of a typical *Homo sapiens* adult, the hobbits were managing some extraordinary things . . . crossing at least two water barriers to reach Flores from mainland Asia." But how exactly *did* they cross the deep, seemingly impassable waters? One scientist speculates that "a giant tsunami . . . swept them out to sea. Survivors clinging to trees could have been washed ashore."[17] Yeah, right. Pure moonshine.

Before the Holocene (only 10 kya), as Jared Diamond wrote in *The Third Chimpanzee,* "unfettered travel was impossible." Early people, he pointed out, "had no social framework for travel beyond immediately neighboring lands." Indeed, other tribes would simply "kill any trespasser. . . . To venture out of one's territory . . . was equivalent to suicide." Human populations, Diamond concluded, are basically sedentary.

Colonizers of new regions are particularly liable to perish.
STEPHEN JONES, *THE CAMBRIDGE
ENCYCLOPEDIA OF HUMAN EVOLUTION*

What a sport it has become, moving the races of men (and the clock of time) around like pawns on a chessboard! How disconcerting, then, that the same scholars, *conversely*, suddenly become strict isolationists, adamantly denying *much more plausible* migrations, to America, for example, by prehistoric (Holocene) Chinese, Egyptians, Phoenicians, proto-Greeks—nations with excellent ships. "Men were on the move around the world much earlier than mainstream scholars would have us believe."* The Vedics of India, a maritime culture, navigated around Africa and across the Atlantic, mining the copper and tin in South America. The Phoenicians left many signs of their foray into Brazil for its mineral wealth. Worldwide legend is full of such traveling wise men in the Stone Age. "And man built ships and sailed over the ocean in all directions, around about the whole world"† (the passage referring to a period around 40 kya).

---

*Joseph, *Ancient America*, 49.
†Oahspe, Synopsis of Sixteen Cycles 1:21.

POLYGENESIS AND THE RACE CARD

Subsapient man was not a traveler, and we needn't take him out of Africa or anywhere else to populate the world. It is much less troublesome and more realistic to take all the early races as indigenous to the lands in which we find their bones (polygenesis).

Carleton Coon, besides positing that the five races developed independently, also theorized that they developed at *different times*. It was this different times that got his approach rejected by the American Anthropological Association. Coon, writing in his memoir *Adventures and Discoveries,* describes how, in 1962, he became, in his own word, the "target" of a passel of do-gooders who accused him of racism when he worked out the relative date when each transition [to the modern form] took place; Coon had Mongoloids and Caucasoids arising around 250 kya, but Africans at a later date.

Coon and his Harvard mentor Hooton discovered that the earliest men of modern type were brachycephalic Caucasians (most early hominids, in contrast, leaned toward long headedness: dolichocephaly). Well, for some reason, their critics jumped to the conclusion that the *first race* to cross the sapiens threshold must therefore be the most advanced. The implication, Coon explained, is that whoever came first is thereby best; but this is a logical fallacy (Coon saw the timing in a different light, simply a matter of different environments).

Hooton had found the first mods to be a Europoid race, an understandable position considering the Ihins came *before* all subsequent mixes. The first AMH specimens, as we have seen many times, resemble the Europoid type—the Ihin. "The parent stock [of sapiens] must be regarded as proto-Caucasoids."[18] But consider this: Since Caucasians have the straight hair of the anthropoid apes and since skulls of anthropoids are brachycephalic, it is the dolichocephalic Africans who are morphologically *furthest* from the simian type. George Frederick Wright's observations put to rest any attempted definition of blacks as relics of the evolutionary past:

> *The European has a small face and a high nose—all*
> *features farther removed from the probable animal*
> *ancestor of man. On the other hand, the European*
> *retains in the strongest degree the hairiness of the animal*
> *ancestor, while the specifically human development of the*
> *red lip is most marked in the negro. The proportions of the*
> *limbs of the negro are also more markedly distinct from*
> *the corresponding proportions in the higher apes than are*
> *those of the European.*
>
> GEORGE FREDERICK WRIGHT,
> *ORIGIN AND ANTIQUITY OF MAN*

Coon, it seems, was assailed by people who were plainly afraid of blowback in the form of "a certain snobbishness as to whose ancestors became sapiens first." A moot point, really, seeing that *Homo sapiens sapiens* (the Ghans) arose "in *all* [e.a.] the divisions of the earth" out of the cross between Ihins and Ihuans.[19] In fact, like the Batek Negritos say, the very first people were created from brown soil; next came beings from white soil (Ihin), therefore they (Batek) were created first. This is what we saw in chapter 2: Negritos were not only a universal race, but were the original occupants of most regions in the world. Coon went on to explain "When I wrote that our Caucasoid ancestors crossed [the sapiens threshold] earlier than the ancestors of some Africans. . . . I knew that I was in for trouble." T. Dobzhansky, leading the charge, stated publicly that Coon's parallel evolution (polygenesis, separate descents), "would require a mystical inner drive that propels evolution." Which is nonsense, since parallel developments are seen everywhere in the record. "The multifaceted phenomenon of man," Jeffrey Goodman noted in *The Genesis Mystery*, "started off in a number of geographic regions at once." And, I might add, not by evolving, but simply by crossbreeding.

> *In reality, the Garden of Eden was worldwide.*
> SIR ARTHUR KEITH, *THE ANTIQUITY OF MAN*

To Robert Braidwood, it seemed "unlikely that only one little region saw the beginning of the [human] drama."[20]

Apparently Coon's opponents were worried that polygenism could be taken advantage of by racial propagandists.* Yet I wonder why they were (and are) not so worried about what racists could do with their *current* reconstructions, like the Smithsonian's and National Geographic's highly publicized models of African hominids looking like a cross between gorilla and Negro. After all, those Miocene apes, "our closest

---

*The more virulent theories of racial differences co-opted parallel evolution, paleontologists like H. F. Osborn claiming that the races were distinct *species,* which had evolved in isolation. From Haeckel to Hitler, this was the framework used to couch the extreme form of white supremacy that led to the Nazi claim of Aryan superiority over the submen of the world.

primary relatives," lived in Africa, and with Africa as the cradle, direct links must be found between apes and black Africans.

We are essentially elaborated African apes.

DONALD JOHANSON AND BLAKE EDGAR,
FROM LUCY TO LANGUAGE

So I'd like to ask: What's so politically correct about apes being the ancestors of African people? Why not deplore the racist H. F. Osborn instead of the humanist Coon? One-time president of the American Museum of Natural History, Osborn believed in the superiority of certain races; he was part of the eugenics movement and linked to the policies that gave rise to Hitlerism. Oh yes, there were close ties between American race scientists and those working in Nazi Germany.[21] Darwin himself had culled (prejudicial) evidence to show that nonwhite races had "more apelike features."[22]

But ever since Hitler, the subject of race has been closed down. Taboo. Throwing out the baby with the bathwater, anthropology—running scared of accusations of racism—not only dropped the idea of race (notwithstanding euphemisms like ethnic groups or even clans), but abandoned the first-rate, unbeatable work of Dixon, Hooton, Coon, and even Weidenreich. It was all too emotional, irrational. And it comes down to threat management—not science.

And it set the stage for revisions, such as the untenable monogenism of an African Garden of Eden, conforming, not to the evidence at hand, but to the pressure of (ill-informed) public opinion and nervous ideas of political correctness. Although anthropology sincerely hoped to purge its doctrines of any hint of racial supremacy, scapegoating Coon (and polygenism) was a disgraceful nonsolution.

The firestorm over supposed racism in anthropology led to egregious manipulation of the record. To put out the fire, it became necessary, for example, to change the date of Africa's Kabwe Man from 40 kyr to 125 kyr—as ammunition against Coon's position that "Africans remained at a *Homo erectus* level . . . at a time when *Homo sapiens* had already appeared in Europe." Kabwe Man was then "adjusted" to cross the *H. sapiens* threshold earlier on. But could his bones really have been so old, with no traces of fossilization? Kabwe (aka Broken Hill, Rhodesian Man) was a tall and strong man with the modern loaf-shaped cranium, modern also in limbs, neck, and dental arch; yet his tibia was not like living Africans, and he abounded in *pre*-Neanderthal (erectoid) traits in his massive brow, skull proportions (only 1,280 cc), and prognathism. His was a long face, large upper jaw, enormous palate, and extremely sloped forehead. Kabwe's tools (Acheulian) were rudimentary, and finally he was a cannibal.

Was Kabwe, this cannibal, the ancestor of all modern men? Was he Africa's phylogenetic link from *H. erectus* (or *H. ergaster*) to *H. sapiens*? Hardly. His Neanderthaloid skull was synchronous with similar types in Europe, such as France's La Chapelle-aux-Saints skull, Greece's Petralona, and Italy's Olmo II. "The Broken Hill Man," thought Marcellin Boule, "is much more like *Homo neanderthalensis* than any other race." Although originally dated 40 kyr, that was revised to 125 to 300 to 500 kyr! But he was found with *extant* mammal bones, and "their appearance is very fresh."[23] In fact, Kabwe may be a mere 33 or 11 or even 6 kyr, "startlingly recent times,"[24] for the bones are not mineralized and his implements hardly differ from those of the modern Bushman.

Franz Weidenreich in *Apes, Giants, and Man* (1946) correctly concluded that "human forms preceding . . . modern man were distributed *all over the entire Old World*" [e.a.], which agrees with the Oahspe verse: "In all parts of the earth there lived ground people,"[25] that is, *H. erectus* types. Weidenreich painstakingly showed how *H. erectus,* at *many* locations, not just Africa, preceded the modern type. Called polycentric evolution (but sometimes parallel evolution or convergence), this approach sees early man grading into the modern type in Europe, Australasia, China, and even the Americas. Not just Africa.

But what does convergence really mean? Does it mean parallel *evo-*

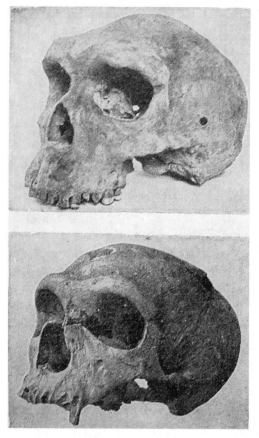

Figure 11.2. Broken Hill (Kabwe) skull (top) compared to La Chapelle-aux-Saints Neanderthal (bottom).

lution? I don't think that's possible. Ultimately, simply, and parsimoniously, it represents the *mating of disparate types* (AMH and Druk), occurring in all parts of the world.

"Convergent evolution is really a special kind of coincidence," offered Richard Dawkins,[26] apparently in an effort to offset the implausibility of evolution happening the very same way in lots of different places. Coincidence. But what exactly does Dawkins's statement mean? Does it mean parallel mutations (the *same* accidental mutations, the same changes) miraculously and coincidentally, occurring in all the separate places where modern man arose? That would be a marvelous feat indeed!

Darwin's own improbable monogenesis forced him to ponder: How on Earth have the same types managed to arise in unrelated parts of the world?* To answer this, evolution wed itself to pseudomigrations. How does one explain, otherwise, the parallels between the Paleolithic stages of India, Europe, and Africa?

Isn't it wiser to allow the same types of early races to have *mixed,* interbred, in each of those different localities? The fact is that Darwin's stubborn monogenism has been pretty much disproven—if only because a single origin forces the different races (Mongoloid, Caucasoid, Australoid, and so on) to evolve or rather diverge too rapidly. Yet now again, monogenism reincarnates in the illogical OOA theory.

THE GREAT EXODUS

Was Au "unique to Africa," as is claimed and required by the OOA Garden of Eden? Since *Australopithecus* presumably evolved to *H. erectus only* in Africa, we should, of course, not be finding any Au elsewhere. And if we do, we must ignore or debunk it.

Rewind about fifty years: Au in Asia was actually common knowledge among paleo students until the latter half of the twentieth century, that is, before OOA became *de rigueur.* "Until recently [1973], it was thought that from forms like Australopithecus, which lived in *Asia* [e.a.], *Pithecanthropus* could have originated."[27] *Au. robustus* features were indeed evident in Java's Sangiran VI and VIII specimens. And near Java, various Au traits were discerned in the amazing Flo (hobbit), who possessed both *H. erectus* and Au features—thick bones, Au-type wrist structure, pelvis, and stature, and very long arms.

> *I would not be surprised if australopithecine remains start turning up outside Africa.*
>
> DEAN FALK, *THE FOSSIL CHRONICLES*

---

*Parallel evolution was the precursor of today's multiregional model, anticipated by Franz Weidenreich in 1939 when he found progression from one type to the next in different parts of the world and with *genetic exchanges* between regions. Indeed, his model called for *universal hybridization,* a continual exchange of genes, which is pretty close to the thesis of this book: man the eternal mixer.

Although Falk acknowledges that "australopithecines were thought to have lived exclusively in Africa," she does remark that Indonesia's "LB1 [hobbit] had arms . . . similar to those of the famous australopithecine Lucy . . . and closely resembled the little, short-legged and small-brained australopithecines. . . . The inside of [LB1's] . . . lower jaws has a little ledge that was typical for australopithecines"; other features "harkened back to early *Homo habilis*." Well, if hobbit branched off from *H. habilis,* that would have been *before* the great (part one) *H. erectus* exodus from Africa. And that's a problem. But here is the solution: Flo's "lineage left Africa before" *H. erectus* evolved, as suggested by anthropology professors Peter Brown, Mike Morwood, and William Jungers, Brown postulating that australopithecines "migrated out of Africa a very long time ago"![28]

But Au were indigenous to each of their locales, their supposed move OOA even greater folly than making travelers out of *H. erectus*.

Isn't it interesting that formerly Asia itself was called Asu-wa—for it was covered over with Asu man, the primeval race, the aborigines of Earth, who abounded in Central Asia (India) during the Pleistocene. In Sri Lanka we have found his descendants, the extremely primitive, race of Nittevo. Were they ape-men? They walked on two legs and had short arms and very hairy legs, indeed a reddish "fur" over most of their bodies. Completely naked, they were a tree- and cave-dwelling people. Asia's most spectacular tree climbers, the Nittevo were at home in the trees, with their powerful grasping hands (and "claws"). With Nittevo, we are almost back where we started—with Asu man, our very first (tree-loving) ancestor. In China, too, Asu man lived alongside *H. erectus,*[29] contradicting the "unique to Africa" idea: Au teeth have been found in China at Hubei and Guangxi provinces. Teeth from China's Longgupo cave, it was thought, "belonged to an ape," this pre–*H. erectus* man dated to 2 myr.

Time line is important, so we must ask: Why did South African Au live at the *same time* as Far Eastern *H. erectus*? Au, according to theory, is supposed to have *evolved* into *H. erectus* before the latter moved

OOA and on to the Far East. But Au led to *H. erectus* "not necessarily"[30] in Africa alone, his remains noted in the Jordan Valley, China, Indonesia, Southeast Asia, and Java, the last referring to G. H. R. Von Koenigswald's 1952 discovery of *Meganthropus,* a creature older than *H. erectus* and possessing some Au traits.

   H. habilis–like industries (more archaic than those of *H. erectus*) are found in Pakistan: these 2-myr chopping tools of the habiline type don't exactly fit with 1-myr or 1.6-myr or even 1.8-myr *H. erectus* as our first migrant OOA to fill an "empty" world. And then there is the Zinj-like Dmanisi Man, in the Caucasus: How was this pre–*H. erectus* (outside of Africa) handled? By coining another wave of migration, of course. I'd get rich collecting a toll for every imagined journey OOA.

> *Be suspicious of a theory if more and more hypotheses are*
> *needed to support it, as new facts become available.*
> FRED HOYLE AND N. C. WICKRAMASINGHE,
> *EVOLUTION FROM SPACE*

   This odd Dmanisi race was christened when six specimens with some Au traits (like upper-arm anatomy) were found in the Caucasus and tentatively assigned to early *H. erectus,* if only because of the time frame. But at 1.8 myr, he was *older* than most African *H. erectus,* such as Kenya's Turkana Boy at 1.6 myr; and Dmanisi's brain was *smaller* than Turkana Boy's (which was 900 cc). Dmanisi was *smaller* (in both height and brain) than *H. erectus,* his brain *H. habilis*–sized at only 600 to 750 cc. Don't such habiline traits prove *other* Gardens of Eden besides Africa?

   Complicating matters, Dmanisi's feet and body were of more modern proportions than *H. habilis.* On the other hand, his prominent canines were almost Zinj-like (robust Au). Described as a hodgepodge, this Dmanisi race had skulls exhibiting a splendid mosaic of features— obviously a mixture of races.

   Discovered in a medieval Georgian town, Dmanisi had huge primitive canines; the point is Dmanisi actually looked *ancestral* to Africa's

H. erectus. Uh-oh. But not to worry: the fire was put out by changing his name to *H. ergaster.* But some of them had traits that *predate H. ergaster* (a predecessor of *H. erectus*), and crazier still, Africa's Turkana Boy was also labeled an *H. ergaster*! What a mess. Immediately after noting these uncomfortable facts, Fagan nonetheless declared that "Dmanisi has strong African associations."[31] On what basis? He did not go on to elaborate.

Dmanisi's mixed and archaic morphology (and its date) was so troublesome to OOA that one scientist said flat out that they ought to put it back in the ground! Said archaeologist Chris Hardaker in *The First Americans,* "If there was ever an equal to Olduvai's Zinj [an Au], the Dmanisi discovery was it." Here was yet another early type, too early, found "aberrantly" outside Africa. Did this pea-brained creature travel from Africa, thousands of miles over new ground (three thousand miles from Olduvai)? Hardaker cries in the wilderness: "How could they have done it without at least some of the Upper Paleolithic [mod] qualities and faculties . . .?"[32] Besides, what were these naked "émigrés" doing in the cold Caucasus, when the presumed route* of travel OOA was along the warm southern rim of Asia?

Overall, scientists regarded Dmanisi's brain as way too small for such an ambitious undertaking, his huge canine teeth indicating a very primitive hominid, indeed more archaic than the first *H. erectus* supposed to have made the great migration. Dmanisi was not inclined or constructed for migration. Instead of giving up OOA, the response was to declare that something even more apelike than *H. erectus* probably made the great exodus OOA: "leaving Africa didn't require a big brain. . . . They took a long hike, and they made it." But their rudimentary tools, of the most primitive type, don't jibe with the tool kit of travelers.

---

*There is nonetheless no archaeological evidence for that journey along the southern rim of the Persian Gulf and Pakistan, according to Brian Sykes. (Sykes, *Seven Daughters,* 284.) Part two routes also look like guesswork: "After reaching Malaysia, a group that would eventually settle Europe branched away," that is, turned back west and "looped northwards to Turkey." (Hooper, "The Scenic Route Out of Africa," 14.)

The last field season at Dmanisi (1991) confirmed these were the oldest hominid bones ever found in Eurasia. Which made analysts wonder if this *Homo georgicus* evolved *separately* from African hominids and if "*Homo erectus* evolved from this primitive Dmanisi stock somewhere in Asia," painting in yet another handy scenario, "and then moved back to Africa. Maybe there were multiple migrations back and forth."[33] Lotta maybes and ad hoc journeys in this highly subjunctive out-of-Africa dictum. Better look to Asia, some thought.

MOTHER ASIA—AND BEYOND

In 1997 a primatologist told the *National Geographic* writer Rick Gore: "The idea that all hominids originated in Africa is a myth created by people working in Africa. Sure, they've found a lot there, but if we'd invested that much time and money in Asia, we would find fossil hominids just as old there, too."[34]

MtDNA analysis, called in to support OOA, came up with a notably small number of mutations distinguishing the living races of men; it was then (gratuitously) assumed that everyone must have descended from a single ancestral root population. But since it is hard to get DNA from fossils older than Neanderthal (ca 35 kya),[35] how can such a judgment be made about people of much earlier times (the modern African dispersal, part two, is said to have begun some 80 kya or even 133 kya)?

Craniometric surveys have found that the little (Bay of Bengal) Andamanese people, though black skinned, actually have more affinities with Asians and Polynesians than with Africans. Blood groups, antigens, and proteins relate these Negritos to Oceanic and Asian populations, not Africans. So how can anyone say the Andamanese or Australians or Javanese are descended of migrants who left Africa? Neither are Australian Aborigines Negroid in type, but a mix of Caucasian and Melanesian genes. "Although dark in skin color, neither the Australians nor the Wadjak [Javanese] skulls are 'Negroid.' We still have a very great deal to learn in these matters," cautioned Robert Braidwood.[36]

Very early art in northwest Australia (at Jinmium) has been dated to 75 kya (twice the age, incidentally, of the earliest known cave paintings in France). This discovery presented a problem to scientists who "have long assumed that *Homo sapiens* lacked the seafaring technology to have reached Australia before about 50,000 years ago." By now you can probably guess what the simple solution was: perhaps "modern humans were on the move much earlier."[37] On the move? Or there all along?

The there-all-along idea (which is polygenism) garners support from burials found at Lake Mungo, Australia, where gracile types are much *older* than the primitive Kow Swamp type. The Lake Mungo Specimen 3, dated 60 kyr, is AMH, "a challenge to current opinions on human origins,"[38] as is WLH50, a very early Australian Caucasoid. Was there high culture here so long ago? (Appendix D charts a dozen Oceanic sites indicative of advanced engineering and architecture in a long-forgotten lost horizon.) But because Australia's *later* Kow Swamp people (10 kyr) were robust and erectoid, theorists postulated—you guessed it—separate waves of migrations. But there were two strains living there all along, not two separate migrations or colonizations.

Maybe Prince Regent River offers a clue to the puzzle of early mods (like Lake Mungo Man) in Australia, people who lent their Caucasian genes to today's Aborigines, for here are singular cave paintings depicting European types, the men bearded, the women delicate; behind them is a snake, which is the Aborigines' symbol of the most remote past: Dreamtime. Of equal interest are Australia's huge limestone pillars near Roper River, attributed to members of a white race—the site boasting streets and polished walls. Are they the same people depicted as bearded Caucasians in the rock art of central Australia near Alice Springs? Might they also be the ancestors of the Murrian people of south and southeast Australia, possessed of a Caucasoid skull form, light skin, beards, and narrow noses? Australian archaeologist Vesna Tenodi is on record stating that much of her country's protohistory has been suppressed by the Australian Archaeological Association (AAA), which "has turned into a political body whose main concern is to please Aborigines. Thanks to the AAA, fossilized human remains were destroyed. These included remains from pre-Aboriginal time, which proved the existence of highly developed pre-Aboriginal races before the arrival of the ancestors of the current Aboriginal tribes."[39]

The archaic whites of Australia were of a similar cast to the Japanese Ainu, both of whom have nothing to do with Africa, but do represent an aboriginal white race of the Far East and beyond. Indeed, Dixon had these proto-Australians in the region *before* proto-Negroids ever wandered OOA. DNA studies seem to corroborate this, finding at least one Lake Mungo lineage that is deeper than African Eve.[40]

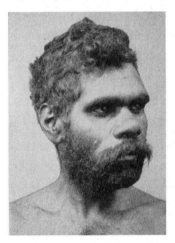

Figure 11.3. Australian with blend of Caucasoid features, note beard.

Before OOA became de rigueur, William Howells reported that the Andaman Islanders "do not at all resemble African Pygmies of the Congo"; they are allied with Africans (by multivariate analysis) "only occasionally. . . . The Andamanese are mysterious."[41] Although these "mysterious" little people do have Bushman-like peppercorn hair and steatopygia, their closest affines reside in Papua, Southeast Asia, the Philippines, Australia, Tasmania, and Sundaland. Cranial morphology links the Andaman Islanders to Australo-Melanesians. Asian (and Oceanic) ancestry is indicated, not African.

The Andaman mtDNA lineage M is common in Asia, not Africa (where lineage I is dominant). Genetic analysis of the Onge people of the Andaman Islands reveals a special change in the Y chromosome, casting

Figure 11.4. A North Andamanese man and his son.

the Onge as actually *ancestral* to the populations of Asia. Indeed, standing OOA on its head, some theories have derived the African Negroes from the East, mankind thought to have "originated . . . somewhere in northern Asia."[42]

William H. Flower, late nineteenth-century surgeon, curator, and anatomist, considered "the south of India as the centre from which the whole of the great Negro race spread." Hooton also speculated, based on blood group B, that Congolese pygmies came from Asia. I present these views only for sake of comparison, not because I agree. Nevertheless it is true that Africa's pygmies are not racially aligned with Black Africans. They may share some blood types (due to intermixings), but the Negrillo is both racially and linguistically distinct from the tall African.

> *The tree of life looks different depending on which genes you choose to analyze.*
>
> GARRY HAMILTON,
> "MOTHER SUPERIOR," *NEW SCIENTIST*

Geneticist Alan Templeton, entering DNA data to the computer in a different way than the Afrocentrist popularizers, got Europe and Asia for the last common ancestor (LCA), his team ultimately concluding that you cannot pinpoint the location of LCA using *living* DNA (the OOA method). "An Asian—not African origin"[43] is indicated by these alternate mtDNA trees.

Before the first Au (Taung Child) was found in South Africa in the 1920s, most scholars did indeed think the Garden of Eden was in Asia—perhaps in the Pamir Plateau.

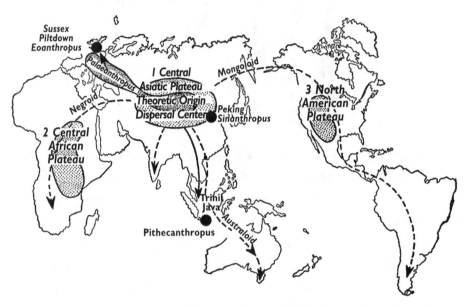

Figure 11.5. Map showing theoretical dispersal of man out of Asia (1929).

The favorite cradle for Egyptologists, ever searching for the origin of the Egyptian kingdoms, has been Asia, not sub-Saharan Africa. Shot from the Stone Age, Egyptian civilization "didn't evolve. . . . Architecture, engineering, medicine, science and well-organized big cities materialized within a century or two—almost as if they had been imported from somewhere else."[44] Somewhere outside Africa. Stephen Oppenheimer and others have identified Southeast Asia as the center of innovations that eventually reached the West, contributing navigation, astronomy and farming to the "historical" cultures.

The people of southern India have been traced to Southeast Asia, not Africa—their languages related to those of Burma and Cambodia.

"Java and . . . all of southeast Asia is a serious rival to the [African] cradle of mankind."[45] Dixon's brachycephalic AMHs came into Europe from the East (Asia Minor, Russia, India), not from Africa. Finally, the disperal of myth cycles (creation legends) seems to have an Austronesian/ Asian source: the fact that the Far Eastern stories are "more complex and internally coherent than the Mesopotamian versions supports the view that the diffusion pattern may have been from East to West rather than the other way round."[46] Indeed, Mesopotamia's monument builders are also traced to the East.

If moderns came out of Africa and spread throughout the world, why should such civilized industries as rice cultivation be evident in China and India[47] *before* they got to Africa? Early rice-growing dates in the Malay Peninsula presage the westward spread of farming into India. Microliths (used on farming scythes?) may be as old as 60 kyr in Sri Lanka, not appearing in the Mediterranean until 10 kya.

The Belgian archaeologist Marcel Otte, in *The Aurignacian in Asia*, does not accept OOA, pointing instead to the Zagros Mountains as the hub of Upper Paleolithic industries. Michael Cremo has also presented arguments for mods arising not in Africa but in Pakistan, Siberia, and Russia.[48] The earliest ivory carving of figurines (a craft developed only by modern humans) is found in Russia and dated 45 kyr. If mods originated in Africa, why are there no such artifacts of ivory south of the Sahara, though they are found in Europe and western Asia? The high art of the French and Spanish caves was, according to Cambridge archaeologist and historian Colin Renfrew, "not a general feature of early *Homo sapiens* in Africa." North Africa's rich heritage of rock art, after all, is of recent date. "Only in France and Spain . . . can the human revolution be generally associated with the appearance of art. Elsewhere at this time there was little or no."[49] Contemporary South African sites were still using Middle Stone Age technologies at this time.[50]

Why is there no *Homo sapiens* explosion in Africa? Instead there are signs of it at such places as Russia, Czechoslovakia, the Americas, Turkey, Japan, Sri Lanka, and Oceania.

NO ONE KNOWS

*Where he [Cro-Magnon] came from or how he came
about we have not the slightest idea.*

ASHLEY MONTAGU, *MAN: HIS FIRST TWO MILLION YEARS*

Those cave artists in Western Europe were, of course, the Cro-Magnon people, the first known edition of true modern man. But "no one knows where the Cro-Magnon came from."[51] Guesses include: "The Cro-Magnon race is thought to have emigrated to North Africa from Europe"[52] (to, not from). In Afghanistan (at Kara Kamar), Coon found remains of Aurignacian Cro-Magnons *older* than the European ones, suggesting again an eastern origin. Arthur Keith, for his part, predicted that when digs are undertaken in Asia, "we may expect to find the earlier history of the Cro-Magnon race," who he thought resembled today's Sikhs.[53]

The unanswered question is: Why do we find their Upper Paleolithic industries (fine tools, artistic works, quarries, and so on) starting *abruptly* in western Europe around 40 or 35,000 BP and somewhat *earlier* to the east? The answer: Because they represent a hybrid (instant) race—the second wave (39 kyr) of Ihuans, the fount of Aurignacian culture. One of the earliest models of European man was found in the 35 kyr mandible at Pestera cu Oase (Southwest Romania); then moving west of the Danube, AMH fossils begin to appear, dating to 32 kya and getting younger the farther west we go.

Now it is plain that for out-of-Africa (part two) to be right, mods *must* be earliest in Africa, having evolved only there. But consider the Ainu of Japan, a very old race, indigenous to the East, though made over to represent an "ancient migration" from Europe, simply because of their (taxonomically) troublesome light color and "Caucasian" features. But how could they be from Europe when Europe was swarming with Neanderthals (not mods) at the same time the Ainu stock took form? No, Europe was the *last* place to be settled by Cro-Magnon AMHs: Ainu migration from Europe? No way, their skulls are Asiatic.

Figure 11.6. A major racial enigma, the Ainu combine Mongoloid and Caucasoid features, not African ones.

Neither do we find the earliest mods of China or Europe or Australia or Indonesia resembling (facially, cranially) any fossil men of Africa. Nor do "the languages of East Africa have [any] demonstrable relation to the languages of Asia."[54] China's sites at Jinniushan, Dali, and Maba all have archaic *H. sapiens* types, which show continuity from China's own (indigenous) erectoid forms. Jinniushan woman, for instance, (dated 250 kya) has a large cranial cap, thin cranial bones, and much reduced supraorbital region, clearly someone in between the old erectoid type and today's Chinese. Nothing African about it. Chinese paleontologists have pointed out this continuity from Peking Man (*Sinanthropus*) to the present people of China, indicating local development only. Indeed, a real problem (for Afrocentrists, anyway) is the persistence of *Sinanthropus*'s shovel-shaped incisors (and several other features) in *modern* Mongoloid populations. This continuity comes under the multiregional (and polygenetic) approach, sometimes called Noah's Sons, which, in opposition to OOA (sometimes called Noah's Ark), holds that mod characteristics arose independently in different parts of the world.

TABLE 11.1. EARLY AMH FEATURES
IN PLACES OUTSIDE AFRICA

Where/Who	Feature	Date
Borneo, Niah Cave/Deep Skull	Mod skull	52 kyr
China/Dali, Maba, Liujiang fossils	Mod features	ca 75 kyr
France/Fontéchevade Man	Mod features	70 kyr
France, Nice, Terra Amata	Borderline between *H. erectus* and *H. sapiens*	400 kyr
Germany/Hamburg	Carvings of mods	200 kyr
Germany/Heidelberg Man	Mod dentition	400 kyr
Germany/Steinheim Man	Mod forehead	300 kyr
Hungary/Vertesszollos Man	1,500 cc brain	100–700 kyr
India/Narmada skull	Mod feature	Mid-Pleistocene
Iran, Hotu Cave/skeletal remains	Mod skulls	100 kyr
Israel, Qafzeh/skeletal remains	AMH	78 kyr
Italy/Castenedolo skulls	Modern	Pliocene
Java/Sangiran 17	Robust *H. sapiens*	700 kya[*]
UK, England/Swanscombe Man	1,325 cc; the first "cast-iron" case of early AMH	250 kyr[†]
UK, Wales, Pontnewydd Cave/teeth	Mod mandible and vertebrae	225 kyr

[*]Cremo, *Forbidden Archaeology*, 494.
[†]England's Galley Hill, Boxgrove, and Foxhall men might also be included here; Hooton, *Up from the Ape*, 359.

Franz Weidenreich, the brilliant German-Jewish anatomist and paleoanthropologist who inspired today's multiregional approach, traced the earliest Javanese *H. erectus* up to Sangiran, Wadjak, and Ngandong, and right on up to the tribal people of today's Java. Continuity. Milford Wolpoff, a leader of today's multiregional school, questions the dating of putative earliest man in Africa and argues persuasively for *H. erectus* populations moving in the modern direction, not only in Africa but also in Europe and Asia, with "occasional interbreeding" (getting along)—versus the OOA replacement (extirpation) model. "I see continuity everywhere," says Wolpoff. He is right. In fact, the independent and worldwide (not just African) appearance of early AMHs can be seen in Java, America, and Europe, putting paid to OOA and its Africa-only origin of mods.

> *South Africa is not the only area yielding earlier origin*
> *dates for modern man. Very early dates for fully modern*
> *man are also coming in from Australia and America.*
>
> JEFFREY GOODMAN, *THE GENESIS MYSTERY*

Fagan has written that the study of African Adams (meaning mods), using Y chromosome, yeilds dates to only 59 kya. Well, if this is right, why do we find AMH features that age or *older* in regions outside Africa? Why does the genetic line of the Malaysian Semang (Negritos) as well as Australia's Negritos go back more than 60 kyr?

HAM, SON OF NOAH

With several different kinds of humans running around early Africa, the question has been posed: What caused this wonderful proliferation of types? The usual answer is that some environmental change created new selective pressures for gene mutation. The climate card. This brings us right back to the unsolved problem of radical variability. And some new answers.

> *There is general recognition that Africans have greater*
> *genetic diversity than other people, but the significance of*
> *that fact remains unclear.*
>
> MARVIN L. LUBENOW, *BONES OF CONTENTION:*
> *A CREATIONIST ASSESSMENT OF HUMAN FOSSILS*

Ham's habits, as we will come to see, explain that "significance." We have mentioned Noah's sons Jaffeth (in Asia) and Shem (in India and the Near East). The third son was Ham (in Africa, see Genesis 10:6–14). Now, with genetics as handmaiden of OOA, mtDNA studies claim that mods have lived in Africa *longer* than anywhere else. This assumption is based on the following observation: As distance and time removed from Africa increases, *diversity* diminishes. Native Americans, for example, have much less variety in their genomes than Africans.

It was this great diversity in Africa that moved theorists to suppose that Africa is both the oldest population and the *founder* population of our species, *Homo sapiens*. A University of Cambridge study, for example (which used data from thousands of skeletons all less than 2,000 years old), determined that the farther a population was located from Africa, the *fewer variations* in skull shape; and this was apparently corroborated by DNA, which said that mods arose in Africa *200,000 or 150,000* BP and spread out (migrated) about 50,000 years ago.

In my view, however, variations and diversity appear simply by *mixing* different races, not by mutations or climate changes. This is where the sons of Ham enter the picture. From Ham's unique history, we will discover why the East African "Hamitic strain" is associated with Caucasoids, and why Saharan and Sudanese tribes are an exact cross between Caucasoids and African Negroes. In the quote that follows, the key word is *interbreeding:* "The entire area from North Africa to the Sudan is one of the major zones of interbreeding in the world. . . . A massive interchange of genes between Caucasoids and Africans has clearly taken place in Africa in the past."[55]

The gloriously mixed Berber have wavy or curly hair, heavy beards, hawk and straight noses, and lightish eyes. The Somalis have black skin and frizzy hair combined with facial features of the European mold. And along the Nile, between Cairo and Khartoum, we find nothing but an intermediate series between the South European type and Negro.

Stretching from the east coast of Africa (the point of Ham's entry from Pan), from Cameroon to Senegal, the people are "extremely black in the west; they tend to become slightly lighter eastward"—and also shorter. Both paleness and shortness are signs of Ihins, in this case, the Hamitic sons of Noah who settled Africa 24 kya. A further example is the Tibbus tribe of the Sahara, a mixture of proto-Negroid and Caspian—tall and very black, yet the hair is not woolly, the nose is aquiline, and prognathism is absent. "All Negros are partly Caucasoid by interbreeding,"[56] which is further indicated by common blood groups and architecture of the teeth.

Of all colors were the tribes of Ham . . . with hair red, and white, and brown, whereby might be traced in after ages the genealogy of nations.

OAHSPE, THE LORDS' FIFTH BOOK 3:3 AND 3:13

The White Lady of Brandberg is a rock painting in southwest Africa, showing Bushmen standing next to white women with European profiles and yellow or red hair. Kalahari cave art in Damarland, according to the world's leading authority on these paintings, Henry Breuil, was unmistakably the work of a mysterious race of white people. But not so mysterious, once Ham, Noah's son in Africa, is factored in.

A British major stationed at Africa's Gulf of Guinea once saw the members of a tribe marching toward the shore as a canoe was approaching, with white-painted natives. He asked what the apparent ritual was about and was told that "it was a custom handed down from the very earliest times, perpetuating the tradition that white men had come once from . . . an island now no longer in existence."[57] Those men were lawgivers and teachers of justice.

Another telling tradition, that of the African Herreros, a Bantu tribe, says that after a great deluge, white men came and mingled among them. Were they the teachers and lawgivers we so often hear about in world traditions, founders who came from across the sea? One great puzzle is how Africa's Ituri forest Efe people have long known of planet Saturn as "the star of nine moons." (Its nine moons were not discovered in the West until 1899.) How did the Efes know this? Well, the Dogon of Africa also claim to have received cosmic information from teachers who gave them knowledge of the stars. Local legends of white gods have long intrigued scholars of Africa, and if oral history be credited, those gods (the euhemerized sacred tribes of Ham) once inhabited their country. (Also deified was Chamha, the solar god of Syria, called Hama in Persia; while the name *Ham* itself is the Zeus of ancient Greece.) But they were Ihins, mortals, the holy sons of Noah, called Ham.

> *One way or another, Caucasoids and their ancestors have played a major role in the genetic history of sub-Saharan Africa.*
>
> JOSEPH THORNDIKE, *MYSTERIES OF THE PAST*

And this is the key to Africa's mysteries: When the Caucasian-like Hamites reached Africa, they ignored the rules given to all the Ihins, and mixed freely with the indigenous black people (Genesis 9:25–27). No, Africans are not the oldest people in the world, but the most varied by reason of admixture.* For the Ihins who settled Africa "broke the law of God more than all other Faithists, being of warm blood—and they mixed greatly with the [African] Ihuans . . . and they ceased to exist as a separate people, because of their amalgamations."[58] For the "warm-blooded" Ham (the name for the Ihins who settled Africa) mingled with aboriginal types. This Hamitic Cro-Magnon Man, as the patriarchs took every opportunity to stress, was the most rebellious, licentious son of Noah. His descendants, as these old stories claim, "married a thousand wives."

---

*Twenty-one kya, the African Ihins were, significantly, the first to go EBA (extinct by amalgamation). The last surviving pure Ihins were in the time of Moses, inhabiting Egypt, Arabia, China, and North America.

My point is this: Greater genetic variation in African populations represents neither deeper time nor more mutations, but crossbreeding only. In fact, this unbridled mixing is exactly the opposite of the "persistent isolation" posited by the mtDNA people to explain the greater genetic diversity in Africa! Diversity (variation) is the face of nothing more complicated than amalgamation.

"Africa is racially the most confusing of all continents."[59] African blood proteins and mtDNA are more varied than elsewhere, yet this is due neither to immensity of time, greater population, environmental instability, nor mutations—but simply to Ham, the mixer. The full impact of this mingling in Africa is evidenced by the fact that here "the Chellean, Acheullean and Mousterian [material] cultures are not sharply defined . . . as in Europe."[60] Admixture is also expressed in the blurring of borders: the indigenous populations of Africa are mostly clinal, that is, blending, overlapping from one region to the next.

Greater variability in Africa? Not according to Stephen Oppenheimer's work, which has shown it is actually the *least* in Africa, from the point of view of tribal legends: Tracing origin myths to their geographical root, Oppenheimer found they spread from Oceania across the Indian Ocean *to* Africa. "In looking for a homeland . . . look for the region with the deepest and greatest diversity of story-types. . . . Africa clearly has the least diversity, Australasia . . . has the most."[61] Even geneticists have gotten a similar result in connection with the path of disease-related molecules: "The greatest diversity in the frequency of gene variants lay outside Africa . . . Oceania . . . had much more variation."[62]

Notwithstanding all I have said, I hasten to add: There *was* a great African diaspora, but it was postflood. Dating no earlier than 24 kya, these were the outward-bound peregrinations of Ham. Yes, waves of men *did* come OOA, but only in the Upper Paleolithic, after the flood, in the Solutrean. Neolithic Negroids in Europe (dolichocephalic, prognathous, African in proportion of limbs) have been uncovered in Switzerland, Lombardy, Brittany, and North Italy. Thus do scholars find an African aspect to French sculptures and implements of the Mesolithic. "Bushman mural art is extraordinarily like that of our [French] caves," Marcellin Boule marveled. "The two centers are united by a long, connected series of works of art, from France to the Cape by way of Spain, North Africa, the Soudan, the Tehad States and the Transvaal. This almost uninterrupted series leads us to regard the African continent as a centre of important migrations . . . stocking southern Europe. . . .The Grimaldi Negroid skeletons [on the Riviera] show many points of resemblance with the Bushman skeletons."[63]

Of all the sons of Noah, Ham was the "travelingest" man. "To Ham I allotted the foundation of the migratory tribes of earth . . . to teach the barbarians."[64] In their path, the Hamite tribes taught mining, metals, writing, crafts, and astronomy, also instructing the Druks and Neanderthals about the world beyond. Thus did Aham become a sacred name and the language of oracles in Arabinia, the name Ham notable also in the original term for Egypt—Aham or Khamet (Hemia to the Greeks).

The civilizing aura of Ham is enshrined in places like Jebel Dukham in Bahrein (with its mound fields*), Abu El 'Idham in Transjordan, Hamman and Hamath in Syria, Hama on the Orontes, the Hamrin basin of lower Mesopotamia, Tehama (along the Red Sea), and the Interahamwe tribe of Africa. Even the African-transmitted cult of ker-ham (the Breton dolmens) was carried from Africa to Europe, the route of these "Ethiopians" threaded through Iberia to northwest Europe.

In addition, the Ogham script, seen on ancient *megalithic* stones of the Celt-Iberian region, recalls the Great Zimbabwe Ruins, constructed of quite similar cut stones, most unusual in Africa, its walls comparable to the forts of Ireland, suggesting that these megalithic structures were built by the same people. In the British Isles, the noble *ham* name, as suffix, lives on at Brixham, Durham, Tottenham, Chatham, Upham, Egham, Birmingham, Bingham, Wickham, Newham, and many other places. Even the English word *hamlet* signifies "settlement" (civilization).

---

*See *Lost History of the Little People:* the sons of Noah built mounds in all the divisions of Earth.

BLACKS DID NOT TURN WHITE

Those émigrés from Africa (part two) who allegedly settled all the world (long before the flood) were, of course, dark skinned and, according to theory, turned white and yellow in Europe and Asia! Different skin colors presumably developed under "genetic selection" after moving out of Africa.

This is certainly one of the most outrageous parts of OOA, that these African migrants lost their dark skin and distinctive hair and turned into Mongoloids and Caucasians, the original African genome lost to "genetic drift." It is the same bizarre and anti-scientific school of thought that claims Neanderthals at higher latitudes "were probably depigmented."[65] Negroes, say our anthropogenecists, turned into Orientals at a late stage of human evolution,* simply by migrating northeast and adapting to the new environment. Wondrously, Africans changed into Mongols in a few thousand years!

As the argument for lightening runs, white skin appears as an adaptation, the pale coloring of northern Europeans due presumably to mutations favoring a dimly lit environment, thus selecting depigmented skin in cloudy periglacial regions. As for the depigmentation of Africa's own lighter-toned pygmies, this evolved because their campsites were in the shade, just as other Negrito populations allegedly got depigmented in forested regions.

That whites or anyone else owe their light complexion to a process of depigmentation is entirely specious. "All of them propagated after their own kind, and do so to this day. And though the blacks might live for thousands of generations . . . in any country in the world, they would never become whites."[66]

Once again, it is race mixing, not evolution or selection, that explains the trait. After white people (the Hamite sons of Noah) entered Africa

---

*Recent genetic research, appearing in prestigious journals, informs us that "European skin turned pale only recently." But whether six or sixty thousand years ago, this is not enough time to evolve from Africans into an entirely different race through natural selection.

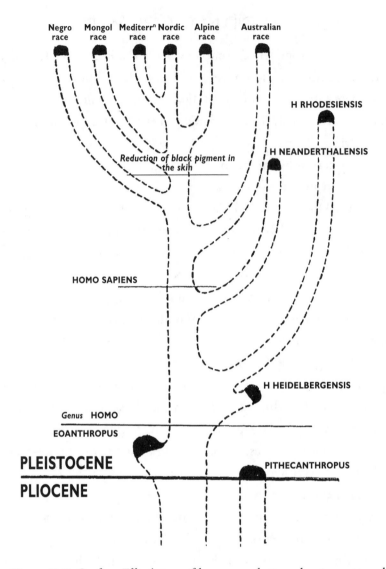

Figure 11.7. Grafton Elliot's tree of human evolution, showing supposed depigmentation of the modern races.

ca 24 kya, new blends quickly arose. Boskop Man, for example, found in the cliff shelters of South Africa, was a splendid Ham hybrid who would have nicely fit the bill as first mods OOA (part two)—only he was too recent! Buried only four and a half feet down, Boskop is dated (unhelpfully) anywhere between 40 and 10 kyr. Although assigned

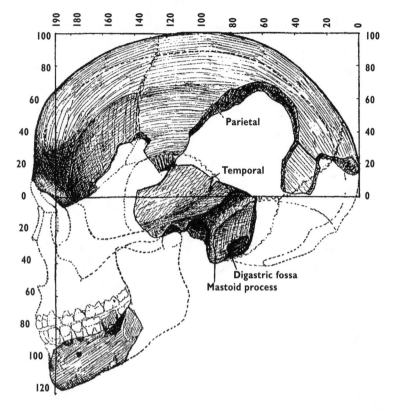

Figure 11.8. Boskop Man skull; Boskop is thought to have had deep tan or yellow-toned skin.

to the Mousterian culture, they were nonetheless "much later in date than the corresponding culture in Europe,"[67] eliminating any African Garden of Eden.

A composite of white Ihin (Ham) and Negroid Ihuan (the latter signaled by his supramastoid crest), Boskop Man was found in 1913 in the Transvaal. His forebrain was 50 percent larger than ours, a total of 1,800 cc! With no prognathism at all (which surprised experts in South African fossils), he was an unusual variety of the modern European type; yet there were primitive traits as well—thick skull wall and very slight chin.

Loren Eiseley was quite taken with these "unique [Boskop] people," whom he overenthusiastically pegged as ultrahuman and the first true men. Boskop was, to Eiseley, "an unknown branch of humanity," possessing delicate and gracefully reduced teeth, fragile jaw, small face, and fine-boned facial structure, all of which indicated "some strange inward hastening of change."[68] But this is nonsense, as is Eiseley's doggerel that Boskop Man was "born by error into a lion country." No error, only an offshoot of those Hamite refugees who mixed so freely with indigenous Africans. Boskop Man's location was on the southeast African *coast*. Isn't this a clue to the landing place of those arrivals from Pan?

Boskop Man's discovery, at first exciting and controversial, was later forgotten. Since 1960, we hardly hear about him. In 2008, John Hawks blogged that Boskop is not any different than a big Bushman or Khoikhoi, with the "large skull of a Khoisan type." Hawks concluded that the uniqueness of Boskop Man "was entirely a figment of anthropologists' imaginations."[69] Just insignificant variations. Once more, the escape clause "variation" conceals the fact of multiple races and mixing thereof.

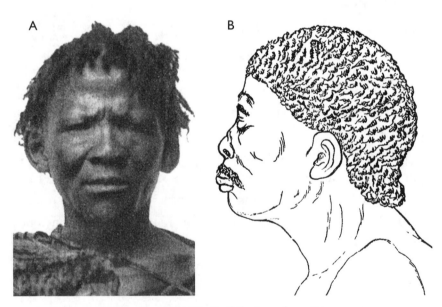

A   B

Figure 11.9. Khoikhois: (A) photo from Dixon,
(B) Swoon/Khoikhoi bust modeled on a living subject.

SEARCH FOR MODS TURNS UP MIXES

Perhaps the weakest point of OOA is the lack of evidence for the first mods in Africa, the people who supposedly left their motherland and did all that colonizing and replacing (part two). This is the crux of OOA: Proponents must prove that Africans were the only ones to cross the *H. sapiens* threshold. The dawn of *H. sapiens* in East Africa, however, is not accompanied by an expected increase in population density, nor is the fossil evidence convincing: Ethiopia's Omo Men, for example, were in reality a mixed bunch, some rather robust; their culture, moreover, remained Mousterian (Neanderthalian). Redated (suspiciously) from recent to 50 to 130 to 195 kyr, Omo Man looks modern enough, but his torus is continuous, and alas, he has no chin. Another Ethiopian candidate, Herto Man, very modern and big brained, was promoted as the (needed) first mod OOA, ignoring, of course, his Neanderthaloid aspects: Acheulian toolkit and thick skulls.

Turning to South Africa, Afrocentrists had another shot at OOA with the fossil men of Klasies River, possessed of modern features, high forehead, reduced brow, and strong chin; but again, all this was mixed in with more archaic and robust features. Held up as the world's first AMHs, Klasies are actually the same age as equivalent types in the Near East (Mt. Carmel).

Located significantly at the tip of Africa, near the beach (port of Hamite landing?), Klasies is contemporary with Border Cave finds (also on the southeast coast). Both represent near modern AMHs, but again, with Mousterian culture. All these near-mod Africans lived at the *same time* as European Neanderthals, whose culture they shared, archaeologically speaking: same flake blades, pigments, fireplaces, limited tool range, little if any use of ivory, bone or shell, and not much hunting.

Klasies's supposedly "complex" behavior consisted of making improved blades, hafting, and using red pigments. So what— Neanderthals used pigments, and the mixed people at Qafzeh in the Mt. Carmel caves had tools and careful burials—at the same time![70]

These AMHs were basically Neanderthal in culture, not just at Klasies, but also in the Levant, the Balkans (Krapina), and the Crimea.

There were no bone/antler tools at Klasies (as used by European Cro-Magnon), which are generally regarded as an index of the true modern type. Also missing here in South Africa are the art and ornament so typical of the *H. sapiens* assemblage. Klasies had neither fishing,* fowling, nor hunting of big animals (the "modern behavior" complex). Their hunting of eland and making of fire have been hyped as modern cognitive behavior, but these hardly comprise the *H. sapiens* explosion we are looking for (and, as we saw earlier, fire use has also been attributed to *H. erectus*).

Afrocentrists strain, making much of little. In fact, a "glaring exception to the record of advanced knowledge," notes Paul Von Ward, "exists in [sub-Saharan] Africa."[71] How then could they be the founders of all that is modern on Earth? Sure early Africans had the gracile limbs, high forehead, and reduced brow of true *H. sapiens,* but these same groups have no chins, others sport robust jaws and teeth. They are just as mixed as Mt. Carmel's proto-Cro-Magnon—and about the same age.

Klasies, let us remember, also had unmistakable cannibalism, as well as sexual dimorphism—the great difference in size between the sexes that is typical of the most archaic hominids: Klasies specimens range from large to quite tiny, the people as a whole showing the most sexual dimorphism ever found in a "modern" population. Well, maybe not so modern, after all. Indeed, with signs of cannibalism and sexual dimorphism, they do not really stand up as a paragon of *H. sapiens*. According to University of Chicago's Richard G. Klein, they are too archaic in frontal arch and mandible to qualify as that paragon.

Klein also doubted the value of Border Cave specimens as first mods—the bones probably not "as old as they say."[72] Swaziland's

---

*Were they fishermen, as some have argued? Shells here may be only washed up.

Border Cave Man, who coexisted with *H. erectus,* shows a whopping 1,500 cc (brain size) as well as a good chin. These people left an engraved object along with rather nice awls, daggers, and beads, hence their "capacity for modern symbolism." Yet there is even earlier symbolic art in India,* and some say that examples of early African "symbolic" works (like covered burials) may be the work of Neanderthals! (Indeed, a few Border Cave specimens have the Neanderthaloid brow.) Even Neanderthals made perforators and discs and prepared pigments. Border Cave, though variously and inconclusively dated at 115, 100, 90, 50, and 30 kyr, may be much more recent (Holocene). It is not clear if the bones (some only four feet down) were contemporary with the sediments in which they were found. South African hominids were found mostly in caves, which are hard to date; these breccias are considered unreliable. Most likely, these specimens represent a post-flood Ham-black mix, judging from their less-than-Negroid features, such as lack of alveolar prognathism. And isn't it interesting that all these best candidates are on South African soil, close to the beach.

South Africa's Florisbad Man is yet another Africa-first candidate but may actually be too recent—his scatter is anywhere from 260 to 130 to 35 kyr[73] or even much younger. Though the Florisbad face is pretty mod, it possesses an extremely broad and flat frontal bone (Neanderthaloid) and very thick skull wall. Florisbad's stone points are Mousterian. Frankly, they can't decide if the Florisbad skull fragment represents a Bushman, a Neanderthal, an archaic sapiens, or Cro-Magnon!

Then, too, there were some kind of early mods at Blombos Cave (again, tellingly, at the tip of South Africa), its date of 20 kya adjusted to 75 kya, thus letting them depart Africa 60 kya, which would be within the theoretical time frame of the modern exodus. But don't bank on the date: the Blombos deposits are mixed with much overlaying of

---

*The cognitive explosion did not take place in Africa, according to other studies: "The chronological and genetic threshold to our modernity . . . at 50,000 years ago [was] in West Eurasia and *after* [e.a.] the out-of-Africa dispersal." (Oppenheimer, *Eve,* 109.)

younger material. Hailed nonetheless as our OOA ancestors, these people ate a varied diet, produced engraved geometric designs, perforated seashells for ornaments, and made delicate bone points: Heating stones to sharpen spearheads is a method called pressure flaking, similar to Europe's 20-kyr Solutrean industry. This technical know-how (even though most of the good points were recovered from a Holocene horizon) earned them the name African Athens—what a hype!

Before OOA came into fashion, Hooton, finding little evidence of Cro-Magnon ancestors in North Africa, saw no "progressiveness and acheivments" comparable to Europe's representational art of the Aurignacian and Solutrean in Africa's Paleolithic west of the Nile.[74]

OCEANIA AND THE OOA THEORY

What light have the Pacific Islanders shed on OOA theory? Has the light even been allowed in? For some reason, historians insist on deriving Oceanic peoples from *elsewhere*. Theory has also decided that Oceania was settled (from Southeast Asia and Sundaland) only in relatively recent time—even though plant (floral) evidence militates against this supposition. The Polynesian races, we might note, seem to combine all types: Caucasoid, Mongoloid, Australoid, and Negroid. So if they came from Indonesia only one thousand years ago, where exactly did they get all those different genes? Dixon fancied the "Caucasic" element in Polynesia arrived there in Neolithic times as some stray branch of Eurasian stock, while J. B. Birdsell fancied a chain of migrations by ancient Negritos across South Asia to the Pacific.

But human life in the Pacific region goes farther back, "more than 30,000 years,"[75] given the ancient argonauts of the Pacific who reached the Solomon Islands 30 kya. Almost matching this date are the remains of 28 kyr starch grains found in the Solomons at Kioly Cave on Buka Island, grains of the cultivated, not wild, taro plant.

"In the Pacific Islands alone," thought Tom Valentine in *The Great Pyramid,* "there is enough evidence of previous civilizations to obliterate the theory that the cradle of civilization was the Middle East" (the statement made in the 1970s, shortly before OOA took the spotlight). Perhaps tens of thousands of years old[76] are the nine feet nine inches high lime-and-mortar cast pillars in the New Caledonian Isle of Pines; here, 400 tumuli (mounds) are three hundred feet in diameter; it was while excavating these mounds, that the pillars were found. Still, historians tell us that the first humans arrived there around 2000 BCE, even though by radiocarbon, the pillars are dated at least 13 kyr.

Today, blondism among these natives is a real clue to the genes of those early Ihins (of Pan) whose distinctive brachycephaly is also evident in the New Caledonian population. Concerning the almost global extent of such "out-of-place" blonds, let us consider the alleged colonization (of New Zealand) dated to a mere thousand years ago: The Maori, though, remember a race of round headed blonds, white little people. In their tradition, very small, fair people were once seen dancing in the sand, people with slight figures and long yellow hair. The Maori said these people had come from Hawaiki, a sunken land; just as Hawaii's (and New Hebrides's) legends recall a fair-haired race of highly civilized men long, long ago.

On Rotumah, too, famed for its amazing massive stone tombs, are found natives of pale skin and notably mild manners. Likewise, some of the Easter Islanders first seen by Dutch sailors in the eighteenth century were yellow- and red-haired people with quite European features. No imagined migrations (or depigmentation) can explain away these *original* genes for light hair, eyes, and skin, belonging to an ancient and lost race.

If Oceania was settled in such recent time, we should not find colossal engineering feats like Rotumah's great stone tombs, or the prehistoric buildings, cyclopean monuments and megaliths, which are found at Caroline, Tonga, Easter, and Malekula Islands—all screaming Mesolithic builders, and about which the natives *know nothing*. Many of the South Sea Islands contain mysterious remains of enormous temples, stone-lined canals and roads, as well as pyramids (the latter at Gilbert, Marshall, Tapiteau, Tahiti, Saipan, Guam, and Tinian)—constructed by an unknown race and built long before man is alleged to have reached Oceania.

We might also ask how the thousands of distinct Oceanic cultures could have evolved in such a short span of time. When Europeans arrived in New Caledonia, for example, the sixty thousand islanders were speaking thirty-seven different languages!* Yet historians say that they wandered over from Southeast Asia (mainland) only recently. They also assume that Papuans came to New Guinea from the Asian mainland, even though a thousand languages are spoken in New Guinea, a land the size of California; in Madang province alone, one hundred seventy-three distinct languages are spoken. I don't think these people came from anywhere, not with such deep linguistic diversity.

Although some do confess that the history of island people such as the Filipino Aeta continues to baffle them, the pro forma solution has them *arriving* through land bridges that once linked their country with the Asian mainland. But all these hypothetical migrations turn their back on (1) the possibility of a lost continent, indeed the motherland of Pan, and (2) the likelihood that these Negritos are aboriginals having evolved in situ—rather than wending their way, by and by, from Africa.

Down under: According to current theory, the first entry (migration) into Australia by our globe-trotting Africans was 50 or 60 kya. "What could have motivated such a trip is almost beyond imagination."[77] Between Sundaland and Australia lay at least fifty miles of open sea. How could travelers from Africa have reached Australia fifteen thousand years *before* reaching the Levant (45 kya), which is so close to Africa? This is, to put it mildly, counterintuitive: How could Australia, the farthest from Africa, have the *earliest* evidence of these migrants? What is more, Australian abos, in their blood chemistry, are as far from

---

*There is also greater diversity of languages in the New World than there should be (see appendix F), which puts the lie to any theory of recent settlement of either the Americas or Oceania.

Africans as possible (and, as we saw, they are racially Australoid with a Caucasoid base).

Australia, besides seriously challenging part two of OOA, also challenges part one, which, you may recall, claims that the earliest wave of *Homo erectus* OOA *stopped* at Indonesia; thus there should be no signs of *H. erectus* farther south, that is, in Australia and New Guinea. But indigenous *H. erectus* types *are* found south and east of Indonesia—at Palau for example, as well as at Australia's Kow Swamp and Coobool Crossing (in New South Wales), where a robust *H. erectus* type mixed with gracile individuals. These were very different contemporary types. So, ad hoc migrations come to the rescue! Instead of taking such coexisting races at face value, evolutionists account for their differences by postulating "two or even three [separate] migrations into Australia. . . . [However] this explanation does not solve the problem," (just shifts it off to the Asian mainland).[78]

It is out of fear (that mulitregionalism could be *interpreted*, used, in racist ways), that the monogenistic model of OOA survives as the chosen wisdom. This was the liberal position (many were Marxist) of the American anthropological establishment—to take Africa as the cradle of mankind, as if that could somehow make up for centuries of black exploitation. Thought to be politically correct, OOA is nonetheless a wrong turn, fed by a weird combination of historical guilt and intellectual sophistry. Marvin L. Lubenow's book, *Bones of Contention: A Creationist Assessment of Human Fossils,* is the one to read for deep insight into the politically motivated OOA theory. In the end we will find that none of this Afrocentrist maneuvering can compensate for the underlying racism of Darwinism itself, which is *still* loath to admit the amalgamation of the races, offering instead the fiction of transmutation of species to account for the different humans in the fossil record.

Who really is the racist? OOA, posing as egalitarian, not only disdains intermarriage (hybridization) but also asserts that the *superior* group, like a conquering army, extirpates all the inferiors in its path (replacement model). Moreover, the Darwinian theory of competition and struggle is (and has been) easily transferred to the idea of racial or national competition, whereby the elimination of unfit races may simply be a part of evolution. Is this not racist? Darwin's own racial bigotry (typical of his time and class) has yet to be exposed, but it is there, hiding in such passages as the one in which he argues for sexual selection, say, of skin color, according to tribal "standards of beauty"; still, he grants his reader that it is "a monstrous supposition that the jet blackness of the negro has been gained through sexual selection,"[79] in other words, that it is a preferential trait.

NEW WORLD NOT SO NEW

Uncensored finds in the Americas support polygenesis, not monogenesis. It certainly upsets out-of-Africa monogenesis to find very early hominids in the New World, of all places. If the textbook version of the peopling of the New World was undertaken by *perfectly modern men* in the very late Pleistocene, say 15 or 20 or even 25 kya, why have crude chopping tools been found in the Americas, dating as early as 42 kyr?[80] Stephen J. Gould himself mentioned inhabitants of our continent who possessed "shortened foreheads, prominent cheeks, deep-set eyes, and slightly apish nose."[81]

This man, *Homo erectus,* as current theory goes, *never crossed into the Americas.*

Nevertheless, the New World has its very own edition of pithecanthropines who always called the American continent their home. What do we make of Utah and Nevada footprints with no arch,[82] a trait belonging to most australopiths, *H. erectus,* and Neanderthals? Consider also a prognathous, beetle-browed specimen from the Southwest dated 27 kyr.[83] Here, the Anasazi people, whose site yields telltale bones with cut marks, had legends of cannibalistic giants (who may well have been large *H. erectus* types), scoring a match with a mysterious tribe of giant red-haired cannibals who once terrorized the Nevada Indians. Indeed, giant mummies have been found in a Lovelock, Nevada, cave—

apparently corroborated by nineteen-inch footprints in sandstone at Carson, Nevada. The Paiutes had a great deal to say about a race of giant "red-headed people eaters."[84]

Neither are Neanderthals supposed to be in the New World, but their tools (fist hatchets) have been unearthed in Abilene, Texas, along with skulls of a dolichocephalic people, similar to specimens in Michigan and Ecuador.[85] All this flies in the face of standard theory, which says mods, of the classically brachycephalic or Mongoloid type, were the very first (and actually only) people to settle America from Siberia.

Roland Dixon, in *The Racial History of Man* (Plate 34), mapped these dolichos in the New World, revealing that long headedness (typical of "archaic culture") is concentrated in two widely separated areas, the largest in the Northeast coastal portions of the continent, including Greenland, and the other at the opposite extreme in southwestern regions. The Southwest, as noted above, also gave us the beetle-browed, hairy, flat-footed type, including some cannibals. Concerning that Northeastern swath, *H. erectus*–type tools (Acheulean), found in New York's Catskill Mountains, are thought to be about 70 kyr—reminding us of similar Acheulian implements discovered in Chile, at the diagonally opposite end of the hemisphere.[86]

The American Southwest also gives us someone less refined than Neanderthal who once inhabited Santa Barbara, California (two skulls). From the coast of California to eastern Texas comes evidence of men with visor brows, depressed nasal roots, and broad noses,[87] exemplified by a specimen from Central California with thick browridge and small brain case, the skull collected by Charles Ostrander in the 1970s.[88]

Earlier in the twentieth century, Nebraska skulls (at Long Hill), were examined by Ales Hrdlicka and proved to be low and receding with strongly marked supraorbital ridges. American physicist William Corliss more recently brought to light signs of Neanderthaloids in Nebraska and Kansas—all typified by prognathism, overhanging brow, and sloping forehead.[89] Very prognathous, too, was Minnesota Woman,

Figure 11.10. Sioux.
Traces of prognathism
are common in the
living races in many
parts of the Americas.

dated 40 kyr, with large teeth (bigger than Neanderthal's) and long arms. She was, however, gracile and large brained: 1,345 cc (a real mix of *H. erectus* and modern people).

Speaking of Kansas's Neanderthaloids and erectoids: There was a report some years ago in *The Kansas City Times* on an apparent race of giants (two skeletons with huge bones) who once lived along the Missouri River. In particular the frontal bone was very low, differing radically from any of the existing races of Indians; too, the torus was continuous and the forehead almost flat, receding back in a dramatically flat slope.

If we look carefully, we find the same archaics south of the border: depicted in Guatemalan ruins is a creature described (rather brutally) by one early observer as a "weak, purposeless, degenerate type, loose-lipped, chinless and imbecile"[90]—perhaps akin to the Solorzano skull cap found near Guadalajara, with its classic *H. erectus* measurements.[91]

Harold Gladwin, in *Men Out of Asia,* also gave us a peek at some primitive skulls found in Ecuador, Brazil, and Chile. These are not one-off hits, for similars appear all over Ecuador at Punin, Paltacalo, and Cuzco, the latter buried one hundred feet below the surface, indicating "the real aborigines,"[92] an early race that may have gone extinct before the Indians. Although the Punin skull was labeled Australoid, I don't think there was a Pleistocene invasion of America by Australoid people, as some have guessed; rather, these are homegrown races, born and raised in the Western Hemisphere, our very own Senor and Senora *Homo erectus.*

As for Brazil, well enough known (since 1970), the Lagoa Santa caves should have settled the matter of early *Homo* in America. Buried among the bones of extinct animals (indicating considerable age), these earliest Brazilians had an almost nonexistent (sloping) forehead and a very wide space between the eyes. With thick skull walls, these people were dolichocephalic, extremely strong, and prognathous; yet they, too, were mixed, boasting a well-developed brain, high skull, and good chin—in short, "a primitive race . . . mixed with other elements."[93] Elsewhere in Brazil, the Sumidouro cave specimens, thick skulled and large browed, not only add to *H. erectus* types in America, but also speak for polygenism, in their near match to equivalent types in Australia (Kow Swamp) and China (Mapa).

Though ignored and dismissed, such types are not rare but known from Ecuador to Tierra del Fuego,[94] where living groups carry forward some of these ancient genes—the subject of the next chapter. In particular, Fuegans can still be encountered with a trace of the old sagittal crest, overhanging brow, and low-sloping vault. Lacking any sign of the tool kit of Old World *H. sapiens* (no rope, needles, lamps, nets, or pots), these people, said Charles Darwin in *Descent,* "possessed hardly any arts and [lived] like wild animals . . . the blood of some more humble creature flows in his veins."

Homo pampaeus in Argentina, announced a century ago by multidiscplinary scientist Florentino Ameghino, possessed a "simian peculiarity of skull"—including very large orbits, so typical of the most archaic races. *H. pampaeus*'s crudely worked and out-sized stones at Monte Hermoso suggest a pithecanthropine of great size. Though Ameghino convincingly argued for an independent origin of South American man, his thesis was shamelessly blackballed by the mighty Smithsonian powers that be. Nevertheless, here in Argentina, the Baradero skeleton was another typical *H. erectus,* with long arms reaching down to the knees. In the same region, the remains of *Homo sinemento* brought to bear one more example of these early blends of AMHs with *H. erectus* types, for the *H. sinemento* people were very prognathous pygmies, dolichocephalic, with no chin at all yet possessed a quite modern dental arch and notably gracile build—an apparent, and very early, blend of Ihin (the AMH little people) with pithecanthropines who "disappeared without leaving descendants."[95]

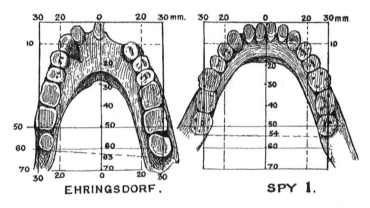

EHRINGSDORF.   SPY 1.

Figure 11.11. Dental arch, after Keith. The more modern shape of dental arch is seen in the Spy I specimen from Belgium.

This presentation of American erectoids is meant to give the reader perhaps a little more confidence in the blacklisted tenet of polygenism: man arising in all divisions of the Earth. Before the 1987 debut of the Afrocentrist Garden of Eden, the parent stock of mankind was thought to be Asian or Near Eastern and of the Caucasian race. Despite the popularization of OOA, a human itinerary out of Africa or out of anywhere else has never been proven. We have now wasted twenty-five years on Afrocentrist monogenism—a real step backward. Almost a hundred years ago, when thinkers were solid and theories were responsible, when things were not so hurried and pressured, Dixon laid out a dependable blueprint: "The whole trend of anthropological investigation can have

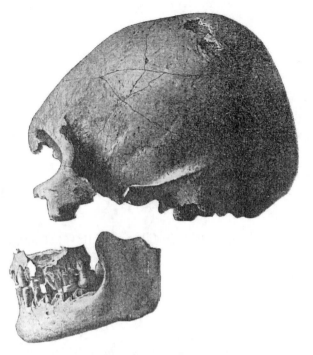

Figure 11.12. Miramar Man, an H. pampaeus
*of Patagonia, shows extreme forehead slope and
ultradolichocephaly but no brow bulge. Miramar Man
was big brained at 1,464 cc and had quite a good
chin—a marvelous hybrid!*

no other outcome than the abandonment of the monogenist position
and the frank acceptance of polygenism."[96] The out-of-Africa theory, as
much a political statement as a piece of science, is a misguided effort to
slough off the Eurocentric image that adheres to modern scholarship.
Are we playing to image now? When image trumps knowledge or truth,
we are headed for trouble.

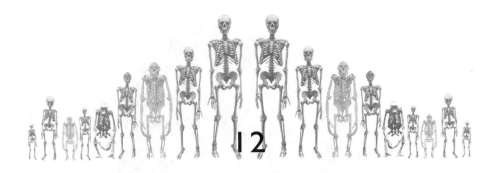

12

DEAD MAN WALKING

Living Evidence of Ancient Races

Man still bears in his bodily frame the indelible stamp of his lowly origin.

CHARLES DARWIN, *THE DESCENT OF MAN*

SWAMPED

Mayan anthropogenesis makes the apt observation that remnants of each World Age survive to the present moment. Yes, early hominid traits still linger—we carry forward even some of the most ancient genes.

Every person alive today has both Druk and Ihin lineage in them.

VERNON WOBSCHALL, OAHSPE STANDARD EDITION

Once crossbreeding is taken to heart, we will be gratefully relieved of the many inconsistencies and difficulties that assail paleoanthropology, including how to explain extinctions; now we can look to EBAs (extinct by amalgamation). How did Neanderthal disappear? Was he wiped out? Is he really extinct? "It is doubtful," thought Ashley Montagu, "whether [Neanderthal] was exterminated anywhere. The truth seems to be that he mixed with whatever populations he encountered and in the course of time was absorbed by such populations."[1] Here we come full circle with "the Neanderthal problem," which is a problem only if we ignore the nookie factor. Montagu made that statement in 1957; it took another fifty-four years for experts to reluctantly concede: "Perhaps they were eliminated by interbreeding with us"[2] (2011). Still, Neanderthals, according to most evolutionists, were "replaced" by *Homo sapiens*. But they bred with *H. sapiens* and disappeared, not really through replacement, but largely through absorption (EBA) into Cro-Magnon groups.

The second wave of Ihuans, the Aurignacians (produced, if you recall, 39 kya) also went EBA, just as they had done in the first round, ca 70 kya—by back breeding. It was, in short, the retrobreeding of Ihuan with Druk, both before and after the flood, that gave us the various Neanderthaloid types. In a way, EBA is similar to the genetic concept of swamping: if Neanderthals crossbred with mods, such as Cro-Magnons, ca 35,000 years ago, their disappearance can be attributed to interbreeding. They were swamped. Hooton was right: "I think they were absorbed."[3] C. Loring Brace was wrong: "Nothing but modern skeletal material is evident since about 35,000 years ago"[4]—which, as we've seen, has been blown out of the water by younger specimens proving EBA, such as the mixed types in Portugal, Romania, Croatia, Czechoslovakia, France, and Spain.

More recent EBAs might include the following cases:

- Dolicocephaly in Europe (swamped by race mixtures with brachycephals from the East) was *not* changed by mutations, but by EBA.
- The Bushman race (except at Kalahari) have been absorbed by the "Cape coloured" people, a mixed people of southern Africa.
- Mandan and other North AmerInd groups, the Wai Wai pygmies

of British Guiana, and many others were absorbed into larger groups, as were most of the Yaghans of Tierra del Fuego and Chono Indians of Chile.

- The Native Hawaiians became so interbred with immigrants to their islands that they have almost vanished as a distinct people.
- The Ainu of Japan are headed in the direction of EBA, driven out, absorbed, or destroyed.
- The original Negritos of Australia, according to J. B. Birdsell, were entirely absorbed by the Murrian people; just as today, the remaining Tasmanians are all of mixed blood.
- The aboriginal stock of New Zealand's Maori were absorbed or destroyed.

NEANDERTHAL: MISSING, PRESUMED DEAD

It was long believed that the coelacanth fish became extinct 70 mya, since their fossil remains do not show up in deposits after that time (Cretaceous). But they are not extinct after all: folks at Comoro Islands, off the South African coast, fish for it and eat it. Likewise did nineteenth-century attempts to dredge up life from the sea, in Norway's fjords and along the Atlantic coast of Britain, reveal that some sea creatures thought to be long extinct still live on the ocean floor. Other "living fossils" include the tuatara (a reptile that, like the coelacanth, was not known since the Cretaceous), *Neopilina* (mollusk, since Devonian), *Lingula* (shellfish, since Ordovician), and *Metasequoia* (redwood, since Miocene).

This subject reminds me of a recently published story about a boy lifted and dropped by a huge bird, the extraordinary scene witnessed by six individuals, including the boy's mother, who said the bird looked like a giant condor, with a wing span about ten feet. Predictably, the experts declared the incident impossible, despite other thunderbird types sighted in Illinois, Pennsylvania, and Tennessee, which seem to resemble the extinct teratorns, made famous by their remains at California's La Brea tar pits.[5]

Numerous studies, moreover, have uncovered evidence of man living at the same time as some of the "bygone" Tertiary beasts: pottery at Tiahuanaco, for example, is adorned with pictures of a *Toxodon*, supposedly extinct for a million years.

As you may have noticed, the extinction date for Neanderthal is a shifty thing, forever moving closer to recent time. At first, the Neanderthal race was said to have *disappeared* around 140 kya; Hooton (before 1950), changed that to 72 kya. Then in 1960, Le Gros Clark made it 50 kya; and by the 1980s it inched up another 10,000 years to 40 kya.[6] That was whittled down to 35 kya soon after; and with new finds at Zafarraya, Spain, it went down another notch, to 33 kya. Then in 1988 Neanderthals' extinction date was made even later—32 kya.[7] No sooner was that established than finds at St. Cesaire, France, and Mladec, Czechoslavakia, knocked it down to 31 kya (these were mosaics, actually, mixed Neanderthal and modern, some with big brows and projecting faces). Even so, it has been further bumped down to 28 kya, after a hard look at Croatia's Vindija cave specimens; and, in the year 2000, experts move the final Neanderthal "extinction" to 27 kya.

But it dropped again to 26 kya, with Predmosti men discovered during renewed digs in the Moravian region; although they try to pass them off as Cro-Magnons, clearly they were mixed with Neanderthals—their dolichocephalic skulls showing orbital ridges and heavy brows. And if they were indeed Neanderthaloid Cro-Magnons (these people did work in ivory), the cannibalism discovered here (bones heavily charred) is a telling sign of Cro-Magnon back breeding and retrogression.

But wait, we now have 24-kyr Neanderthaloids, with the 1998 Portugal finds, thousands of years after the supposed Neanderthal extinction, not to mention 18-kyr Neanderthals at Girona, Spain, and at Amud I (Galilee) and yet other Neanderthals said to date less than 6000 BP![8] How reliable are our concepts of or dates for racial extinctions?

Flores's controversially archaic (but chronologically recent) hobbit was found to be alive and well just 12 kya, which made some observers think that sightings of Yeti-like creatures might indeed have a grain of truth. Maybe cryptozoology could come in from the cold after all. The cryptid Sasquatch/Yeti is described as having long arms reaching to the knees—like hobbits and the Asuans of old, and indeed like most early men: Au, *H. habilis, H. erectus,* and Neanderthal. Long arms, even today, are retained by some groups of *H. sapiens pygmaeus,* such as the Andamanese, Malaysian Negritos, Aetas, and Veddas; in India, a visitor to Bengal in 1824 reported seeing short-statured wild tribes, their arms extremely long, reddish hair on their skin.

WHATEVER HAPPENED TO *HOMO ERECTUS?*

In every major race, both erectus and sapiens forms were represented.

CARLETON COON, *ADVENTURES AND DISCOVERIES*

Just as Mr. and Mrs. Neanderthal lasted a lot longer into the Mesolithic (and perhaps the Neolithic) than is supposed, Mr. and Mrs. Erectus* also survived so late in time that they seem to have coexisted not only with the Neanderthals but even with Cro-Magnons.[9]

The western Pacific (Micronesia's Caroline Islands) holds a rich vein of prehistoric cave remains bearing an extinct population of little people: twenty-five miniature humans, with erectoid features, who lived as recently as *three thousand* years ago on the rock islands of Palau.

Excavated by a National Geographic team in July 2006, these little Palauans, together with their hobbit cousins on Flores, have been the center of a hot scientific debate since their discovery. The tiny Palau specimens, like hobbit, had small faces, large teeth, and a wide gap between their eyes (rather like the feral, completely wild, and extant Orang Pendek of Sumatra). Fairly chinless like hobbit but larger brained, the little people of Palau have caused much head scratching;

---

*Such as 12-kyr Ngandong (Java) and Kow Swamp (Australia) erectoids.

anthropologist Lee Berger and his team were perplexed by Palau man's (supposedly surprising) mixture of traits, some of which (reduced chin, vertical depth of jaw) scientists thought of as prehuman.

But what is so *perplexing* about the *blend* of traits, which pervades every single branch of the family tree? While our little people of Palau do indeed have a hard time fitting into a smooth evolutionary picture, they go right along with the rest of mankind in demonstrating a *mixed* gene pool. A wonderful hodgepodge of traits. Call it what you like—crossbreeding, interbreeding, hybridization, or amalgamation—*such racial mixing has been going on since the beginning.*

If these erectoid Palauans were here as recently as 1400 BP, when did the Druk or *H. erectus* race actually die out? If we go shopping for *H. erectus* dates, there is an ample selection to choose from: The experts tell us that they went extinct 120 kya, or was it 400 kya? Gone from Africa, others say, 500 kya; or was that 300 kya? In more recent books we are likely to hear that they vanished only 60 kya, which agrees with Le Gros Clark's date of *H. erectus* in the Far East less than 60 kya. *H. erectus* is getting younger all the time, like Ndandong Man of Java, with recent evaluation of teeth done of course *since* the upsetting hobbit find (in 2004). Other Javanese skullcaps of *H. erectus* are dated as young as 28 kyr—a far cry from his supposed extinction 500 kya. But old habits die hard and even current books (say, Brian Switek's 2010 book *Written in Stone*) put the *H. erectus* die-off at 200 kya. Something is definitely wrong with these dates. Jeffrey Goodman's chart on page 165 of *The Genesis Mystery* lists *H. erectus* specimens dated as recent as 10 kya!

Histories drawn from Oahspe (Druks persisting up to the time of Zarathustra, 8 kya, when they intermarried with Ihuan's, and the long-nosed ground people, hoodas/Druks, living in Arabinia until around 5 kya[10]) seem to score a match with findings in Australia, where recent *H. erectus* groups date anywhere from 40 to 6 kya: the Mossgiel cranium from New South Wales, the Cossack skull from western Australia, the Cohuna cranium of Victoria, the Talgai cranium of Queensland, and, finally, the Kow Swamp specimens of Victoria.[11] These robust Kow

Swamp people are only 10 kyr, notwithstanding their certifiably archaic appearance: receding skulls, continuous torus, massive jaw and brow, big teeth, large erectoid face. A spot of controversy here: Since their femurs are perfectly *modern* and since the continent of Australia supposedly has never had but *H. sapiens* types upon it, therefore the Kow people *must be H. sapiens* if only because the out-of-Africa theory says *H. erectus* never made it to the nether continent, which was "colonized" only later, ca 60 kya, by fully human types!

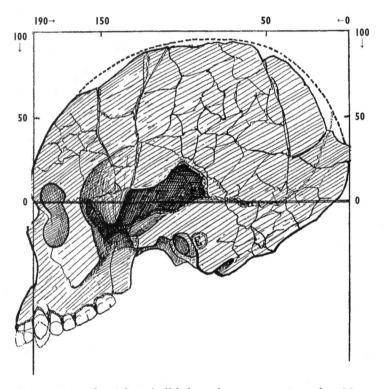

Figure 12.1. The Talgai skull belonged to a recent Australian H. erectus, *dated to about 12 kya, the low-vaulted cranium discovered at Talgai near Brisbane. Talgai Man is similar to today's Aborigines in the face (prognathous), dentition (canines), forehead (sloping), and palate. But the Talgai skulls are above the average dimensions of mod abos, though the glabella and eyebrow region are very thick and the palate, according to Keith, "anthropoid." Mixed genes are the only solution to all such "mysterious" combinations.*

But that can't be right. Those Kow crania had both *H. erectus* and early *H. sapiens* features: the forehead was flat like mods, but it was sloped. The lower face was prognathous and the cranial vault thick; yet the skull is basically modern, with straight sides. So theorists pulled the isolation card: invoking the separateness of the island continent, evolutionists tried to explain (away) Kow as "peripheral isolates"—a remnant population.

That didn't work, though, when the same kind of people were found two thousand miles away on the west coast of Australia: the Cossack material. This archaic morphology was not a regional freak but continental in distribution. And when they tried to explain *that* away by nonevolutionary factors, such as inbreeding, cranial deformation, anemia, malnutrition, "the admission that one or more of these [nonevolutionary] factors could produce a *Homo erectus*–like morphology is also an admission that the concept of human evolution [itself] is not needed to explain that morphology."[12] Or any morphology.

DEAD MAN WALKING

Archaic traits in *living* races puts paid to evolution and extinction, indicating instead the steady absorption of bygone groups into the mainstream of the more modern phenotype. Some members of *all* of today's races, thought George Frederick Wright, show archaic traits. For example:

- France still has a remnant of the large Cro-Magnon who supposedly died out about 11 kya: the tallish, ruggedly handsome, dark-skinned people of the Dordogne Valley, possessing the peculiar Cro-Magnon skull form.
- The platycephalic (low-domed) skull of early man is still evident in some people of Holland and England.[13] Skull shapes in mid-Wales also show some Stone Age measurements. Some Lapps and Finns have the (Neanderthaloid) occipital bun. The dolichocephaly of many North Europeans is also an archaic leftover.

- Marcellin Boule found the Paleolithic Grimaldi skull type among some modern people of Italy, "explained by the facts of atavism."
- According to Franz Weidenreich, Java's *Pithecanthropus* and *H. soloensis* agree with today's Australian Aborigines in details such as strong browridges, receding foreheads, and angular skulls. Australian and Tiwi Aborigines show a low cranial vault with some keeling at the top of the skull (sagittal crest); Tasmanian and Australian abos are similar to 13-kyr Keilor Man of Victoria.
- Weidenreich also noted a certain continuity that bridges Peking Man (Sinanthropus) with today's Mongolian populations, who also share a short tibia with Neanderthal. Peking Man's upper incisors were shovel shaped, as found in almost all of today's Mongoloid peoples. According to Chinese archaeologists, Peking Man's skull has some characteristics shared by modern Chinese people, particularly the low nose bone and flat cheeks.
- Inuits and Aleuts also compare with Peking Man in thick mandibular torus and robust cheekbones. Peking Man's bony outgrowth at the inside of the lower jaw is found in most of today's Lapps and Inuit. Laplanders and Greenlanders also have a heavy overarching brow and the nonoverlapping, edge-to-edge teeth bite of earlier races. A vestigial sagittal crest (bony ridge along the top of the head, as in Zinj) is found among some AmerInds, Inuit, Chinese, and Australians.
- Mincopies (Andamanese) bones are pretty thick with conspicuous muscular impressions; they share with the Bushman and Khoikhoi the archaic trait of steatopygia (bulked up hips and buttocks).
- Bushman nails are considerably curved like Lucy and some habilines who had curved hands and feet, which helped in tree climbing. Other unusual or archaic traits among the Bushmen include labia majora and knob nipples, not to mention the primeval wide-set eyes.
- The Mbuti of Zaire are pygmies who "in stature, brain capacity, and even way of life, are comparable to *Homo habilis*." Yet the Mbuti "are modern men in every sense except that they do not watch television

documentaries nor receive grants from science-funding bodies."[14]

- The Ona Indians of Tierra del Fuego are of primitive appearance in their low sloping cranial vault. Fuegans and Patagonians (recent skulls) show the jutting browridge and some sagittal keeling.

Figure 12.2. Modern man of Dordogne retains his Cro-Magnon heritage. It is also thought that today's Basque people are morphologically the same as 20-kyr Cro-Magnon Man.

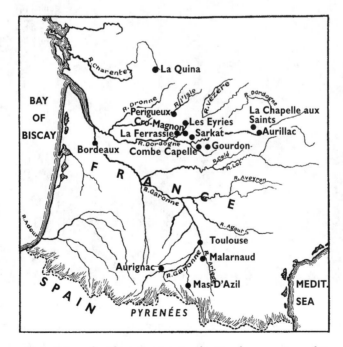

Figure 12.3. Map of prehistoric sites in the Dordogne region of France.

Figure 12.4. (A) Tierra del Fuego people drawn by Captain Fitzroy, Darwin's superior on the HMS Beagle. Three of these Fuegans were brought back to England and presented to the Queen. (B) Fur-clad Ona Indians of Tierra del Fuego in the 1890s.

Figure 12.5. An Ona Indian.

Figure 12.6. This woman, Trugannie, was one of the last of the full-blooded Tasmanians (1875).

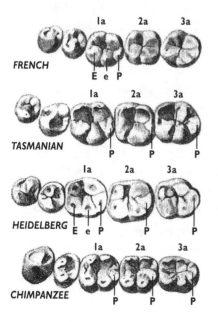

Figure 12.7. Tasmanian molar, compared to lower molars of chimpanzee, Heidelberg Man, and a modern Frenchman.

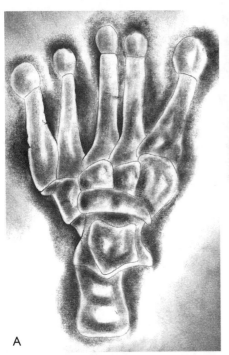

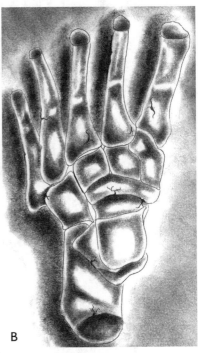

A

B

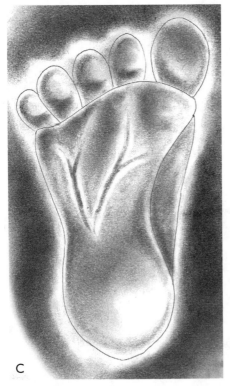

C

Figure 12.8. Splayed big toe as seen in: (A) Olduvai Au child (B) modern Bushman (C) Italian Neanderthal. A feature of the archaic body plan, the divergent big toe is extant in the gene pool of H. sapiens: *Bushman, Negrillos, Veddas, and Malaysian Negritos possess the big toe projecting sideways. The feature was suited to the tree-climbing, branch-grabbing early races, such as the australopiths (Au. afarensis, Sterkfontein, OH 8, etc.). Next in line to inherit the divergent digits was* H. habilis; *then, Neanderthal; then Chancelade man of the Magdelenian—all seen with the divergent great toe.*

Thomas Henry Huxley felt that those recently discovered Neanderthals (at Feldhofer, 1856) were little more than an extension back in time from the living Australian Aborigines, the latter regarded in Victorian times as the most brutish of living humans. Australian and Tiwi prognathism, large teeth, and depressed nasal root appear to be holdovers from Neanderthal.

Speculation about extant Neanderthaloids did not end with the Victorians. A recent issue of *The Barnes Review* highlights the work of Stan Gooch, who said that Jews can trace "integral components" of their heritage to the Neanderthals who, he maintains, were not destroyed by the Cro-Magnon culture but were incorporated into it through inter-breeding.[15] Indeed, at the end of the Paleolithic, the Ghans back-bred with the Druks, accounting for Neanderthal traits in Neolithic mods of the Middle East and beyond.

Gooch went on to speak of a Neanderthal race *still* living in Central Asia today, contradicting out-of-Africa enthusiasts who assure us that Neanderthals went extinct long ago, without issue, and with no par-ticular relationship to moderns, for they were "completely replaced."[16]

A Trip to the Museum of Natural History

Figure 12.9. Shirley and Max. Cartoon by Marvin E. Herring.

Modern Asian features have also been compared to late archaic hominids, the beaky Neanderthal face, for example, still notable among the living people of India and peripheral Mongoloids such as the Nagas of Assam. Some anatomists have remarked that the splayed big toes and nasal cavity of today's Bushman are the same as Neanderthal's, as is the molar formation of the Bantus. In addition, Congoid people typically have the distinctive Neanderthal occipital bun (protuberance at back of head),[17] which is also found among the Finns and Basques. It seems that other aspects of the Neanderthal cranial type persist in today's Europeans, some of whom "bear the traces of their remote Neanderthal ancestry . . . in their rather heavy browridges, deep-sunk eye sockets, receding foreheads and weakly-developed chin,"[18] seen especially among Nordics, as noted by Hooton and Sir William Turner, anatomist. According to Arthur Keith, "traits which may be called . . . primitive are of frequent occurrence . . . still among South Europeans."[19] Indeed, the likelihood of extensive hybridization in the Mesolithic makes it less surprising that some Central Europeans today have in their posterior skull, femur, and nose a structure reminiscent of Neanderthals.

HOMO FERUS

The first amalgamation that got the ball rolling, the Ihin-Asu match of old, was only the beginning: many forbidden matings were to follow, and some of the resulting hybrids did not entirely disappear. Roman historians alluded to "wild races"—were they Neanderthals? And what about today's acclaimed Sasquatch of the Northwest or Yeti of the Himalayas? The abominable snowmen, a hirsute race of humanlike giants, are neither bears nor apes, their faces coarse but manlike, with sad expressions. Some analysts think Yeti might be a relict *Sinanthropus* (Peking Man), having survived in the inaccessible Himalayas and in parts of Central Asia. Jim Dennon portrayed Bigfoot as a subhuman creature, 82 percent beast—the largest hominid. Carleton Coon wrote in his autobiography, *Adventures and Discoveries,* "I am sure that both these bipedal primates [Yeti and Sasquatch] exist." In China, Coon heard stories about wild men, covered with hair, seen in the Yellow River Valley; there have been similar reports from Kiang-Si province.

Cryptozoology is an active and valid field, probing cryptids like the Almas of Central Asia (Gooch's Neanderthals?) and their cousins in the Caucasus, Azerbaijan's white-haired Kaptar. Cryptozoologist Loren Coleman has followed these unknowns among AmerInd, Asian, and African tribes, folk traditions about them going back thousands of years. Usually described as possessing erectoid-like features, such as short legs and long arms, cryptids have also been depicted with bull necks, fanged canines, and heavy browridges, answering in every detail to the (supposedly extinct) hominids of the Paleolithic.

We see hundreds of species in our modern world who are in fact survivors of previous Earth epochs.

VINE DELORIA JR., *RED EARTH, WHITE LIES*

White skinned and black haired is the rumored "ape man" of Africa's Gold Coast—reputedly fourteen feet tall,[20] (these giants perhaps related to the unknown makers of the enormous biface tools found in southern Morocco). On the short end of Africa's spectrum is the man-beast of Tanzania, called Agogwe, a russet-furred manimal, who authorities speculate might be a survival of *Homo erectus,* not unlike cryptids seen in Zaire, Sudan, and Senegal. When these odd "little men" called Agogwe*—reported by many credible witnesses and also an intrinsic part of tribal lore—first came to the attention of academia, they were considered some sort of missing link.

Agog, in Panic, meant "without sense or reason" (as in the later name M'agog). The suffix *-we* in Agog-we indicates "race of people," such as America's Ongwe, Africa's Zimbabwe, Tchokwe, and Mpongwe, and the Arawe people of Melanesia.

Twice as Tall?

Hand axes from sites in Ethiopia, Morocco, and Syria (at Ain Fritassa and Sasnych) are "far too large" (ten pounds and more) to be hand-held by people of today's world; the bifaces from Agadir, Morocco, weigh in at 17 pounds, and their use requires the finger spread of a 13-foot tall man. Ordinary flint tools of the same type weigh less than a pound. They certainly suggest a bygone race of much greater size. "Mortals became twice the size of men today . . . an imperfect giant stalk . . . and the time of generation of mortals had risen. Many lived to be 300 years old."* Theopompus knew of people twice as big as normal men, and they lived double our age. Explorer and author David Hatcher Childress, on Nan Madol in the Pacific, learned of a human femur found in the jungle that was twice as big as a normal man's.

---

*Oahspe, Synopsis of Sixteen Cycles 2:13.

But the only thing "missing" is a frank understanding of crossbreeding: sexual congress enjoyed by the disobedient Ihuans with the ground people, the most common type of exogamous unions documented in these chapters. Africa, thought Loren Eiseley, "like other great land areas, has its uneasy amalgams."[21] It was here, in 1888, that Henry Morton Stanley encountered pygmies with "fell" over their body, almost feathery; just as the first white men to meet Africa's Twa people in the nineteenth century said their whole body was covered by thick hair, almost like felt. To this day there are pygmy tribes among whom infants are born with reddish lanugo.

> *The early progenitors of man were no doubt once covered with hair.*
>
> CHARLES DARWIN, THE DESCENT OF MAN

Other furry tribes include the hairy Ainu of Japan. Caucasoid and modern though they are, the Ainu yet retain the extreme hairiness of the archaic type. (They themselves must be Darwinists, for they claim their furriness came from their ancestor, the bear!) One story out of China recalls the Chou Dynasty's brain trust (twelfth century BCE), one member of which was so hairy that no one could see his skin (see figure 12.10).

We are aware of some strange creatures of this type quite a bit closer to home. Jacko for example, seen by a huntsman in British Columbia, was a pint-sized fellow who "resembles a human being" but is covered with glossy hair and has extraordinarily strong arms.[22] Bernard Heuvelmans (Mr. Cryptozoology) reported on another wild type known as *didi* in Venezuela and British Guiana, who answers to the description of Jacko; short and powerful, with human features and reddish brown fur, didi resembles, in turn, the Sisemite of Central America, who are covered in black hair, which grows almost to the ground.

BATS, BATAKS, AND BATUTUTS

First man, Asu, with his long arms and hairy body, left a few genes to the atavistic Bat people of Brazil, as well as to other groups of absolutely wild men, *Homo ferus*—all considered dead ends on the human family tree. But a few did survive.

In the iconography of our forebears, such beings with hairy bodies, deformed bones, snarling fanged mouths and claws are not symbols of anything or the mythic fantasy of unschooled people; rather are they remembrances of the monstrosities that were not corrected until Apollo's time, eighteen thousand years ago.[23] A few probably survived to a much later date, though deeply ensconced in the hidden places of Earth.

A. F. R. Wollaston described a relict people he encountered during his daring expedition to New Guinea's interior a hundred years

Figure 12.10. A European conception of a wild man. Note his short stature and shaggy coat of hair, probably akin to the hairy dwarves (farfadets) featured in the folklore of France and the classical Irish grogach, hairy and naked (unlike the clothed and civilized leprechauns).

ago: small, short-legged, broad-faced, short-skulled, and very hairy, the description reminding us a bit of the Bogenah tribe still living in the heart of Panama's Guaymi country: very short, extremely primitive, and rather like the early races with their strong, bowed legs, long arms, large hands and feet, low and receding foreheads, flat and bridgeless noses. Other explorers of Panama have documented "an animal which appeared . . . to be a cross between a [human] . . . and a gigantic ape . . . [it] walked erect, weighed possibly 300 pounds and was covered with long black hair. . . . The Indians from Ecuador to Nicaragua assert that [such] creatures inhabit isolated jungle-covered mountains."[24] Among these little known races are Ecuador's *shiru* (four and a half feet tall and fur covered) and the little *dwendis* of Belize, no taller than four feet, well proportioned but hairy with rather long arms and yellowish faces. Loren Coleman has documented and illustrated various relict pygmy types: hairy and swift, like Mexico's alux, sehite, and sedapa, most with thick black or red fur and an "ancient look to their face"; very capable tree climbers, they are naked and give off strange vocalizations.

Even though Asu man is long extinct, his genes live on in all of us, though most notably in remote groups who, not intermarrying with outsiders, kept recycling some of that ancestor's genes. Remnants of hairy, brutal, and cannibalistic people still exist in the interior of South America, and they are greatly dreaded. Known to the Spanish as Cabelludos (hairy ones), to the Portuguese they are Morcegos (bats), based on their custom of hiding during the day and hunting by night. The Indians call them Tatus, or armadillos, from their Druk-like way of *burrowing in the ground,* in holes about twelve feet in diameter, roofed over with branches.*

Pliny the Elder, in the first century, furnished details about the

---

*The Bushmen sometimes sleep in hollows, shallow depressions in the sand lined with tufts of soft grass, like the scooped nests of shore birds on a beach; here the people can lie curled just below the surface of the plain to let the cold night wind blow across the veld and pass over them.

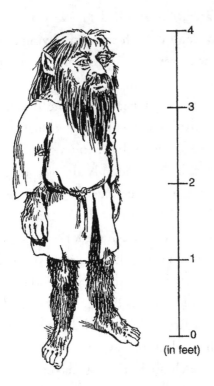

Figure 12.11. Alux of Mexico. Courtesy of Loren Coleman, author of The Field Guide to Bigfoot, Yeti, and Other Mystery Primates Worldwide *(1999).*

"dreadful dwellers" of the Sahara, troglodytes who had no language but "batlike squeaks" and lived in holes in the ground—as the Matmata of Tunisia apparently still do. The Yaks also burrowed in the ground "like beasts in the forest." Relict creatures answering this description may still be around. Some Negritos in Malaysia know of a group called Batak, said to live around the headwaters of the Plus, who are cannibals and dwell in "burrows in the ground."

The 1889 *Proceedings of the Royal Geographical Society* reported on a wild man, perfectly nude, captured in Madagascar while asleep in the branch of a tree. He resisted capture violently, a powerfully built man, his body thickly covered with long black hair. Possessed of a strange gait, he traveled very fast, sometimes on all fours. After a few weeks in captivity, he began to communicate; it seems he had a father and brother in the forest, and when the authorities went to capture them, they "jumped from tree to tree like monkeys." The poor captured man died five months later.

Here in Madagascar near the erstwhile home of the noble Vazimba (white pygmies), wild men, small of stature (called *kalonoro*) are known to inhabit the woods around the Marojejy Mountains, deep in the forest. Frightening and bizarre (*hafahafa*), they have long flowing beards and glowing eyes. Whence came these awful little men? We learn from Hindu scriptures "there was born into the world evil offspring" prior to the flood.[25] And while their height and beards speak of Ihin (Vazimba) genes, their fierceness puts them with the tribes of Druks.

In Sumatra live manimals in the upland jungles of the Kerinci Seblat.

> *[I]t was not a man. It was not an orang-utan.*
>
> MR. OOSTINGH, QUOTED IN *FORBIDDEN ARCHAEOLOGY:*
> *THE HIDDEN HISTORY OF THE HUMAN RACE*

Hundreds of witnesses have seen this actually gentle creature called Orang Pendek; he does not inspire fear among the natives. Little and broad shouldered, with dark fur and short legs, he is so powerfully built that (villagers say) he can uproot small trees. He is described as having a gorilla-like torso; yet the face is human. Analysis of Orang Pendek hair finds nothing but human DNA.

Such erectoids are known not only in Sumatra but in Indonesia, Malaysia, southern Asia, and Oceania—always in the most remote rain forests. Such were the hairy Ebu Gogo on Flores, four and a half feet tall, still alive in the seventeenth century, maybe even up to the twentieth century, though no longer seen. The Burmese Batutut (aka Nguoi Rung) are four feet high with long hair, their counterparts seen also in Vietnam and Laos. A related creature is known in Yunnan province, China: four feet high, hairy, with a human body.

Are these *Homo erectus?* Even today, some natives of New Caledonia show *Homo erectus* traits, especially in teeth, browridge, and sloping forehead. The pithecanthropine forehead and supraorbitals have not been entirely lost.

AT THE END OF THE DAY

"Perhaps generations of students of human evolution, including myself, have been flailing about in the dark. . . . The theories are more statements about us and ideology than about the past," confessed David Pilbeam, Harvard anthropologist, in his 1978 review of Richard Leakey's *Origin*. The official explainers of man have "explained" (or rationalized) human evolution six ways from Sunday; but it doesn't fit together into a coherent whole. So many out sequence specimens (appendix E) make a good chunk of the fossil record an *anomaly*—these inconsistencies whitewashed by rhetoric and forests of obfuscating jargon.

Today, evolution stands supreme not because it is true but because we live in an age that expects a scientific explanation for everything.

> *The . . . world has been bamboozled into believing*
> *that evolution has been proved. . . . The prognosis for*
> *Darwinism is now very poor. . . . Evolutionary theory has*
> *reached an impasse.*
>
> FRED HOYLE AND N. C. WICKRAMASINGHE,
> *EVOLUTION FROM SPACE*

Synthesis is desperately needed: *Thesis,* before the nineteenth century, was biblical genesis; *antithesis,* with Darwin, was descent with modification. Synthesis must surely follow, for the winds of change are a'blowing. But the tug-of-war between evolution and design has no end in sight, because both sides are right—and wrong. Science and religion are actually inseparable. Science at war with religion is man at war with himself.

Until the seventeenth century, it was widely held (largely due to religious ideas) that Earth was at the center of the universe. Until the twenty-first century, men believed (largely due to antireligious ideas) that they were descended from apes. Today, methinks, people want the real deal, not the rigged consensus of official opinion. Science today is unconscionably politicized, theory-driven, funds-hungry, and agenda-bound; no longer a quest for truth but for a place in the academic (corporate or governmental) sun.

Evolution has enjoyed its monopoly long enough; like all monopolies, it is a ticking bomb. A professional giant with feet of clay, it is one of the greatest scientific snow jobs, or in Richard Milton's words, "the most pervasive myth of the 20th century."[26] We've been mushroomed—kept in the dark and fed crap. Our experts have been on a hunt for something that doesn't exist and never did (the missing link). The link between man and ape hangs in the air. The value of natural selection hangs in the air. The cause of racial variability hangs in the air. The origin of mind hangs in the air. The identity of the common ancestor hangs in the air. The causes of extinction hang in the air.

Evolution is a bubble, along with many other scientific bubbles, getting ready to burst—my next book! The AMH genotype, they say, has hardly changed in the last 80 kyr, and indeed, many agree that the physical evolution of man is finished. The brain is in stasis, actually decreasing in size, shrinking for at least the past 10 kyr. If it is true that we are no longer evolving, perhaps it is also true that we never did evolve in the first place. In any event, now we must adapt in a different way: If evolutionary theory was based on advantages achieved by *particular* (competing) groups, now is the time, having entered the universal age, to look for that which is advantageous *for all* (really, the reverse of competition).

If we are still a work in progress, it is not physical; it is spiritual and ethical. Ahead of us lies the challenge of learning how to assimilate to a larger entity than family or tribe or nation. The whole world is our pasture and the forecast is amalgamation—and a new chosen people—as foreseen in the Kosmon prophecies: "In one, tallness; in another, shortness; in one, sound teeth and bones, and well-formed limbs; in another, sagacity; in one, a dense population and well-tilled lands; in another, plain food and long life; and in kosmon, man shall go abroad into all countries, one nation with another; and they shall profit by wisdom, to bring forth *a new race* [e.a.] with all the glories selected from the whole. . . . In this era, I do not come to an exclusive people, but to the combination of all peoples commingled together as one people. Hence, I have called this, the Kosmon Era. From this time forward, My chosen shall be of the amalgamated races, who choose Me. And these shall become the best, most perfect of all peoples on the earth. And they shall not consider race or color, but health and nobleness as to the mortal part; and as to spirit: peace, love, wisdom and good works, and one Great Spirit only."[27]

Every possible combination of human traits has been examined in this book, enabling us to look at the fossil record without the obligatory evolutionary lens. Let the chips fall as they may. The picture that emerges of its own accord is one of amalgamation. And with it, with this amalgamation, comes the final blow to racialism, which, as I see it, includes *all* forms of excessive, demonstrative, group pride. It is in trust and collaboration and reaching out that man of the future will be tested. The "master race" is yet to come.

> *To no one race can the palm be arrogantly assigned, rather*
> *to the product of the blending of types. . . . In the history*
> *of mankind . . . amalgamations between two or more of*
> *the great fundamental types have occurred . . . mark[ing]*
> *the progress of the race. . . . May we not in this see a*
> *prospect of a still nobler growth of all that makes for the*
> *best in man?*
>
> ROLAND DIXON,
> *THE RACIAL HISTORY OF MAN*

Realistically, man for his part, tends to be more cyclic than evolutionary; our history on this planet has been rise and fall, rise and fall. But this trajectory, overall, *is* progressive.

Neither have I given progress to a stone, a tree, nor to any animal, but to man only have I given progress.

OAHSPE, BOOK OF FRAGAPATTI 11.18

What is progress, what is improvement? Instead of *physical* evolution or speciation, we are in the throes of cultural and moral evolution as well as racial amalgamation, moving toward the destined harmony and oneness of the Human Race. The present uptick in amalgamation of the races is the signal that heralds this Dawn of universal man. Apart from technological advances and creature comforts, mankind's true evolution is in Goodness, similar to what anthropologist G. H. Williamson once called the Kingdom of Good Judgment.

Adam Sedgwick, Darwin's geology mentor (who much to Darwin's chagrin did not buy the "cold atheistical materialism" that drove evolutionism), saw "a moral and metaphysical part of nature" linked to the physical. Today, the sciences ignore this link: "Were it possible (which, thank God, it is not), to break it [the link], humanity would suffer a damage that might brutalize it, and sink the human race into a lower grade of degradation than any into which it has fallen."[28] In quest of an objective science of man, evolution has given us raw bones, leaving out the best and most distinctive aspects of our lineage.

Evolution is just one of many secular "isms" that have been popularized in our day to take the place of the unseen powers. The cold eye of science can tell us the how; the warm heart of faith can open our understanding to the why, the purpose. Just as progress brought us from an age of superstition and dogma to an age of science and reason, the next phase promises to be one of insight—the quickening of human understanding. The coming years are tasked with revealing, little by little, much of the spirit side that is entirely natural. In his outstanding book, *The Genesis Mystery,* Jeffrey Goodman noted that "Nobel laureate in physics Eugene Wigner, analyzing the implications of quantum mechanics, believes that man has a nonmaterial mind that can influence matter. These striking concepts cannot be ignored in our search for man's origin."

Darwinism has made its case against the uniqueness of man, evolutionists on every hand parroting that we are not special. But we are special. Add to the animal, vegetable, and mineral kingdom, the kingdom of man: a stunning cross between beast (corpor) and angel (spirit). Johann Friedrich Blumenbach, the father of physical anthropology, and others, like Cuvier who inaugurated the study of paleontology, did indeed give man the dignity of a separate order of being.

> *Distinct and different from all other animals . . . he has a force or soul. . . . This great gift has been bestowed on no other form of life . . . [for] man is a separate and distinct creation, possessing a divine force. It is impossible that he can have come out of, or evolved from some animal not having that force.*
>
> JAMES CHURCHWARD,
> THE LOST CONTINENT OF MU

> *Higher forms of humanity await future emergence. Our race is evolving to a godlike state. . . . Everything in the cosmos is the work or play of God the Creator-spirit.*
>
> JOHN WHITE, ENLIGHTENMENT 101

Evolutionism has made man's critical attainment hunting and technology, focusing on what tools they made, what plants they chomped, what animals they exploited, what predators they avoided. Fine, but this is only the beginning of insight.

Mary Leakey, writing of her life work in the "trenches," said she was impelled mostly by curiosity. Let us extend that sentiment to curiosity about both this world and the next. Both the seen and the unseen. Without the two, man's story cannot be fully told. There will always be that *gap*. Oh, things can be scientized, but only up to a point. And haven't we picked over the bones long enough? After all, there are no new human species to find—just new combinations, owing to incessant race mixing. Maybe it is time to climb out of the pit, fold the tent, and give the fossils a rest. Rather than excavating more bones, we might excavate our minds, for the lost spirit that is so deeply buried. Evolution in its present form is stuck in the pit, the trench of materialism. Nor will the light of day shine on a philosophy that craftily uses assumptions to prove *itself*.

We might have to retrace our steps. We westerners, especially in the English-speaking world, have too many theories. We need greater discernment, even wisdom, not more theories. We need to choose, to decide, whether it is chance or purpose that governs this world. This is not a matter that will, at the end of the day, be settled with more theories or experiments or computer runs even with proof. It will be settled with understanding. Norman Macbeth said: "The effort devoted to explaining should be diverted to contemplating."[29] What is man evolving toward? Rather than competition for life and survival, organisms in the real world help each other, accomplishing together what cannot be done separately. The same is true for us: man alone is weak. Where we came from has a lot to do with where we are going—toward a peaceful, humble, reciprocating, trustworthy coexistence. This is the evolution I believe in.

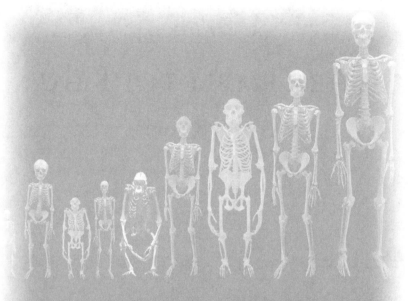

APPENDICES

THE TEMPLES OF KHU

A Word Study

It was prophesied, before the submersion of Pan in the Pacific that "the temples of Khu . . . shall sink to rise no more . . . in the deluge."

OAHSPE, SYNOPSIS OF SIXTEEN CYCLES 3.28

Although the fawn-colored little people of Africa's Kalahari Desert are called Bushman (by the Dutch) and San (by the Khoikhoi), their own name for themselves has the telltale Khu-: !Kung or Khu'ai—which also happens to denote their *tablier egyptien,* a rare anatomical feature they hold in common with the ancient Egyptians. In days past memory, refugees from Pan (and the Khu cult) did indeed inhabit more northerly portions of Africa. And it is here we find remnants of that cult in the Egyptian rite of Khu, which propitiates the dead. Translated from the Book of the Dead, Khu denotes "the light wherewith souls are clothed" or "intelligence of the spirit." The like-named Khusar is found to be the ancient god* of Byblos, the first to teach the arts of divination.

Khub means "the lord" in Khoikhoi, and Ku is a Hawaiian creator deity. *K'uh* also means "deity" in Mayan, and Kumpara means Creator among the South American Jivaro. In North America Kumush is the god of the Modocs (AmerInd) and Kumastambo is the name of the Creator among the Mohave Indians. The Blackfoot culture hero is Kutoyis. There are also the Kutenai Indians in British Columbia and the Kusseta Indians of Oklahoma.

Gospel led the Egyptians to name their temples of light Khuit (Gizeh) and Khufu (Kheops). Wadi Kubbaniya is the site of Egypt's oldest farming community, and Kurru is an Egyptian site with tombs of ancient kings. This tradition is echoed in Java's pyramid site Sukuh, as well as in the Americas, where the Mayan pyramid of Kukulkan (at Chichen Itza) is considered the house of the Great Spirit who is named Hunab-Ku (or Kukumatz). Its door is decorated with short, bearded figures, suggesting the sacred tribes of the little people. The Mayan word for home (on the mounds) is *ku*. Their white god came from the "heart of the sea," the former land "where whites and blacks dwelt together in peace." The Maya's great benefactor* Kukulcan is said to have arrived from the West, founding a new culture, and these survivors landed at Kuhualcan. Kucican is a Mexican ruin in Yucatan with ancient causeways. At Palenque, the royal mother of the land is Zac-Kuk.

The Kuna Indians are in the San Blas (white Indian) area of Panama, their curious pictographs similar to Easter Island *rongorongo* script, which, in turn, has been compared to Chinese ideograms; interesting, then, that *ku-wan* means "ancient pictures" in Chinese. Ku'u-ku'u is the name of a king's son on Easter Island, where Kuihi-kuaha is a magical invocation, taken from the names of the two gods of the original Pacific homeland, the latter sometimes called Ku-mari Nadu (Land of the Sons of God).

Because the name *Khu* was carried so reverently from Pan in Noe's (Noah's) great diaspora, we are not surprised to find it also in situ, at surviving remnants of that lost land such as Kusai Island in the South Seas, as well as in adjacent territories, like Japan with its carvings on the Kusabi rock (at Yonaguni, see prologue). Other Japanese *ku*'s appear in Kumamato on Kyushu and Kume Island off Okinawa, or embedded in names such as Okuninushu, a Japanese semi-divinity, "Ruler of the Land" and in Kun ("King"). Reminding us of Egypt's rite of Khu is Japan's Mizuko Kuyo ceremony.

---

*Caribbean Arawaks say their benefactors were named Kurumany and Kulimina, those who created a new type of man and woman.

Indeed, the Hawaiians also practice a ritual called Ku, honoring the phases of the moon; Hawaiians name Ku as a divinity and, by extension, kuhina are ministers of the chiefdom. There are majestic remains at Kukii in Hawaii. The name of the Hawaiian creation story is Ku-mu-lipo, while Kumu-Honua is the name of the first couple. In New Zealand, Kupe was the first king of the Maori, whose legendary hero is Venuku; the Maori word for chieftain is kura, rather similar to the Quechua kuraca.

Also analogous to Egypt's Khu rite of the dead is Siam's khu'an, or guardian spirit. Here in Thailand the holy district of Khu Mueang appears to be eponymously named, as is Du-ku, the sacred mountain of Turkey, as well as Khuzistan, the great religious center of southwest Persia (Elam), where Kudur Mabug was king of the Elamites; just as Kuresh (Cyrus) was the first Achemenid king of Persia. Khunik Cave, in Iran, also comes to mind, as well as the Mandeans of Iraq who spoke of an earlier ideal world called Kushta. Include here the Khushana Empire of long-distance traders. Other Iraqi centers of peace and learning include Baghdad's Khujut Rabua (with ancient batteries), and Mt. Kurkura, landing place of the ark in Chaldean legend. Consider also Kuyunjik, where royal archives were kept at Nineveh, and Kuwait, and even the holy Kuran (variant of Koran or Qur'an).

The oldest people (legendary) of western India were known as the Kurus; the Kuru were an ancient people living along the Ganges, in the epics; the Kuruba people were also an ancient tribe. Kubera is a flying hero of the *Ramayana*. A ku-named center of learning in India is mentioned in ancient Hindu scriptures, which gives Lemuria (another name for Pan or Mu) as the place where the sons of God, or Kumaras, first taught humanity the path of spirit; their leader was Sanat Kumara. That fabled continent was Kanya Kumari, said to have been submerged.

The semantic range of *ku* is considerable. It is not unusual to find Panic words (especially place names) transferred with devotion to the naming of new lands and peoples (postdiluvial settlements); as such, Ku appears in:

- Sankhu, India: the subcontinent also has its Kurumba Negritos, its hill-tribe Kuki people of Assam, the Kumi tribe, and the Kulu people of the Himalayas.
- Pan-Ku, in Chinese, was the first man and Creator who gave Earth its form; Khara Khutuul, Mongolia, and Kuen Lun mountains are other Asiatic *ku* names.
- Southeast Asia and beyond: Kulen Plateau, Cambodia; Ku Chi, Vietnam; Kuala Lumpur; Kuching, Borneo; Mt. Lawu-Kukusan, Java; the relict Kubu tribe of Indonesia; the Orang Kuba people of Malaysia; and Kumawa Mountains in New Guinea. On the north coast of New Guinea, Kulabob is the name of the culture hero in their migration legend, these villagers speaking a language similar to Polynesian.
- In the ancient Egyptian language, khu means "mountain." Kukuyu means "grass" among Kenya's Ki-ku-yu people (with the same beadwork as Plains Indians). Other African *ku*'s: Kufra Oasis of Libya; Kumasi, Ghana; Kumbi, Mali; Kuru, Ethiopia; and Kurawi people of Sudan. In South Africa are the Ku-ruman Hills, while Kuhistan (mountainous land) is a province in Afghanistan (as is Kunar) with 14-kyr cave drawings. Afghanistan's nomads are the Kuuchi people. Khumbaba is lord of the Cedar Forest, in the Gilgamesh epic of the flood (Babylonia).
- Elsewhere in the Old World are: Kubra in the Red Sea area; Khuzestan, east of Turkey; Kura Valley, near the Caspian (Kura Depression in the Ba-ku region of the Caucasus); Kusamo, Finland; Kukuteni people of southern Ukraine; Kukes, Albania; and Kuman people of southeast Europe. *Kulak* means "landowner" in Slavic.
- Cuba is phonetically Kuba; Curacao is phonetically Kuracao; Cuzco (Ecuador) is phonetically Kusko; Cuenca is phonetically Kuenca. Also in South America are the Kukurital people of Brazil, the Ma-ku Indians of Brazil, the Kuikuru people of Brazil, and the Kubeo Indians of Colombia.

APPENDIX B

GHAN

A Word Study

The era of the fighting Ghans, beginning some 18 kya, brings us to the time of the first kings and conquerors. As such, they were victors: *ganar* means "victor, win" in Spanish (like the English word *gain*). When I realized that these same kings and conquerors gave rise to the divine monarchs of the ancient world, it was easy to see how kings became gods: It was politics, as usual. Empowered by deification, the Ghan rulers gave their name to the Siamese god Waizganthos as well as the Mexican god of maize, the all-nourishing Ghanan. In Mother India (where the Ganges River itself may have been named after these immortal emperors), the Gandarvas are formidable angel-warriors of Sanskrit fame, whose job is to convey messages between gods and mortals. Another Hindu (Gonds) god, this one protector of crops, is Ghansyam.

The tall and stately Ghans (half Ihuan, half Ihin) of the forgotten past lent their genes to the royal families of Africa, Sumeria, Peru, and Polynesia. Very grand, indeed, were their monuments, their society stratified into castes (the world now carefully divided into the rich and the poor). In the Turkic, Indic, Tartar, and Mongolian languages, we can trace these great lords philologically, which is to say, *ghan* means "ruler" (it is actually a title) in the aforementioned languages and is a variant of *khan* and *kagan*: Genghis Khan and his son Kublai Khan are the most notable examples. In Russian, pa-khan means "boss" and ka-gan means "ruler."

Dagan was a Sumerian king name (and Akkadian god), while Sumugan was the Sumerian "king of the mountain" (strangely reminiscent of the Khingan Mountains in Mongolia) and god of the steppes. In fact GaniEden Ganudia is one name of ancient Babylonia itself, where Ghan/gan was designated "lord of the land," and in Sumerian, *gana* meant land (Gana-ugigga was, for example, a territory of ancient Lagash). Ghanem, thereafter, became a (Syrian) surname, while Ghani is a distinguished name in Afghanistan.

Damghan, on the Iranian plateau, is where the Persian King Darius was murdered. Gurgan, a province in Iran, contains countless mounds, while Magan is an ancient port city east of the Persian Gulf. Ganj Dareh is also in Iran.

Afghanistan lies in the heart of this vast khan-ruled empire, dating beyond the Neolithic, and in that terribly ancient land are sanctuaries such as Yorgan Tepe (near the Zab River) and districts such as Arghandab, Oruzgan, and Yamgan. Ferghana in Uzbekistan is known for its petroglyphs* of helmeted men (warriors?), similar to figures drawn in Siberia and India. Spilling into Pakistan are the valley of Koghan and the ancient kingdom called Gandhara (at its height under the Kushan kings). Gandhara once had its center in Peshawar, which literally means "city of man." Yes, ghan, as the paragon of manhood (perfected *Homo sapiens sapiens*), indicates the race itself, mankind. Isn't the Chinese word for mankind *egan*? In Panic, *dan'gan* meant a man of light, which happens to be Na'ganwag in Algonquin. The Arabic book of songs, Kitab al-Aghani, may have the same root.

Numerous clans and tribes adopted the prestigious appellation: the Dolgan of Siberia, the Pangan of Malaysia, the Yaghan of Tierra del Fuego. In Europe, the original clan name of the early Celts was Eoghan (which became Ewan, then the ubiquitous Evan of the Welsh). Here in the British Isles, Gwazig-Gan was the name of a branch of the legendary Korred. Even names like Korrigan and Morrigan (the war goddesses

---

*Related perhaps to Cro-Magnon (Ghan) cave art in *Gan*dia, Spain.

of the Celts) and numerous modern surnames—Mulligan, Finnegan, Morgan, Logan, Hogan, Gaughan, Reagan, Grogan, Fagan, Coogan, and Monaghan—all seem to draw on the original *gan*.

Ancestral Indo-European names of the Slavic world include Ganovce, Slovakia, and the Kurgan (mound) cult of South Russia; in Sweden these tumuli are called gangriften.

I wonder, too, if ghan is embedded in Wigan and Bugan (heroes of the Filipino Ifugao flood legend), or at Filipino sites like Vigan, Kiangan, and Ilagan, or in Pagan, Burma (with its stately ruins), Changan (one of China's first cities), Ramat Gan (Israel), Kalibangan (India), and Gangoji (Japan). Africa's Ganda people, Ghana, Uganda, Baganda, and Tanganyika may also be part of the picture.

America, too, had its fair share of ghans, usually suffixed, starting with the enchanting Ongweeghans and following through with the Iroquois prophet Ganyodiyo and statesman Deganawida. Consider also Ganogwioeon, war chief of Seneca and the Cherokee Ganadi ("the Great Hunter"), as well as the Apache "Mountain of the Ghans," Waukegan (Illinois), Michigan, Gansagi (an old Cherokee settlement), as well as Apatzin-gan (Mexico), Ganchavita (Colombia), Navagandi and Portogandi (villages in Darien, Panama, with white Indians), Chorotegan (Nicaragua), and Manicugan (Canada), along with North American groups like the Piegan, Unagan, Okanogan, Sheboygan, Mohegan, and, last but not least, the splendid queen of it all, with her yellow hair long and hanging, Minne-gan-ewashaka.

APPENDIX C

IHUAN

A Word Study

Although the name *Ihuan* is lost, the Andean Huancas are one of many tribes in the New World whose name betrays I-huan identity. The Huanac, a tribe of Peru, were great warriors, like the Ihuans—*huan, wan*, or *uan* being phonetically interchangeable. Tawantinsuyu was the name of the Inca Empire (similar to Ta-wan-anna, the title of the Hittite queen); Guanape is an island off Peru. The Chirigwano people are in Bolivia, and the Timukwana people in Florida.

In the Old World, the suffix *-wan* can be found in many place names like Taiwan, Nihewan (Basin China), Palawan (Philippines), Silwan (Palestine), Issaowan (Algeria), Parwan (Afghanistan), and Helwan (Egypt). It appears in Africa in Botswana, Wanyoro, and Rwanda. There are the Bantu people of Tswana and Wanyamwezi, and the Congolese Wangatta, the Chesowanja of Kenya. *Wan* crops up in the names of various peoples, such as the Ikhuan, an Arabian tribal group (the name means "brothers"), and the Guanches of Canary Island, of classic Ihuan stock. All these tribal names bespeak Ihuan (variant of Iwan) ancestry, including the Berawan tribe in Borneo (near the Niah Caves) and the Paiwan people of Formosa.

The name *Ihuan* (Iwan) was most widely retained in the Americas: Saskatchewan, Mattawan, Tiahuanaco (Bolivia), Huanuco (Andes), and

Huancayo (Peruvian Andes). Chavin de Huantar is a megalithic ruin in Peru. The pre-Incan Huari Huanca people worked in gold trepanning. In Brazil are the Tahuamanu (blue-eyed giants); in Mexico: Cuanalan, Tehuantepec, Huanta de Jimenez (Oaxaca), Guanajuato; in Honduras: Aguan; in Nicaragua: Guanacaste; in Antilles: Guanahani, and in Cuba: Guantanamo and Guanabo.

Among the Guajiros of Venezuela, *wa-yu* means "man"; "wa-" is seen again in the Wabanaki, Wanana, and Wai Wai Indians of British Guiana. I think its original broad meaning was man or mankind,* setting the tribes of Ihuans apart from the hordes of Druks and half-breeds. In the Americas are Sacsahuaman, Marcahuasi, and Huaura River in Peru, and Lake Chiguana in Bolivia. In Mexico: Chihuahua, Huasteca (people), and the Nahua (Aztec) language. In North America, tribal names were routinely suffixed with *-wa*: Ojibwa, Chippewa, Kiowa, Iowa, Etowa, Tonkawa, Ottowa, and Chautauqwa, even Hiawatha, also known as Wahta: The Creator said—I name thee Eawahtah for you are spirit and flesh evenly balanced. In some places, *wa/hua* retains the meaning of noble, or even the sacred part of man, as in Watuinewa, the Yaghan god of life, and in *wakan* (a Dakota word for "sacred" and the Omaha great spirit: Wakonda; equal to *wak'a*, meaning "shrine" in Cuzco) and *huaca* (sacred relics), as well as in Culhuacan (Toltec), and Teotihuacan (Aztec), which are sacred sites, like Huaca del Sol; (*huaceros* are looters of sacred relics). In Arabic, *iwan* means "sacred space."

In Asia the creator god (of the Bhils, India) is Bhagwan, while among the Sulawesi, Wangko is the moon god and Wangi, the sky god. Australia's Wandjina figures seem to embody the ancestral race. In the Maori language of New Zealand, *ahua* means "man," while Hu-ia is the legendary hero of these people. I suspect that the *hu* in "human" is also traceable to this ancient root: as seen in the Mexican culture hero, Hunaphu as well as the Hue Hue, Mexico's first men of the modern type and the Mayan word for grandee: hunak. Hurakan is a Mayan god, while among the Kubeo Indians of Colombia, Humena Hinku means "my little

---

*Guanenh was the Guanch word for "people"; in Chukchi wanga means "I, me."

Figure C.1. The prophet Hiawatha, his portrait showing what pure North American Ihuans looked like. "I will leave one race of Ihuans on the earth . . . the North American Indian. . . . And I will raise up prophets among them . . . and they shall build unto the Great Spirit." Illustration and quote (Book of Wars 21:11) from Oahspe.

spirit." In North America are the Huron (Iroquoian) and Hupa Indians (California). Huluppu is the tree of immortality in Mesopotamian myth, while Hu is the Celtic god. It may be possible to trace Hue (Vietnam), Huahine (Society Islands), Oahu and the Menehune (in Hawaii), the Hua people of New Guinea, the Huva people (Madagascar), and the Hutus of central Africa to the same root.

In China are: Huang River, Guangzhou province, Dunhuang, Guangxi Zhuang, Hunan, Jiahu, and the sacred Mount Hua. In the

Chinese language, hua-li means "beautiful," while huang signifies "supreme" or "sovereign." Shi Huang Ti was the name of the first historical Chinese emperor (the one who buried all books of knowledge). Huang-hai province is in Korea; Tunhuang in Turkestan; Bengawan River in Java; Okinawa in Japan; Sumbawa in Indonesia; Massawa, Red Sea; Jarawa people (Andaman).

EVIDENCE OF HIGH CIVILIZATION IN ANCIENT OCEANIA

Where	Evidence
Easter Island	Rongorongo script, ruins of a city offshore, megalithic statues, ancient canals, stone houses
Panape	Ancient ruins, temple architecture, cyclopean masonry: gigantic blocks up to fifteen tons, basalt "log" cabins, underwater columns, cyclopean enclosures, underground passages
Marianas	Late monolithic monuments weighing thirty tons, red marble columns, pyramids
Raratonga (Cook Island)	Ancient road of tight-fitting basalt, ruins underwater leading to another island, a highway system radiating from the temple complex
Malden (Line Island)	Pyramids, megaliths, ruins of forty stone temples, continuation of highway system
Tongatabu	Huge stone arch
Raiatea	Step pyramids, huge statues, and massive platforms
Between Maui and Oahu (Hawaii)	Submerged city ruins (top secret)
East of Tahiti	On sea bottom: great ruins, hieroglyphics on column
Fiji	Monolith with undeciphered inscriptions
Samoan Islands	Truncated and step pyramids of (imported) red stone
Melanesia	Stone buildings and monuments

APPENDIX E

SOME ANACHRONISMS

Can we blame all these out sequences on "archaeologic serendipity" (as does Ashley Montagu[1])? Or would it be more truthful to admit, with Jeffrey Goodman, that "such turnabouts . . . stand in direct opposition to the continuous developmental processes Darwinians espouse"?[2]

COMPARISON OF
EARLY AND LATE HOMINIDS

Where	Earlier Hominid/Remains	Traits Compared to Later Specimen
Australia	Lake Mungo remains	AMH, with smaller faces, jaws, brows than Kow Swamp H. erectus
China Jinniushan hominid	Less archaic skull than later Dali Man	
France	Fontéchevade Man	More modern skull, face, 1,470 cc, lighter build than later Neanderthal
Iraq, Shanidar	H. neanderthalensis	Less prognathous than later Neanderthals
Israel	Qafzeh Cave remains	AMH, 20 kyr older than Amud Neanderthal
Java	Sangiran/H. erectus	Much smaller and lighter skulls than Ngandong
Java	Solo Man/H. erectus	Temporal bone more mod than Neanderthal

Where	Earlier Hominid/Remains	Traits Compared to Later Specimen
Kenya	ER 1470/*H. rudolfensis*	More advanced than *H. habilis* and *H. erectus*, bigger brain than younger Au, flatter face, weaker brow than ER 1813
Kenya	ER 3883/*H. ergaster*	Less robust and more delicate skulls than *H. erectus* (such as OH9)
Kenya	Kanapoi hominid/*Au. anamensis*	More modern foot and humerus than later Sterkfontein, Olduvai, and Lucy Au
Kenya	Turkana Boy and ER 3733/ *H. ergaster*	Less robust than later *H. erectus*
Spain, Atapuerca	*H. antecessor*	1.2-myr skull with 1,390 cc larger than 800-kyr skull with 1,000 cc
UK, Swanscombe	Swanscombe Man/*H. erectus*	More *H. sapiens* features than later Neanderthal

NEW WORLD
NOT SO NEW

Ancient America has some surprises in store for conventional paleontology.

According to the flamboyant out-of-Africa theory (OOA), AMHs from Africa arrived in distant Tierra del Fuego (tip of South America), thanks to the hypothetical land of Beringia, which accommodated the crossing: the lowered sea level apparently allowed Paleo-Indians to walk across the Bering Strait. Fitting into the competitive evolutionary model, these migrants from the forbidding steppe of Siberia came to the New World "in search of a new homeland," having been pushed out by other bands.[1]

Just as OOA is fraught with a surfeit of riddles and implausibles, the New World's settlement out of *Asia* can be equally troublesome. According to Howells, for example, American Indian skeletal affinities with Asians are actually minimal. In addition, the dispersal pattern through the Americas, rather than the expected southerly trek, was actually north and south out of a Central American hub. The Bering crossing itself was relatively recent, and of minor significance; we can see this by looking at the paucity of dialects among Inuit, having occupied their home for the shortest amount of time. This is in stark contrast to no less than twelve linguistic stocks in native Oregon and sixteen in California (the area no larger than France), indicating a much greater antiquity than the Inuit. Humans in South America also lived long before the alleged crossing of the Bering land bridge some 12 or 15 or even 25 kya.

"Human beings," observed A. H. Verrill in *Old Civilizations of the New World,* "may have crossed from Asia via Bering Strait; . . . others reached the New World by way of Greenland . . . or came from Oceania or mid-Pacific archipelagoes." Inuit is slightly related to the language of eastern Siberia, but "otherwise, the AmerInd languages show a wide diversity . . . and *no* [*sic*] clear relationship to any Old World tongues."[2] Neither Carib, Arawakan, Quechuan, or Mayan linguistic families have any relationship to Asiatic languages.

Nor are Native Americans, morphologically, necessarily Mongoloid, that is, the Siberian type, the latter being a purely brachycephalic people; but many parts of North and South America are heavily dolichocephalic, even the Inuit. Non-Mongoloid were the white Eskimos along Davis Strait.

Mexico's Maya are not Mongoloid either: Carleton Coon noted that their earwax type makes Maya non-Mongoloid, as well as their darker skin and bigger noses. A 10-kyr skeleton in Brazil shows a different

Figure F.1. Quechua man (from Joan Parisi Wilcox, Masters of the Living Energy*).*

skull type than both modern Indians and the people of northern Asia who are thought to be their closest relatives.

> *A significant population of archaic Caucasoid peoples inhabited North America for at least thirty thousand years. This . . . was not achieved by crossing Beringia from Siberia into Alaska.*
>
> PATRICK CHOUINARD,
> FORGOTTEN WORLDS

Those early Caucasians in America were Ihins and Ihin blends, and they are the same people who built large cities, conducting extensive commerce, most notably in the Ohio Valley and the Midwest, some trade items traveling two thousand miles cross country. These civilizers were craniometrically Alpine types, like the Sumerians. They "surveyed the way for canals, they found the square and the arch, they led the Ihuans (Indians) to the mines." The Menomoni Indians of Wisconsin told, in their sacred histories, of ancient mines being worked by people of fair coloring. "Patiently they taught and instructed the Ihuans in all things and industries."[3] According to the Kayapo Indians of Brazil, their history began when a radiantly white and gentle culture-bringer and teacher taught them the arts of living: he was cleverer than anyone else and so instructed the people in hitherto unknown matters.

There were great boats with cloth sails and beams in North Guatama 18 kya, plus mining, agriculture, and architecture, as taught by the Ihins, the sacred people. Of special note are the arts of engraving, acquired shamanistically by these sons of Noah.

> And god sent his angels down to man, to inspire him in the workmanship of images and *engravings*. . . . And the Lord spake unto the Ihin saying . . . go provide me a stone and. . . . And the Ihins prepared a stone . . . [and] made tablets.
>
> OAHSPE, THE LORDS' FOURTH BOOK 1.5

Carvings and pictures all over the Americas of extinct megafauna place these skilled artisans here before the Mesolithic. The 40-kyr Holly Oak pendant, found in Delaware, for example, has a carving of a mammoth.

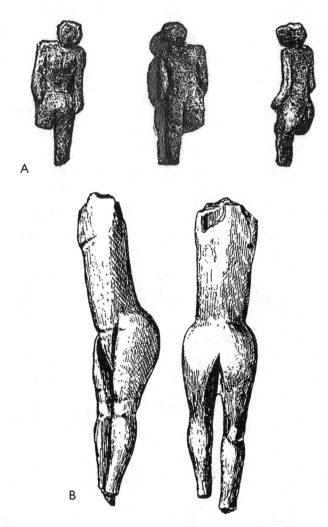

Figure F.2. (A) Nampa image. This skillfully formed miniature, only one and a half inches high, was brought to world attention by Charles Francis Adams in the 1880s. Found in Idaho during the boring of an artesian well (320 feet deep), the figure represents the human form. It has been compared to Europe's Aurignacian sculptures, like (B) this "Venus" from France.

TABLE F.I. SAMPLE OF AMERICAN CULTURES DATED BEFORE THE BERING CROSSING

Time BP	Place	Find
18,000	Lake Superior	Copper mines, implements
18,000	Esmeraldas, Ecuador	Weapons, seals, jewels
18,000–27,000	Tiahuanaco, Bolivia	Gateway of the Sun
20,000–24,000	Peru	Habitation site
22,000	Yuha Desert, California	Fully mod skeleton
27,000	Texas and Southwest sites	Basket Makers
27,000–30,000	Old Crow River, NE Canada	Worked caribou bone tool, domesticated dog
30,000–65,000	Sheguiandah, Ontario	Artifacts, tools
30,000	Sao Raimundo Nonato, Brazil	Animal drawings on rockface
30,000	Otavalo, Ecuador	Human remains*
30,000–35,000	Monte Verde, Chile	Stone and wood artifacts, shelters
37,000–60,000	Alberta, Canada	Fully mod child's bones, human occupation
37,000	Lewisville, Texas	Many hearths
38,000	West central Illinois	Chert tools†
39,000	Pendejo Cave, New Mexico	Fingerprint impressions
40,000	Lime Creek, Nebraska	Bone tools‡
40,000	Mesoamerica	Settlements§
50,000–80,000	Calico desert and Haverty near Los Angeles	Human occupation, hearth, tools¶
50,000	Managua, Nicaragua	AMH footprints, writing\\
50,000	bank of Savannah River	Hearth and stone tools
70,000	California sites	Grinding tools**

Time BP	Place	Find
72,000	Nicaragua, San Diego, Yukon	Artifacts[††]
100,000	Mississippi	Mod pelvis

*Corliss, *Ancient Man*, 668, 387, 472, 642.
[†]Ibid.
[‡]De Camp, *Lost Continents*, 147.
[§]Adams, *Prehistoric Mesoamerica*, 22.
[¶]Corliss, *Ancient Man*.
[\\]Corliss, *The Unexplained*, 11–12.
[**]Goodman, *Genesis*, 269.
[††]Childress, *Lost Cities of North and Central America*, 13.

I believe that Giloff was the original name of the impressive temple ruins at Cocle, Panama, which were found lying under volcanic ash.[4] The awesome site sports a series of stone images and huge *engraved* columns in rows. The main temple is "a great cathedral." Several of its statues wear beards and have eyes set far apart.

> In the Middle Kingdom (Panama) . . . was the temple of Giloff, with a thousand columns . . . and within Giloff dwelt the oracle of the Creator for two thousand years.
>
> OAHSPE, FIRST BOOK OF GOD 25.5

The Akuria people, living today in the Cocle region of Panama, were unknown to the outer world until visited by Alpheus H. Verrill in the early 1920s. The Akuria are described as rather like little Caucasians, the women about *four feet tall* (Ihin genes). Their noses are thin and well bridged, chins well developed, foreheads broad.

Now the classic Bering hypothesis put a cork on the antiquity of man in the New World: ca 1350 BP, even though nineteenth-century scholars thought Native Americans had been here for many more thousands of years.

> *No modern humans lived outside Africa until about 59,000 years ago.*
>
> BRIAN FAGAN, *CRO-MAGNON*

Valsequillo, controversially, was dubbed Mexico's Garden of Eden after fancy spearheads and fine stone tools were excavated—obviously artifacts of modern man, a sophisticated culture. Matching the Old World's Cro-Magnon culture, Valsequillo was dated anywhere from 35 kya to 250 kya!*

There were the *Huehue* (old ones) of Mexico, shown with modern features and beards (a Caucasian trait). These Huehue people predate the Maya, as do the carvers of the Loltun caverns, their style "completely different" from Mayan work, "some other race."[5] The Maya say the first man was molded from white and yellow maize; they were little, pale, bearded, and gentle people.

Even though some genetic studies have concluded that all AmerInds "evolved" from an Asian stock that lived 30 kya in Siberia,[6] North America's Mandan and Pomo Indians speak of their white (not Oriental), bearded forebears. Michigan natives had a curious idol with a beard like a European, though the natives are beardless; there was a special ritual for honoring this statue. Indeed, many tribes recall a white race before their time. Long before contact with the Europeans, the medicine men of the Croatan Indians (North Carolina) communed with the spirit of the first white man, who belonged to a race that inhabited North America even *before* the red man. The Shawnee also knew that white people had lived in the valley before them. The Maidu (California Indian) creation myth claims that the first people whom Earth Starter made were "as white as snow." A culture bringer of the Iroquois, named Ioskeha, was also a white one. And among the Algonquins, it was universally held that their ancestors were Abenakis, literally, "our white ancestors."[7]

---

*The 35-kya date was based on volcanic ash beds and radiocarbon. New dating techniques (uranium method) multiplied that date sevenfold! But the uranium could have migrated in or out of the sample, either way (in or out) giving a falsely old date.

NOTES

INTRODUCTION

1. Gould, "Dorothy, It's Really Oz," 59.
2. Gregory, "Fight the Good Fight," 3.
3. Coyne, *Why Evolution Is True,* xiii, xvii.
4. Mayr, *What Evolution Is,* 9.
5. Jones, *The Cambridge Encyclopedia of Human Evolution,* 239.
6. Feliks, "The Pleistocene Coalition: Exploring a New Paradigm," 5.
7. Lubenow, *Bones of Contention,* 202.
8. Shermer, "Unweaving the Heart," 38.
9. Ibid.
10. Diamond quoted in Mayr, *What Evolution Is,* x.
11. Cronin, *The Ant and the Peacock,* 48, 49, 52.
12. Johnson, *Darwin on Trial,* 121, 178.
13. Ibid., 121.
14. Morley, *A Path of Light,* 2.
15. White, "Enlightenment 101."
16. Seiglie, *Creation or Evolution?* 26.
17. George Frederick White quoted in Jones, *The Cambridge Encyclopedia of Human Evolution,* 330.
18. Velikovsky, *Worlds in Collision,* 304–5.
19. Eldredge, *Darwin: Discovering the Tree of Life,* 112.
20. Ridley, *Evolution,* 413.
21. Time-Life, *The Human Dawn,* 50.

PROLOGUE. DIASPORA

1. Campbell, *Primitive Mythology,* 15, 146, 274.

2. Drake, *Gods and Spacemen of the Ancient East,* 148.

3. White, *Pole Shift,* 313; Charroux, *The Mysteries of the Andes,* 69.

4. Charroux, *Masters of the World,* 67, 50.

5. Oahspe, The Lords' First Book 1:25, 2:4.

6. Kolosimo, *Timeless Earth,* 131.

7. Bahn, *100 Great Archaeological Discoveries,* 64.

8. Schwartz, *The Mysteries of Easter Island,* 157.

9. Gladwin, *Men Out of Asia,* 234.

10. Ellis, *Polynesian Researches,* 123.

11. Brinton, *The Myths of the New World,* 195.

12. Corliss, *Ancient Man,* 234.

13. Kolosimo, *Not of this World,* 153, 199.

14. Chouinard, *Forgotten Worlds,* 241–42.

15. Kolosimo, *Timeless Earth,* 54.

16. Oppenheimer, *Eden of the East,* 66.

17. Ellis, *Polynesian Researches,* 121.

18. Oppenheimer, *Eden of the East,* 477.

19. Chouinard, *Forgotten Worlds,* 156.

20. Kolosimo, *Timeless Earth,* 57.

21. Goodman, *The Genesis Mystery,* 209.

22. Ibid., 175.

23. Gladwin, *Men Out of Asia,* 44.

24. Ellis, *Polynesian Researches,* 122.

25. Jones, *The Lost Data on the Chariots of the Elohim,* 239.

26. Goodman, *The Genesis Mystery,* 220.

27. Von Ward, *We've Never Been Alone,* 174.

28. Wilkins, *Mysteries of Ancient South America,* 117.

29. Honore, *In Quest of the White God,* 188.

30. Brinton, *The Myths of the New World,* 226.

31. Dixon, *The Racial History of Man,* 451.

32. Brinton, *The Myths of the New World,* 200.

33. Radin quoted in Gladwin, *Men Out of Asia,* 343.

34. Ellis, *Polynesian,* 122.

35. Campbell, *Primitive Mythology,* 465.

36. Hancock, *Fingerprints of the Gods,* 194.

37. Goodman, *The Genesis Mystery,* 185.

38. Corliss, *Ancient Man,* 234.

39. Drake, *Gods and Spacemen,* 88.

40. Hitchcock, *American Antiquities* (1894), 209–11.

41. Chouinard, *Forgotten Worlds,* 38.

42. Stuart, *Discovering Man's Past in the Americas,* 191.

43. Kolosimo, *Timeless Earth,* 248.

CHAPTER 1. OUR KNOWLEDGE IS SKELETAL

1. Baroly, *Prehistory from Australopithecus to Mammoth Hunters,* 34.

2. Oahspe, Book of Apollo 13:3.

3. Switek, *Written in Stone,* 20.

4. Oahspe, Book of Sethantes 8:8.

5. Oxnard, *The Order of Man,* 8, 316.

6. DeSalle and Tattersall, *Human Origins,* 119.

7. Johanson quoted in Johanson and Edey, *Lucy,* 349, and in Johanson, "Face-to-Face with Lucy's Family," 101.

8. Reader, *Missing Links,* 226.

9. Isaak quoted in Nardo, ed., *Evolution,* 148.

10. Huse, *The Collapse of Evolution,* 44.

11. Goodman, *The Genesis Mystery,* 32, 41.

12. Glausiusz, "A Trail of Ancient Genes," 71.

13. Federer, *America's God and Country,* 1996.

14. Lubenow, *Bones of Contention,* 168.

15. Reader, *Missing,* 8.

16. Eiseley, *The Immense Journey,* 84.

17. Macbeth, *Darwin Retried,* 14.

18. Ridley, *Evolution,* 4.

19. Eiseley, *The Immense Journey,* 34.

20. Owen Lovejoy, quoted in Johanson and Edey, *Lucy,* 330.

21. Fagan, *World Prehistory,* 53.

22. Schulter, *The Ecology of Adaptive Radiation.*

23. Ridley, *Evolution,* 199.

24. Day, *Guide to Fossil Man,* 81.

25. Eiseley, *Darwin's Century*, 222.

26. Shreeve, *The Neandertal Enigma*, 198.

27. Fagan, *Cro-Magnon*, 24.

28. Ibid., 67.

29. Hall, "Last of the Neanderthals," 50.

30. Montagu, *Man: His First Two Million Years*, 40.

31. Shreeve, *Enigma*, 276.

32. Kolosimo, *Timeless Earth*, 180.

33. Oahspe, Lords' Fourth Book 4:10.

34. Kolosimo, *Timeless Earth*, 55.

35. Oahspe, First Book of the First Lords 3:14.

36. Ibid., Synopsis of Sixteen Cycles 2:5.

37. Ibid., The Lords' Second Book 3:8.

38. Ibid., The Lords' Fifth Book 3:18.

39. Kolosimo, *Not of This World*, 89.

40. Higgins, *Anacalypsis*, 142.

41. Oahspe, The Lords' Fourth Book 4:21.

42. Berlitz, *Mysteries from Forgotten Worlds*, 148.

43. Oahspe, The Lords' Fourth Book 2:16.

44. Ibid., Book of Divinity 13:13.

45. Gore, "The Dawn of Humans."

46. Oahspe, Synopsis of Sixteen Cycles 3:16, Book of Aph 1:10.

47. Montagu, *Man*, 60.

48. Bahn quoted in Jones, *The Cambridge Encyclopedia of Human Evolution*, 330.

49. Dingir, "The Origin of Hubris," 15.

50. Darwin quoted in Himmelfarb, *Darwin and the Darwinian Revolution*, 343.

51. Oahspe, The Lords' Fifth Book 4:24.

52. Goodman, *The Genesis Mystery*.

53. Weikart, "Dehumanizing," 34.

54. Diamond, *The Third Chimpanzee*, 36.

55. Mayr, *Evolution*, 134.

56. Lemonick and Dorfman, "Up from the Apes," 57.

57. Nowak, "Symbiosis," 35.

CHAPTER 2. RETURN OF THE HOBBIT

1. Gorner, "Tiny-human Find Becomes Huge News," A10.

2. Mayell, "Hobbit-Like Human Ancestor Found in Asia."

3. Lovejoy, quoted in Johanson and Edey, *Lucy.*

4. Winters, "Hobbit Wars Heat Up."

5. Gorner, "Tiny-human Find Becomes Huge News," A10; Winters, "Hobbit Wars Heat Up"; Wong, "Hobbit Hullabaloo," 22–24.

6. Wills, *The Darwinian Tourist,* 264.

7. Neimark, "Meet the New Human Family," 54.

8. Coon, *Adventures and Discoveries,* 343.

9. "Bone Collection Backs Up Hobbit Theory," 9.

10. Mayell, "Hobbit-Like Human Ancestor Found in Asia," http://news.nationalgeographic.com/news/2004/10/1027_041027_homo_floresiensis.html.

11. Eldredge, *Darwin,* 228.

12. Oahspe, Synopsis of Sixteen Cycles 1:22.

13. Hooton, *Up from the Ape,* 625.

14. Dixon, *The Racial History of Man,* 230.

15. Day, *Guide to Fossil Man,* 405.

16. Leakey, "Skull 1470," 819.

17. Ibid., 820–28.

18. Tattersall and Schwartz, *Extinct Humans,* 116.

19. Fix, *The Bone Pedlars,* 136.

20. Oxnard, *The Order of Man,* 3.

21. Dixon, *The Racial History of Man,* 503.

22. Cloud, *Oasis in Space,* 488.

23. Schwartz, *Sudden Origins,* 21, 26, 378.

24. Oahspe, Book of Apollo 13:5.

25. Corliss, *Ancient Man,* 319.

26. Trinkaus and Shipman, *The Neandertals,* 310.

27. Shreeve, *The Neandertal Enigma,* 122.

28. Lubenow, *Bones,* 69.

29. www.stewartsynopsis.com.

30. Urban, "Antiquarian Researches," 182.

31. Clark, *Indian Legends of the Northern Rockies.*

32. de la Vega, *Comentarios reales de los Incas, 1609–17.*

33. Oahspe, footnote to Lords' Fifth Book 5:7.

34. Rosenberg, *The Myth of the Twentieth Century.*

35. Wright, *Origin and Antiquity of Man*, 113.

36. Dixon, *The Racial History of Man*, 40.

37. Mack, "Mexico's Little People," 39.

38. Wollaston, *Pygmies and Papuans*, 313.

39. Layard, *Stone Men of Malekula*, 9.

40. Coon, *The Living Races of Man*, 176.

41. Ibid., 206.

42. Gladwin, *Men Out of Asia*, 89.

43. Wright, *Origin and Antiquity of Man*, 119.

44. Dixon, *The Racial History of Man*, references in these paragraphs drawn from 28, 32, 35, 49, 58, 76, 310, 400, 403, 409, 462, 476, 479, 494.

45. Kolosimo, *Not of This World*, 53.

46. Oahspe, God's Book of Eskra 9:15.

47. Hooton, *Up from the Ape*, 465, 592.

48. Goodman, *The Genesis Mystery*, 20.

49. Norman, *Gods and Devils from Outer Space*, 122.

50. Hooton, *Up from the Ape*, 369.

51. Hrdlicka et al., *Early Man in South America*, 306.

CHAPTER 3. CHEEK BY JOWL

1. Lewin, *In the Age of Mankind*, 85.

2. Darwin quoted in Ridley, *Evolution*, 14–15.

3. Feliks, "Ardi"; Leakey and Lewin, *Origins*, 86.

4. Lubenow, *Bones of Contention*, 118.

5. Sagan, *The Dragons of Eden*, 94.

6. Goodman, *The Genesis Mystery*, 166.

7. Nardo, *Evolution*, 215, 218.

8. Oahspe, Book of Sethantes 10:3–4.

9. Goodman, *The Genesis Mystery*, 174.

10. Kleiner, "Early Toolmakers Cast Off Their Rock-banger Image," 7.

11. Le Gros Clark, *The Fossil Evidence for Human Evolution*, 169.

12. Brace, *The Stages of Human Evolution*, 34, 55.

13. Hrdlicka et al., *Early Man in South America,* 155.

14. Boule, *Fossil Men,* 242.

15. Day, *Guide to Fossil Man,* 114.

16. Montagu, *Man: His First Two Million Years,* 70.

17. Oahspe, Book of Divinity 13:17.

18. Ibid., The Lords' Fifth Book 5:12–14.

19. Montagu, *Man,* 137–38; Mithen, *After the Ice,* 43, 113; Dorothy Garrod, first excavator at Mt. Carmel, judged them agriculturalists.

20. Hooton, *Up from the Ape,* 534.

21. Neimark, "Meet the New Human Family," 51.

22. Ibid., "Pinkie Pokes Holes in Human Evolution," 55.

23. Tiffany, "Editor's Comments," 5–14.

24. Oahspe, Book of Divinity 11:18.

25. Ibid., Book of Wars 19:2.

26. Boule, *Fossil Men,* 243.

27. Oahspe, Synopsis of Sixteen Cycles 1:23–25, 10th cycle.

28. Ibid., The Lords' Third Book 2:6 and 2:8.

29. Lewin, *The Origin of Modern Humans,* 69.

30. Hooton, *Up from the Ape,* 384.

31. Roach, "Neandertals, Modern Humans May Have Interbred, Skull Study Suggests," http://news.nationalgeographic.com/news/2007/01/070116 -neanderthals.html.

32. Biello, "Need for Speed?" 31.

33. Trinkaus and Shipman, *The Neandertals,* 392.

34. Dixon, *The Racial History of Man,* 103.

35. Boule, *Fossil Men,* 6, 332.

36. de Quatrefages, *The Pygmies,* 32, 49, 80.

37. Evans, *The Negritos of Malaysia,* 9.

38. Coon, *The Living Races of Man,* 196.

39. Flower, *The Pygmy Races of Men,* 6–7.

40. Hooton, *Up from the Ape,* 356, 400, 612.

41. Dixon, *The Racial History of Man,* 463.

42. Kolosimo, *Not of This World,* 91; Kolosimo, *Timeless Earth,* 204.

43. Landsburg and Landsburg, *In Search of Ancient Mysteries,* 54, 67.

44. Hooton, *Up from the Ape,* 535–37.

45. Oahspe, The Lords' Fifth Book 3:14.

46. Ibid., Book of Aph 13:3.

47. Ibid., First Book of the First Lords 45:24.

48. Deloria, *Red Earth, White Lies,* 212.

49. Pike, *Morals and Dogma of the Ancient and Accepted Scottish Rite of Free Masonry,* 599.

50. Oahspe, The Lords' Second Book 1:18.

51. Ibid., First Book of the First Lords 1:14.

52. Ibid., The Lords' First Book 1:19.

53. Leakey, "Skull 1470," 822.

CHAPTER 4. "I DO NOT BELIEVE I EVER WAS A FISH"

1. Darwin, *The Autobiography of Charles Darwin,* 120.

2. Milton, *Shattering the Myths of Darwinism,* 140.

3. Eiseley, *The Immense Journey,* 34.

4. Disraeli quoted in Hornyanszky and Tasi, *Nature's I.Q.,* 24.

5. Coyne, *Why Evolution Is True,* 170.

6. Eldredge, *Darwin,* 173.

7. Koestler, *The Ghost in the Machine,* 148.

8. Mayr, *What Evolution Is,* 199, 140.

9. Keith, *The Antiquity of Man,* 436.

10. Leakey, *The Origin of Humankind,* 93.

11. Oahspe, Book of Apollo 5:8–9.

12. Cremo, *Human Devolution,* 415.

13. Fix, *The Bone Pedlars,* 288, 293.

14. Oahspe, Book of Divinity 14:3.

15. Oahspe, The Lords' Second Book 1:4.

16. Ibid., The Lords' Third Book 2:6.

17. Ibid., Synopsis of Sixteen Cycles 1:21.

18. Ibid., Book of Osiris 9:15.

19. Ibid., First Book of the First Lords 3:21; see also verse 13 in Second Book of the Lords, chap. 2.

20. Ibid., First Book of the First Lords 1:9.

21. Ibid., The Lords' Third Book 2:3.

22. David Lack, in Milton, *Shattering the Myths of Darwinism,* 147.

23. Teresi, "Q and A," 69.

24. Taylor, *The Great Evolution Mystery,* 174.

CHAPTER 5. THE MATING GAME

1. Hooton, *Up from the Ape,* 337.

2. Hall, "Last of the Neanderthals," 52.

3. Boule, *Fossil Men,* 243.

4. Goodman, *The Genesis Mystery,* 146.

5. Keith, *Ancient Types of Man,* 78.

6. Leakey quoted in Brass, *The Antiquity of Man,* 104.

7. Day, *Guide to Fossil Man,* 125.

8. Brace, *The Stages of Human Evolution,* 95.

9. Montagu, *Man: His First Two Million Years,* 71.

10. Gamble quoted in Gore, "The Dawn of Humans," 102.

11. Johanson and Edgar, *From Lucy to Language.*

12. Hall, "Last of the Neanderthals," 41.

13. Sykes, *The Seven Daughters of Eve,* 126.

14. Lemonick and Dorfman, "Apes," 58.

15. Klein and Edgar, *The Dawn of Human Culture.*

16. Oppenheimer, *The Real Eve.*

17. Hall, "Last of the Neanderthals," 42.

18. Folger, "Scientists Get Inside the Mind (and Genes) of the Neanderthals," 29.

19. Stix, "Traces of a Distant Past," 60–61.

20. Than, "Neanderthals Didn't Mate with Modern Humans, Study Says."

21. DeSalle and Tattersall, *Human Origins.*

22. Wade, "Signs of Neanderthals Mating with Humans," 1–2.

23. Ibid.

24. Neimark, "Meet the New Human Family," 50, 58.

25. Daily Mail reporter, "First Modern Humans Protected Themselves Against Disease After Leaving Africa by 'Interbreeding with Neanderthals.'"

26. "Editorial Note," *Smithsonian,* 8.

27. Hooton, *Up from the Ape,* 538, 596.

28. Hall, "Last of the Neanderthals," 33.

29. Neimark, "Meet the New Human Family," 48.

30. Weidenreich, *Apes, Giants, and Man,* 2–3.

31. Trinkaus and Shipman, *The Neandertals,* 314.

32. Klein and Edgar, *The Dawn of Human Culture,* 86.

33. Howells, *Getting Here,* 87.

34. Leakey and Lewin, *Origins,* 100.

35. Lubenow, *Bones of Contention,* 112–13.

36. Gore, "The Dawn of Humans," 1.

37. Quoted in Taylor, *The Great Evolution Mystery,* 144.

38. Lemonick and Dorfman, "Apes," 52.

39. Gould, *The Panda's Thumb,* 170.

40. J. B. S. Haldane quoted in Lovtrup, *Darwinism,* 308.

41. Fischman, "Part Ape, Part Human," 123–33.

42. W. J. Sollas of Oxford, quoted in Johanson and Edey, *Lucy,* 46.

43. Neimark, "Meet the New Human Family," 55.

44. Lewin, *In the Age of Mankind,* 37.

45. Hooton, *Up from the Ape,* 400.

46. Darwin, *The Descent of Man,* 200–201.

47. Thorndike, *Mysteries of the Past,* 263.

48. Tattersall, *The Fossil Trail,* 219.

49. DeSalle and Tattersall, *Human Origins,* 130.

50. Diamond, *The Third Chimpanzee,* 53.

51. Fagan, *Cro-Magnon,* 139.

52. Braidwood, *Prehistoric Men,* 27.

53. Cannell, "Sexual Selection in Archaic Populations."

54. Verrill, *Old Civilizations of the New World,* 43, 59.

55. Edey and Johanson, *Blueprint,* 351.

56. Fagan, *World Prehistory,* 65–67.

57. Ibid., 75–76.

58. Klein and Edgar, *The Dawn of Human Culture,* 96.

59. Schwartz, *Sudden Origins,* 104.

60. Quoted in Johnson, *Darwin on Trial,* 165.

CHAPTER 6. THE TIMEKEEPERS

1. Ridley, *Evolution,* 14.

2. William Whewell quoted in Higgins, *Anacalypsis,* 316.

3. Cited in Wright, *Origin and Antiquity of Man,* 34.

4. Keith, *The Antiquity of Man;* Andrews, *Meet Your Ancestors,* 19.

5. Oxnard, *The Order of Man;* Braidwood, *Prehistoric Men,* 10.

6. Le Gros Clark, *Fossil Evidence,* 126.

7. Tattersall and Schwartz, *Extinct Humans,* 225.

8. Ager, "The Nature of the Fossil Record," 132.

9. Fischman, "Part Ape, Part Human," 133.

10. Edey and Johanson, *Blueprint,* 348–50.

11. Gorner, "Tiny Human," A-10.

12. Edey and Johanson, *Blueprint,* 348–50.

13. Andrew Hill, quoted in Howells, *Getting Here,* 78.

14. Fagan, *Cro-Magnon,* 12.

15. Leakey, *The Origin of Humankind,* xii; Johanson and Edey, *Lucy,* 20; Trinkaus and Shipman, *The Neandertals,* 412, 414.

16. Braidwood, *Prehistoric Men,* 23.

17. Alexander Marshack, "Use of Symbols Antedates Neanderthal Man," 22.

18. Thorndike, *Mysteries,* 100; Morris, *Scientific Creationism,* 175.

19. Day, *Guide,* 93.

20. Lubenow, *Bones of Contention,* 163–64.

21. Ibid., 112–13.

22. Keith, *The Antiquity of Man,* 366.

23. Eiseley, *The Immense Journey,* 85.

24. Keith, *Ancient Types of Man,* 125.

25. Wilson, *The Chariots Still Crash,* 70.

26. Charroux, *One Hundred Thousand Years of Man's Unknown History,* 29.

27. Minkel, "Food for Symbolic Thought," 16.

28. Leakey, "Skull 1470," 819–29.

29. Lemonick and Dorfman, "Up from the Apes."

30. Johanson, "Face-to-Face with Lucy's Family," 113.

31. Wilson, *The Chariots Still Crash,* 64–65.

32. Huse, *The Collapse of Evolution,* 66; Morris, *Scientific Creationism,* 142–43.

33. Teresi, "Q & A," 71.

34. Oxnard, *The Order of Man,* 6.

35. DeSalle and Tattersall, *Human Origins,* 136.

36. Robbins, "How Old Is the Earth, Really?" 42.

37. Lewin, *The Origin of Modern Humans,* 17.

38. Hamilton, "Mother Superior," 28.

39. Charroux, *One Hundred Thousand Years,* 38.

40. Lubenow, *Bones of Contention,* chapter 23.

41. Sarfati, *Refuting Compromise,* 314–15.

42. Howells quoted in Lubenow, *Bones of Contention,* 279–80.

43. Childress, *Lost Cities of North and Central America,* 251.

44. Winchester, *Krakatoa,* 94.

45. *Discover,* October 2005, 40.

46. Johanson, "Face-to-Face with Lucy's Family," 114.

47. Ibid.

48. Leakey, *The Origin of Humankind,* 99.

49. Howells, *Getting Here,* 131; Leakey and Lewin, *Origins,* 91.

50. Robbins, "How Old Is the Earth, Really?" 43.

51. Oahspe, Book of Sethantes 16:3–5.

52. Ibid., Book of Ah'shong 1:6–9.

53. Ibid., Book of Divinity 13:16.

54. Ibid.

55. Wilford, *The Riddle of the Dinosaur,* 306.

56. Denton, *Evolution: A Theory in Crisis,* 251.

57. Oahspe, The Lords' Fifth Book 5:8.

58. Ibid., Book of Divinity 13:17.

59. Wells, "Dipping into the Gene Pool," xxviii.

60. Fagan, *Cro-Magnon,* xvi, 93.

61. Oahspe, God's Book of Eskra 9:15.

62. Ibid., Synopsis of Sixteen Cycles 1:9–19.

63. Diamond, *The Third Chimpanzee;* Shreeve, *The Neandertal Enigma,* 68.

64. Fagan, *Cro-Magnon,* 64, 113, 116.

65. Hall, "Last of the Neanderthals," 42.

66. Klein and Edgar, *The Dawn of Human Culture,* 198.

67. Goodman, *American Genesis,* 201.

68. Oahspe, The Lords' Third Book 1:4 and 3:16.

69. Eldredge, *Darwin,* 15; Diamond, *The Third Chimpanzee,* 237.

70. Oahspe, First Book of the First Lords 2:10.

71. Ibid., First Book of the First Lords 2:10 and 4:2–4; Book of Sethantes 8:17.

72. Von Ward, *We've Never Been Alone,* 157–66.

73. http://jqjacobs.net/blog/gobekli_tepe.html; www.andrewcollins.com/page/articles/Gobekli_Tepe_interview.htm.

74. Oahspe, Synopsis of Sixteen Cycles 3:9 and 1:21; The Lords' Third Book 3:17.

75. Ibid., Synopsis of Sixteen Cycles 3:9.

76. Keith, *Ancient,* 111.

77. Shreeve, *The Neandertal Enigma,* 276.

78. Joseph, ed., *Discovering the Mysteries of Ancient America,* 132.

79. Oahspe, The Lords' Second Book 1:9, The Lords' Third Book 3:17.

80. Oahspe, The Lords' Fifth Book 5:7 and 5:13.

81. Brackenridge, "On the Population and Tumuli of the Aborigines of North America."

82. Oahspe, Synopsis of Sixteen Cycles 1:21.

83. Schoch, *Forgotten Civilization,* 223.

CHAPTER 7. THE SPARK

1. DeSalle and Tattersall, *Human Origins,* 203.

2. Quoted in Lovtrup, *Darwinism,* 226.

3. Ibid., 227.

4. Wills, *The Darwinian Tourist,* 297.

5. Shapiro, *Origins,* 247, repeating comments made.

6. Gould, *The Panda's Thumb,* 54–57.

7. Fix, *The Bone Pedlars,* xvi.

8. Hornyanszky and Tasi, *Nature's I.Q.,* 135.

9. Tattersall and Schwartz, *Extinct Humans,* 245.

10. Shermer, *In Darwin's Shadow,* 183.

11. Wright, *Origin and Antiquity of Man,* 378, 388.

12. Oahspe, Book of Divinity 11:13.

13. Ibid., Book of Saphah: Port-Pan Algonquin.

14. Charroux, *One Hundred Thousand Years,* 97.

15. Cremo, *Human Devolution,* 127.

16. Oahspe, The Lords' Fourth Book 4:22.

17. Oahspe, Book of Inspiration, Chapter 6.

18. Von Ward, *We've Never Been Alone,* 3, 31–32, 97, 104, 108.

19. Kolosimo, *This World,* 58.

20. Oahspe, First Book of the First Lords 1:1.

21. Ibid., The Lords' Second Book 1:6–7.

22. Sedgwick quoted in Himmelfarb, *Darwin and the Darwinian Revolution,* 291, 325.

23. Sedgwick quoted in Leakey and Lewin, *Origins,* 23.

24. Oahspe, Book of Wars 29:32–33 and First Book of the First Lords.

25. Ginzburg, *Legends of the Jews,* 124.

26. Oahspe, The Lords' Second Book 1:17–18.

27. Ibid., Synopsis of Sixteen Cycles, chapter 2.

28. Ibid., The Lords' Second Book 1:21 and 3:5–7.

29. Trench, *Temple of the Stars,* 100.

30. Oahspe, The Lords' Second Book 1:24.

31. Ibid., Book of Divinity 11.13; First Book of the First Lords 1:20; The Lords' First Book 1:28.

32. Oahspe, First Book of the First Lords 1:20; The Lords' Second Book 1:17.

33. Ibid., Book of Sethantes 5:2.

34. Norman, *Gods and Devils from Outer Space,* 140.

35. Dixon, *The Racial History of Man,* 16.

36. Oahspe, The Lords' Fifth Book 5:12.

37. Ibid., The Lords' Fourth Book 4:3–4.

38. Time-Life, eds., *The Human Dawn,* 51.

39. Lubenow, *Bones of Contention,* 112–13.

40. Dixon, *The Racial History of Man,* 504.

41. Wallace quoted in Moore, *The Post-Darwinian Controversies,* 185.

42. Dixon, *The Racial History of Man,* 503, 521.

43. Dennon, "Removing Oahspe's Enigmas."

44. Oahspe, Book of Cosmogony and Prophecy 5:13.

45. Diamond, *The Third Chimpanzee,* 236.

46. Oahspe, First Book of the First Lords 1:3 and 1:8.

47. Ibid., Book of Divinity 11:13.

48. Ibid., First Book of the First Lords 1:3, 2:6

49. Ibid., Book of Divinity 13:10–11.

50. Ibid., Book of Jehovih 7:21.

51. Ibid., Book of Divinity 14:1.

52. Darwin quoted in Himmelfarb, *Darwin and the Darwinian Revolution,* 151.

53. Diamond, *The Third Chimpanzee,* 180.

54. Switek, *Written in Stone,* 268.

CHAPTER 8. "THAT MYSTERY OF MYSTERIES"

1. White, *Enlightenment 101.*
2. Behe, *Darwin's Black Box,* 46.
3. Richard Dawkins as quoted in "What's so heavenly about the God particle?" 50.
4. Milton, *Shattering the Myths of Darwinism,* preface.
5. Gödel quoted in Dennett, *Darwin's Dangerous Idea,* 444.
6. Rudwick, *The Meaning of Fossils,* 188.
7. Quoted in Tattersall and Schwartz, *Extinct Humans,* 88.
8. Eldredge, *Darwin,* 122; Keith, *The Antiquity of Man,* 57.
9. Jones, *The Cambridge Encyclopedia of Human Evolution,* 281.
10. Lovtrup, *Darwinism,* 132; see also Oahspe, Book of Inspiration, chap. 6, verse 20.
11. Oxnard, *The Order of Man,* 323.
12. Flower, *Pygmy.*
13. Weidenreich, *Apes, Giants, and Man,* 13.
14. Oahsp, Book of Inspiration 7:15.

CHAPTER 9. MUTANTS, MONSTERS, AND MORPHOGENESIS

1. Hitching, *The Neck of the Giraffe,* 104.
2. Johanson, "Lucy's Family," 112.
3. Cited in Gottlieb, "It Ain't Necessarily So," 88.
4. Hooton, *Up from the Ape,* 197.
5. Hamilton, "Mother Superior," 26.
6. Denton, *Evolution,* 304.
7. Eiseley, *Darwin's Century,* 51.
8. Fix, *The Bone Pedlars,* 318.
9. Darwin, *Origin of Species,* 1958 ed., 146.
10. Quoted in Tattersall and Schwartz, *Extinct Humans,* 88.
11. Hitching, *The Neck of the Giraffe,* 87.
12. Huse, *The Collapse of Evolution,* 122.
13. Taylor, *The Great Evolution Mystery,* 232.
14. Koestler, *The Ghost in the Machine,* 154–55.
15. Time-Life, eds., *The Human Dawn,* 56.

16. Keith, *The Antiquity of Man,* 190.

17. Tattersall and Schwartz, *Extinct Humans,* 176, 205.

18. Howells, *Getting Here,* 201.

19. Eiseley, *The Immense Journey,* 65.

20. Hooton, *Up from the Ape,* 196–97.

21. Johanson and Edey, *Lucy,* 326.

22. Lemonick and Dorfman, "Apes," 53.

23. According to Johanson and Edey, *Lucy,* 306.

24. Howells, *Getting Here,* 73.

25. Gore, "The Dawn of Humans," May 1997.

26. Mayr, *What Evolution Is,* 243, 248.

27. Wright, *Origin and Antiquity of Man,* 432.

28. Hooton, *Up from the Ape,* 142.

29. Sagan, *The Dragons of Eden,* 95.

30. Klein and Edgar, *The Dawn of Human Culture,* 56.

31. Gorman, "Cooking Up Bigger Brains," 86.

32. Quoted in Himmelfarb, *Darwinian,* 259.

33. Keith, *The Antiquity of Man,* 727.

34. Mayr, *What Evolution Is,* 245.

35. Brace, *Stages,* 62, 65, 69, 90, 99, 101.

36. Hooton, *Up from the Ape,* 400.

37. Fix, *The Bone Pedlars,* 118.

38. Day, *Guide to Fossil Man,* 368.

39. Tattersall and Schwartz, *Extinct Humans,* 210.

40. Oahspe, The Lords' Second Book 2:23–24; Jones, *Encyclopedia,* 344.

41. Oahspe, The Lords' Third Book 2:6.

42. Klein and Edgar, *The Dawn of Human Culture,* 20.

43. Shreeve, *The Neandertal Enigma,* 16.

44. Taylor, *The Great Evolution Mystery,* 190.

45. South African anthropologist John Robinson in his "classic" 1963 paper quoted in Leakey, *Origin,* 62.

46. Oahspe, First Book of the First Lords 3:8.

47. Linden, "A Curious Kinship: Apes and Humans," 26–27.

48. Mayr, *What Evolution Is,* 253.

49. Palmer, *Origins,* 20.

50. Ibid., 61.

51. Fagan, *Cro-Magnon,* 100.

52. Diamond, *The Third Chimpanzee,* 364; Richard Klein quoted in Hall, "Last of the Neanderthals," 42.

53. MacLeod and Graham-Rowe, "Every Primate's Guide to Schmoozing," 10.

54. Hall, "Last of the Neanderthals," 54.

55. Ibid., 50.

56. Lewin, *Origin,* 33; Fagan, *Prehistory,* 60.

57. DeSalle and Tattersall, *Human Origins,* 88.

58. Tattersall and Schwartz, *Extinct Humans,* 245

59. Hoyle and Wickramasinghe, *Evolution from Space,* 102–3.

60. Denton, *Evolution,* 43, 324, 351.

61. Fisher and Gould quoted in Ridley, *Evolution,* 29, 121.

62. Monad quoted in Ridley, *Evolution,* 413.

63. Hayward, *Creation and Evolution,* 28.

64. Barone, "Not So Fast, Einstein," 12.

65. Dixon, *The Racial History of Man,* 504.

66. Biello, "Need for Speed?"

67. Edey and Johanson, *Blueprint,* 126.

68. Huse, *The Collapse of Evolution,* 150.

69. Behe, *Darwin's Black Box,* 113–15.

70. Denton, *Evolution,* 149.

71. Behe, *Darwin's Black Box,* 47, 126.

72. Hornyanszky and Tasi, *Nature's I.Q.,* 40.

73. Von Daniken, *In Search of Ancient Gods,* 204.

74. Charroux, *Masters of the World,* 66.

75. Sagan, *The Dragons of Eden,* 127.

76. DeSalle and Tattersall, *Human Origins,* 110.

77. Von Ward, *We've Never Been Alone,* 110.

78. Nardo, *Evolution,* 211.

79. Charroux, *The Gods Unknown,* 240.

80. Ginzburg, *Legends of the Jews.*

81. Oahspe, First Book of the First Lords 4:19.

82. Ibid., First Book of the First Lords 2:6–7 and 2:21.

83. Ibid., First Book of the First Lords 2:9.

84. Ibid., Book of Wars 21:2, 21:7, and 21:9.

85. Ibid., Book of Apollo 14:4.

86. Ibid., Synopsis of Sixteen Cycles: 16th cycle.

87. Spence, *The Myths of Mexico and Peru*, 234.

88. Brinton, *The Myths of the New World*, 24.

89. Reader, *Missing Links*, 232.

CHAPTER 10. WHEN THE WORLD WAS YOUNG

1. Gish, *Evolution: The Fossils Say No!*

2. Morris, *Scientific Creationism*, 92.

3. Taylor, *The Great Evolution Mystery*, 88.

4. Johnson, *Trial*, 43. See also Martinez, *The Time of the Quickening*, chapter 4.

5. Dennett, *Darwin's Dangerous Idea*, 149.

6. Behe, *Darwin's Black Box*, 249.

7. Oahspe, Book of Cosmogony and Prophecy 4:18–19.

8. Ibid., Book of Divinity 15:6.

9. Ibid., Book of Cosmogony 8:4–5.

10. Ibid., Book of Jehovih 6.6, Book of Cosmogony and Prophecy 8:4–5.

11. Milton, *Shattering the Myths of Darwinism*, 83, 23.

12. Quoted in Thomson, *The Watch on the Heath*, 205.

13. Krauss, "How the Higgs Boson Posits a New Story of Our Creation," 5.

14. Behe, *Darwin's Black Box*, 168, 173.

15. Hitching, *The Neck of the Giraffe*, 67.

16. Hoyle and Wickramasinghe, *Evolution from Space*, 148.

17. Wright, *Origin and Antiquity of Man*, 412.

18. Quoted in Lovtrup, *Darwinism*, 226.

19. Oahspe, The Lords' Fifth Book 1:11–12.

20. Ibid., Book of Jehovih 4:14, 5:7, 6:6, and 6:11; Book of Cosmogony and Prophecy 5:8; Book of Knowledge 4:28.

21. Oppenheimer, *The Real Eve*, 205.

22. Eiseley, *The Immense Journey*, 84.

23. Ingman and Gyllesten, "Mitochondrial Genome Variation," 1600–606.

24. Sykes, *Daughters*, 124.

25. Oppenheimer, *The Real Eve*, xxi.

26. Fagan, *Cro-Magnon*, chapter 5.

27. Goodman, *The Genesis Mystery*, 218.

28. Himmelfarb, *Darwinian*, 316.

29. Robbins, "How Old Is the Earth, Really?" 69.

30. Wright, *Origin and Antiquity of Man,* 478–79, appendix.

31. Milton, *Shattering the Myths of Darwinism,* 67.

32. Oahspe, Book of Jehovih 5:14.

33. Ibid., The Lords' Fourth Book 4:23–24.

34. Henderson, "Blame for Decline of Bison Misplaced, Scientists Find."

35. Oahspe, Book of Sethantes 16:4.

36. Ibid., Book of Divinity 11:12; Book of Wars 21:7; First Book of the First Lords 4:7–9.

37. Norman, *Gods and Devils from Outer Space,* 127.

38. Oahspe, The Lords' Second Book 2:5; First Book of the First Lords 2:8; Book of Sue; The Lords' Third Book 2:6.

39. Trench, *Temple of the Stars,* 98.

40. Oahspe, Book of Cosmogony and Prophecy 5:11.

41. Diamond, *The Third Chimpanzee,* 218.

42. Johanson and Edgar, *From Lucy to Language,* 356.

43. Brace, *The Stages of Human Evolution,* 100.

44. Oahspe, Synopsis of Sixteen Cycles 1:22.

45. Oahspe, Book of Jehovih 5:8.

46. Himmelfarb, *Darwinian,* 344.

47. Cited by Coon, *Adventures and Discoveries,* 358.

48. Thorndike, *Mysteries,* 227.

49. Edey and Johanson, *Blueprint,* 277.

50. Abrams, "Brain Map Shows You Think Like a Worm."

51. Hoyle and Wickramasinghe, *Evolution from Space,* 114.

52. Keith, *The Antiquity of Man,* 456, 719, 726–77.

53. Thorndike, *Mysteries of the Past,* 224.

54. Dixon, *The Racial History of Man,* 503.

55. Kolosimo, *Timeless Earth,* 8.

56. Hazen, *The Scientific Quest for Life's Origin,* 141.

CHAPTER 11. NOT OUT OF AFRICA

1. Klein and Edgar, *The Dawn of Human Culture,* 97.

2. Goodman, *The Genesis Mystery,* 135.

3. Johanson and Edey, *Lucy,* 121.

4. Klein and Edgar, *The Dawn of Human Culture,* 119.

5. Lewin, *The Origin of Modern Humans,* 104.

6. Reported in Cremo, *Human Devolution,* 89.

7. Stix, "Traces of a Distant Past," 60; Shreeve, *The Neandertal Enigma,* 67.

8. Brig, "Letter to the Editor," 9.

9. Trench, *Temple of the Stars,* 99–100.

10. Wills, *The Darwinian Tourist,* 260.

11. Gore, "The Dawn of Humans," 90.

12. Leakey and Lewin, *Origins,* 120.

13. Gore, "The Dawn of Humans," 101.

14. Fagan, *Cro-Magnon,* 90, 97, 104.

15. Nardo, *Evolution,* 214.

16. Wong, "Hobbit Hullabaloo," 22.

17. Neimark, "Meet the New Human Family," 53–54.

18. Thorndike, *Mysteries of the Past,* 225–26.

19. Oahspe, The Lords' Third Book 3:15.

20. Braidwood, *Prehistoric Men,* 166.

21. Bowler, *Evolution: The History of an Idea,* 297.

22. Mellars, *The Neanderthal Legacy,* 211.

23. Boule, *Fossil Men,* 482–83.

24. Goodman, *The Genesis Mystery,* 157.

25. Oahspe, The Lords' First Book 1:74.

26. Dawkins, *The Blind Watchmaker,* 274.

27. Kolosimo, *Not of This World,* 114.

28. Falk, *The Fossil Chronicles,* 162, 184–85.

29. Cremo and Thompson, *Forbidden Archaeology,* 566–69.

30. Le Gros Clark, *The Fossil Evidence for Human Evolution,* 166, 173.

31. Fagan, *Cro-Magnon,* 35.

32. Hardaker, *The First Americans,* 258–59.

33. Gore, "Georgian Skull Find."

34. Gore, "The Dawn of Humans," 101.

35. DeSalle and Tattersall, *Human Origins,* 133.

36. Braidwood, *Prehistoric Men,* 27.

37. Gore, "The Dawn of Humans," 108.

38. Adcock, et al., "Mitochondrial DNA Sequences in Ancient Australians," 537–42.

39. Tenodi, "Problems in Australian Art and Archaeology," 15–16.

40. Lubenow, *Bones of Contention*, 268.

41. Howells, *Getting Here*, 174.

42. de Quatrefages, *The Pygmies*, 185–86.

43. Lewin, *The Origin of Modern Humans*, 104.

44. Landsburg and Landsburg, *In Search of Ancient Mysteries*, 90.

45. Coon, *The Living Races of Man*, 656; see also Mithen, *After the Ice*, 314.

46. Oppenheimer, *Eden of the East*, 373.

47. Mithen, *After the Ice*, 361–63.

48. Cremo and Thompson, *Forbidden Archaeology*, 183.

49. Renfrew, *Prehistory*, 69, 72, 85.

50. Brass, *The Antiquity of Man*, 186.

51. Goodman, *The Genesis Mystery*, 127.

52. Reader's Digest Association, eds., *Man and Beast*, 300.

53. Keith, *The Antiquity of Man*, 71.

54. Coon, *The Living Races of Man*, 140.

55. Thorndike, *Mysteries of the Past*, 225.

56. Thorndike, *Mysteries of the Past*, 225–26.

57. Kolosimo, *Not of This World*, 29.

58. Oahspe, The Lords' Fifth Book 5:13.

59. Coon, *The Living Races of Man*, 84–85.

60. Keith, *The Antiquity of Man*, 362.

61. Oppenheimer, *Eden of the East*, 398.

62. Lehrman, "From Race to DNA," 23–24.

63. Boule, *Fossil Men*, 280, 306.

64. Oahspe, The Lords' Fifth Book 3:3.

65. Palmer, *Origins*, 29.

66. Oahspe, Book of Cosmogony and Prophecy 5:12.

67. Keith, *The Antiquity of Man*, 366.

68. Eiseley, *The Immense Journey*, 133–38.

69. Hawk, "The 'Amazing' Boskops," http://johnhawks.net/weblog/reviews/brain/paleo/lynch-granger-big-brain-boskops-2008.html.

70. Davis et al., *Of Pandas and People*, 111; Tattersall and Schwartz, *Extinct*, 414.

71. Von Ward, *We've Never Been Alone*, 175.

72. Klein and Edgar, *The Dawn of Human Culture*, 229.

73. According to Day, *Guide to Fossil Man,* 325; Goodman, *The Genesis Mystery,* 83.

74. Hooton, *Up from the Ape,* 381.

75. Bahn, *100 Great Archaeological Discoveries,* 192.

76. Charles Berlitz, *World of the Incredible but True,* 76.

77. Leakey and Lewin, *Origins,* 142.

78. Lubenow, *Bones of Contention,* 119, 226.

79. Passage cited in Himmelfarb, *Darwinian,* 344.

80. Stuart, *Discovering Man's Past in the Americas,* 36.

81. Gould, *The Panda's Thumb,* 165.

82. Steiger, *Mysteries of Time and Space,* 20.

83. Gladwin, *Men Out of Asia,* 89, 158.

84. Norman, *Gods and Devils,* 126–27, 139.

85. De Camp, *Lost Continents,* 148.

86. Keith, *The Antiquity of Man,* 484.

87. Gladwin, *Men Out of Asia,* 59.

88. Steen-McIntyre, "The Enigmatic Ostrander Skull," 17.

89. Corliss, *Ancient Man,* 672–74, 677.

90. Corliss, *The Unexplained,* 8.

91. Steen-McIntyre, "The Enigmatic Ostrander Skull," 17.

92. Wright, *Origin and Antiquity of Man,* chapter 7.

93. Dixon, *The Racial History of Man,* 458–59.

94. Hrdlicka et al., *Early Man in South America,* 178, 155.

95. All Argentina references are from Hrdlicka et al., *Early Man in South America,* 273–76, 292.

96. Dixon, *The Racial History of Man,* 503.

CHAPTER 12. DEAD MAN WALKING

1. Montagu, *Man: His First Two Million Years,* 68–69.

2. Neimark, "Meet the New Human Family," 58.

3. Hooton, *Up from the Ape,* 338, 695.

4. Brace, *The Stages of Human Evolution,* 95.

5. Childress, *Lost Cities,* 358.

6. Diamond, *The Third Chimpanzee,* 41; Shanidar is dated 60–40 kyr.

7. Lewin, *In the Age of Mankind,* 126.

8. Lubenow, *Bones of Contention,* 81–82.

9. Goodman, *The Genesis Mystery,* 170.

10. Oahspe, Book of Divinity 11:16 and Book of Wars 21:7.

11. Lubenow, *Bones of Contention,* 348–49.

12. Ibid., 265–66.

13. Keith, *The Antiquity of Man,* 202.

14. Milton, *Shattering the Myths of Darwinism,* 207.

15. Tiffany, "Editor's Comments," 5–14. See also for example, Oahspe, The Lords' Second Book 3:10; The Lords' Fourth Book 4:4; Book of Fragapatti 39:1.

16. Sykes, *Seven Daughters,* 129.

17. Coon, *The Living Races of Man,* 12.

18. Montagu, *Man: His First Two Million Years,* 69.

19. Keith, *The Antiquity of Man,* 100.

20. Coleman, *The Field Guide to Bigfoot, Yeti, and Other Mystery Primates Worldwide,* 98.

21. Eiseley, *The Immense Journey,* 133.

22. Steiger, *Monsters among Us,* 72.

23. See Apollo's mission to correct the monstrosities in Oahspe, Book of Apollo, Chapter 5 and 2:5.

24. Marsh, *White Indians of Darien,* 20–21.

25. Oahspe, The Lords' Fifth Book 1:14.

26. Milton, *Shattering the Myths of Darwinism,* 271.

27. Oahspe, Book of Es 8:29 and 20:37–39.

28. Higgins, *Anacalypsis,* 258.

29. Macbeth, *Darwin Retried,* 147.

APPENDIX E: SOME ANACHRONISMS

1. Montagu, *Man: His First Two Million Years,* 48.

2. Goodman, *The Genesis Mystery,* 179.

APPENDIX F: NEW WORLD NOT SO NEW

1. Time-Life eds., *The Human Dawn,* 49.

2. De Camp, *Lost Continents,* 105.

3. Oahspe, The Lords' Third Book 1.

4. Wilkins, *Mysteries of Ancient South America,* 98.

5. Berlitz, *Mysteries from Forgotten Worlds,* 151–53.

6. Palmer, *Origins,* 249.

7. Brinton, *The Myths of the New World,* 188.

GLOSSARY

anthropogenesis: the origin of man

Aurignacian: period of European culture from ca 40 to 28 kya; time of the Great Leap Forward

bipedal: the two-legged gait typical of hominids

brachycephalic: round headed, as were the little people; Mongoloids are typically brachycephalic

Cambrian Period: 600 to 500 mya, many forms of life arise but in the seas only

Cretaceous Period: 135 to 65 mya, after the Jurassic, rise of small mammals and birds

dolichocephalic: long headed from front to back (as viewed from above)

genome: the sum total of our, or any animal's, genetic makeup, the DNA package of a species

genus: in scientific classification, a well-defined group (or taxon) with a unique set of traits; for example, in *Homo sapiens*, *Homo* is the genus, and *sapiens* is the species

Holocene Epoch: Recent time, starting around the same time as the Neolithic, about 10 or 11 kya

hominid: term used to designate all manlike fossils discovered

Magdalenian Period: 18 to 11 kya, just before the Neolithic, applies to culture level in Europe, which is best known for its impressive (Cro-Magnon) cave art

Mesolithic: 20 to 10 kya; lit., Middle Stone Age

Miocene Epoch: 25 to 13 mya, before the Pliocene

monogenesis: theory of single (continent) as the cradle of man (versus polygenesis)

morphogenesis: how structures came about; origin of anatomical features

morphology: form and structure of living things; observations on skeletal anatomy

Mousterian: about 70 to 40 kya; mostly in reference to European Neanderthal culture, tools, and environment

Neolithic: literally "new stone"; about 10 to 5 kya, the prehistoric period before the Bronze Age

Paleolithic: literally "old stone"; about 600 to 15 kya, divided into Lower, Middle, and Upper

Panic: the language of Pan (also called Lemuria or Mu)

phenotype: sum total of observable inherited traits (term used in genetics)

phylogeny: an evolutionary sequence; the idea that all creatures have descended over time from a common ancestor, assuming that one species has evolved lineally to the other

pithecanthropines: hominids of the *Homo erectus* type; first specimens were discovered in Asia

Pleistocene Epoch: literally "most recent"; geological time 2-plus mya to 10 kya

Pliocene Epoch: geological time 12 to 2 mya (some say began only 5 mya)

Glossary   529

polygenesis: multiple Gardens of Eden; theory of man's advent in several different places

prognathous: seen in profile, bulging of midface or alveolar (mouth) region

Solutrean: about 25 to 18 kya; culture in Europe in Late Paleolithic (Cro-Magnon)

species: in scientific classification (taxonomy), species is a unique type of plant or animal (e.g., in *Homo sapiens*, *sapiens* is the species name, while *Homo* is the genus). Below the level of species is subspecies, which is the same as race. Species is a breeding group.

Tertiary Period: 65 to ca 3 mya, sometimes called the age of mammals

torus: bar of bone (eyebrow area)

BIBLIOGRAPHY

Abrams, Michael. "Brain Map Shows You Think Like a Worm." *Discover,* January 2011, 36.

Adams, Richard E. W. *Prehistoric Mesoamerica.* Norman, Okla.: University of Oklahoma Press, 1996.

Adcock, Gregory J., Elizabeth S. Dennis, Simon Easteal, Gavin A. Huttley, Lars S. Jermiin, W. James Peacock, and Alan Thorne. "Mitochondrial DNA Sequences in Ancient Australians: Implications for modern human origins." *Proceedings of the National Academy of Sciences of the USA* 98, no. 2 (January 16, 2001): 537–42.

Ager, Derek V. "The Nature of the Fossil Record." *Proceedings of the Geologists' Association* 87, no. 2 (1976): 131–159.

Andrews, Roy Chapman. *Meet Your Ancestors.* New York: Viking Press, 1945.

Bahn, Paul G. *100 Great Archaeological Discoveries.* New York: Barnes & Noble Books, 1995.

Barloy, Jean-Jacques. *Prehistory from Australopithecus to Mammoth Hunters.* Paris: Barron's Educational Series, 1986.

Barone, Jennifer. "Not So Fast, Einstein." *Discover,* April 2007, 12.

Barzun, Jacques. *Darwin, Marx, Wagner: Critique of a Heritage.* Chicago: University of Chicago Press, 1981.

Behe, Michael J. *Darwin's Black Box.* New York: The Free Press, 1996.

Berlitz, Charles. *Incredible But True.* New York: Stonesong, 1991.

———. *Mysteries from Forgotten Worlds.* New York: Dell, 1972.

Bernheimer, Richard. *Wild Men in the Middle Age.* Cambridge, Mass.: Harvard University Press, 1952.

Biello, David. "Need for Speed?" *Scientific American,* June 2008, 31.

"Bone Collection Backs Up Hobbit Theory." *New Scientist,* October 15, 2005, 9.

Boule, Marcellin. *Fossil Men: A Textbook of Human Paleontology.* Trans. Jesse Elliot Ritchie and James Ritchie. Edinburgh: Oliver and Boyd, 1923.

Bowler, Peter J. *Evolution: The History of an Idea.* Berkeley: University of California Press, 2003.

Brace, C. Loring. *The Stages of Human Evolution.* Englewood Cliffs, N.J.: Prentice-Hall, 1979.

Brackenridge, H. H. "On the Population and Tumuli of the Aborigines of North America." A letter to Thomas Jefferson read before the American Philosophical Society, October 1, 1818.

Braidwood, Robert J. *Prehistoric Men.* New York: William Morrow, 1967.

Brass, Michael. *The Antiquity of Man.* Baltimore, Md.: AmErica House, 2002.

Brig. "Letter to the Editor." *Atlantis Rising* 67 (January/February 2008): 9.

Brinton, Daniel G. *The Myths of the New World.* Blauvelt, N.Y.: Multimedia Publishing, 1976.

Broom, Robert. *The Coming of Man.* London: H. F. & G. Witherby, 1933.

Campbell, Joseph. *Primitive Mythology.* Vol. 1 of *The Masks of God.* New York: Penguin Books, 1987.

Cannell, Alan. "Sexual Selection in Archaic Populations." *Pleistocene Coalition News* 2, no. 4 (July/August 2010): 2–3.

Caudill, Edward. *Darwinian Myths.* Knoxville: University of Tennessee Press, 1997.

Charroux, Robert. *The Gods Unknown.* New York: Berkley Publishing, 1972.

———. *Masters of the World.* New York: Berkley Books, 1967.

———. *The Mysteries of the Andes.* New York: Avon Books, 1974

———. *One Hundred Thousand Years of Man's Unknown History.* New York: Berkley Publishing, 1963.

Childe, V. Gordon. *The Most Ancient East.* Hertford, UK: Stephen Austin & Sons, 1928.

Childress, David Hatcher. *Lost Cities of North and Central America.* Stelle, Ill.: Adventures Unlimited Press, 1993.

Chouinard, Patrick. *Forgotten Worlds.* Rochester, Vt.: Inner Traditions, 2012.

Churchward, James. *The Children of Mu.* New York: Paperback Library, 1931.

———. *The Lost Continent of Mu.* New York: Paperback Library, 1931,

Clark, Ella E. *Indian Legends of the Northern Rockies*. Norman: University of Oklahoma Press, 1966.

Cloud, Preston. *Oasis in Space*. New York: W. W. Norton, 1988.

Coleman, Loren, and Patrick Huyghe. *The Field Guide to Bigfoot, Yeti, and Other Mystery Primates Worldwide*. New York: Avon Books, 1999.

Coon, Carleton. *Adventures and Discoveries*. Englewood Cliffs, N.J.: Prentice Hall, 1981.

———. *The Living Races of Man*. New York: Alfred A. Knopf, 1965.

———. *The Origin of Races*. New York: Alfred A. Knopf, 1962.

Coppens, Philip. *The Lost Civilization Enigma*. Franklin Lakes, N.J.: New Page Books, 2013.

Corliss, William. *Ancient Man*. Glen Arms, Md.: The Sourcebook Project, 1978.

———. *The Unexplained*. New York: Bantam Books, 1976.

Cottrell, Leonard, ed. *The Concise Encyclopedia of Archaeology*. New York: Hawthorne Books, 1960.

Coyne, Jerry A. *Why Evolution Is True*. New York: Viking Press, 2009.

Cremo, Michael. *Human Devolution*. Los Angeles: Torchlight Publishing, 2003.

Cremo, Michael, and Richard Thompson. *Forbidden Archaeology*. Los Angeles: Bhaktivedanta Book Publishing, 1998.

Cronin, Helena. *The Ant and the Peacock: Altruism and Sexual Selection from Darwin to Today*. Cambridge, UK: Cambridge University Press, 1991.

Curry, Andrew. "Ancient Excrement." *Archaeology*, July/August 2008, 42–45.

Daily Mail reporter. "First Modern Humans Protected Themselves Against Disease After Leaving Africa by 'Interbreeding with Neanderthals'," June 17, 2011. www.dailymail.co.uk/sciencetech/article-2004705.

Darwin, Charles. *The Autobiography of Charles Darwin*. Ed. Nora Barlow. New York: Harcourt Brace, 1958.

———. *The Descent of Man*. New York: Penguin Books, 2004.

———. *The Origin of Species*. Cambridge, Mass: Harvard University Press, 1964.

Davis, Percival and Dean H. Kenyon. *Of Pandas and People: The Central Question of Biological Origins*. 2nd ed. Dallas: Haughton Publishing, 1993.

Dawkins, Richard. *The Blind Watchmaker*. New York: W. W. Norton, 1996.

Day, Michael H. *Guide to Fossil Man*. Chicago: University of Chicago Press, 1988.

De Camp, L. Sprague. *Lost Continents*. New York: Ballantine Books, 1975.

Deloria, Vine, Jr. *Red Earth, White Lies*. New York: Scribner, 1995.

Dennett, Daniel C. *Darwin's Dangerous Idea*. New York: Simon & Schuster, 1995.

Dennon, Jim. "Removing Oahspe's Enigmas." Privately published (1979).

Denton, Michael. *Evolution: A Theory in Crisis*. Bethesda, Md.: Adler & Adler, 1986.

de Quatrefages, Armand. *The Pygmies*. New York: D. Appleton, 1895.

de la Vega, Garcilaso. *Comentarios reales de los Incas, 1609–17.*

DeSalle, Rob, and Ian Tattersall. *Human Origins*. Bronx, N.Y.: Texas A&M Press, 2008.

Diamond, Jared. *The Third Chimpanzee*. New York: HarperPerennial, 1992.

Dingir, Ishtar Babilu. "The Origin of Hubris." *Pleistocene Coalition News* 2, no. 4 (July/August 2010): 14–16.

Dixon, Roland. *The Racial History of Man*. New York: Charles Scribner's Sons, 1923.

Drake, W. Raymond. *Gods and Spacemen of the Ancient East*. New York: Signet Book, 1968.

———. *Gods and Spacemen of the Ancient Past*. New York: Signet Book, 1974.

Edey, Maitland, and Donald Johanson. *Blueprint: Solving the Mystery of Evolution*. New York: Penguin Books, 1990.

Eiseley, Loren. *Darwin's Century*. Garden City, N.Y.: Anchor Books, 1961.

———. *The Immense Journey*. New York: Time Reading Program, 1962.

Eldredge, Niles. *Darwin: Discovering the Tree of Life*. New York: W. W. Norton, 2005.

Ellis, William. *Polynesian Researches*. Rutland, Vt.: Charles E. Tuttle, 1969.

Evans, Ivor H. N. *The Negritos of Malaysia*. Cambridge, Mass.: Cambridge University Press, 1937.

Fagan, Brian M. *Cro-Magnon: How the Ice Age Gave Birth to the First Modern Humans*. New York: Bloomsbury Press, 2010.

———. *World Prehistory: A Brief Introduction*. New York: Longman, 1999.

Falk, Dean. *The Fossil Chronicles*. Berkeley: University of California Press, 2011.

Feder, Kenneth L. *Frauds, Myths and Mysteries*. Mountain View, Calif.: Mayfield Publishing, 1990.

Federer, William J. *America's God and Country: Encyclopedia of Quotations*. Coppell, Tex.: Fame Publishing, 1996.

Feliks, John. "Ardi." *Pleistocene Coalition News* 2, no. 1 (January/February 2010): 1–3.

———. "The Pleistocene Coalition: Exploring a New Paradigm." *Pleistocene Coalition News* 2, no. 5 (September/October 2010): 5–7.

Fell, Barry. *America B.C.* New York: Demeter Press Book, 1977.

———. *Saga America.* New York: Times Books, 1980.

"Fight the Good Fight." (Editorial) *New Scientist,* October 2005, 3.

Filler, Aaron G. *The Upright Ape.* Franklin Lakes, N.J.: New Page Books, 2007.

Fischman, Josh. "Part Ape, Part Human." *National Geographic,* August 2011, 120–33.

Fix, William R. *The Bone Pedlars.* New York: Macmillan, 1984.

Flindt, Max H., and Otto O. Binder. *Mankind: Child of the Stars.* Huntsville, Ark.: Ozark Mountain Publishing, 1999.

Flower, William H. *The Pygmy Races of Men.* London: Royal Institute Lecture, 1888. Reprinted in *Essays on Museums,* 1898.

Fodor, Jerry, and Massimo Piattelli-Palmarini. *What Darwin Got Wrong.* New York: Farrar, Straus & Giroux, 2010.

Fodor, Nandor. *An Encyclopedia of Psychic Science.* Secaucus, N.J.: The Citadel Press, 1966.

Folger, Tim. "Scientists Get Inside the Mind (and Genes) of the Neanderthals." *Discover,* January 2007, 29.

Frobenius, Leo. *The Childhood of Man.* London: Seeley & Co., 1909.

Gamble, Clive. *Timewalkers.* Cambridge, Mass.: Harvard University Press, 1994.

Ginzburg, Louis. *Legends of the Jews.* Philadelphia: Publications Society, 2003.

Gish, Duane T. *Evolution: The Fossils Say No!* San Diego: Creation Life Publishers, 1978.

Gladwin, Harold S. *Men Out of Asia.* New York: McGraw-Hill, 1947.

Glausiusz, Josie. "A Trail of Ancient Genes." *Discover,* October 2006, 71.

Goodman, Jeffrey. *American Genesis.* New York: Summit Books, 1981.

———. *The Genesis Mystery.* New York: Times Books, 1983.

Gore, Rick. "The Dawn of Humans." *National Geographic,* May 1997, 84–109.

———. "The Dawn of Humans." *National Geographic,* July 1997, 96–102.

———. "Georgian Skull Find." *National Geographic,* August 2002.

———. "Tracking the First of Our Kind." *National Geographic,* September 1997, 92–99.

Gorman, Rachel Moeller. "Cooking Up Bigger Brains." *Scientific American,* January 2008, 86.

Gorner, Peter. "Tiny-human Find Becomes Huge News." *Orlando Sentinel,* October 28, 2004, A10.

Gottlieb, Anthony. "It Ain't Necessarily So." *The New Yorker,* September 17, 2012, 84–88.

Gould, Stephen Jay. *The Panda's Thumb.* New York: W. W. Norton, 1980.

———. "Dorothy, It's Really Oz." *Time,* August 23, 1999, 59.

Gregory, Jane. "Fight the Good Fight." *New Scientist* 2283 (March 2001): 3

Hall, Stephen S. "Last of the Neanderthals." *National Geographic,* October 2008, 36–59.

Hamilton, Garry. "Mother Superior." *New Scientist,* September 3, 2005, 26–29.

Hancock, Graham. *Fingerprints of the Gods: Evidence of Earth's Lost Civilization.* New York: Three Rivers Press, 1995.

Hardaker, Christopher. *The First Americans.* Franklin Lakes, N.J.: New Page Books, 2007.

Harrell, Eben. "Scientists Discover an Ancient Human Relative." *TIME: Science & Space,* March 24, 2010.

Hayward, Alan. *Creation and Evolution.* Minneapolis: Bethany House, 1985.

Hazen, Robert M. *The Scientific Quest for Life's Origin.* Washington, D.C.: Joseph Henry Press, 2005.

Henderson, Diedtra. "Blame for Decline of Bison Misplaced, Scientists Find." *Orlando Sentinel,* November 28, 2004. www.time.com/time/health/article/0,8599,1974903,00.html.

Hibben, Frank. *The Lost Americans.* Rev. New York: T. Y. Crowell, 1968. First published 1946.

Higgins, Godfrey. *Anacalypsis.* Chesapeake, N.Y.: ECA Associates, 1991. First published London, 1836.

Himmelfarb, Gertrude. *Darwin and the Darwinian Revolution.* Garden City, N.Y.: Doubleday, 1959.

Hitchcock, Romyn. *American Antiquities.* 1894.

Hitching, Francis. *The Neck of the Giraffe.* New Haven, Conn.: Ticknor and Fields, 1982.

Honore, Pierre. *In Quest of the White God.* New York: G. P. Putnam & Sons, 1964.

Hooper, Rowan. "The Scenic Route Out of Africa." *New Scientist,* May 21, 2005, 14.

Hooton, Earnest A. *Up from the Ape.* New York: Macmillan, 1963.

Hornyanszky, Balazs, and Istvan Tasi. *Nature's I.Q.* Badger, Calif.: Torchlight Publishing, 2009.

Howells, William. *Getting Here.* Washington, D.C.: The Compass Press, 1993.

Hoyle, Fred, and N. C. Wickramasinghe. *Evolution from Space.* New York: Simon & Schuster, 1984.

Hrdlicka, Ales, William Henry Holmes, Bailey Willis, Frederic Eugene Wright, and Clarence Norman Fenner. *Early Man in South America.* Bulletin 52. Washington, D.C.: Smithsonian Institution, 1912.

Huse, Scott. *The Collapse of Evolution.* Grand Rapids, Mich.: Baker Books, 1997.

Hyman, Stanley Edgar. *The Tangled Bank.* New York: Athenaeum Press, 1962.

Ingman, M., and U. Gyllesten. "Mitochondrial Genome Variation and Evolutionary History of Australian and New Guinean Aborigines." *Genome Research* 13, no. 7 (2003): 1600–6.

Johanson, Donald. "Face-to-Face with Lucy's Family." *National Geographic,* March 1996, 96–117.

Johanson, Donald, and Maitland Edey. *Lucy.* New York: Simon & Schuster, 1981.

Johanson, Donald, and Blake Edgar. *From Lucy to Language.* New York: Simon & Schuster, 1996.

Johnson, Phillip. *Darwin on Trial.* Downers Grove, Ill.: Intervarsity Press, 1991.

Jones, J. S. Book review. *Nature* 345 (May 31, 1990): 395

Jones, Martha Helene. *The Lost Data on the Chariots of the Elohim.* Arkansas: self-published on lulu.com, 2008. butsuri.homelinux.net/MHJ.

Jones, Stephen, ed. *The Cambridge Encyclopedia of Human Evolution.* New York: Cambridge University Press, 1994.

Joseph, Frank, ed. *Discovering the Mysteries of Ancient America.* Franklin Lakes, N.J.: New Page Books, 2006.

Keith, Arthur. *Ancient Types of Man.* London, New York: Harper & Brothers, 1911.

———. *The Antiquity of Man.* London: Williams & Norgate, 1929.

Klein, Richard G., and Blake Edgar. *The Dawn of Human Culture.* New York: John Wiley, 2002.

Kleiner, Kurt. "Early Toolmakers Cast Off Their Rock-banger Image." *New Scientist* 186, no. 2494 (April 2005): 7.

Koestler, Arthur. *The Ghost in the Machine.* Chicago: Henry Regnery, 1967.

Kolosimo, Peter. *Not of This World.* New York: Bantam Books, 1973.

———. *Timeless Earth.* New York: Bantam Books, 1975.

Krauss, Lawrence M. "How the Higgs Boson Posits a New Story of Our Creation." *Newsweek,* July 16, 2012, 5.

Kreisberg, Glenn, ed. *Mysteries of the Ancient Past.* Rochester, Vt.: Bear & Co., 2012.

Kruglinksi, Susan. "Are We All Asians?" *Discover,* May 2006, 13.

Kuhn, Herbert. *On the Track of Prehistoric Man.* New York: Random House, 1955.

Landsburg, Alan, and Sally Landsburg. *In Search of Ancient Mysteries.* New York: Bantam Books, 1974.

Layard, John. *Stone Men of Malekula.* London: Chatto and Windus, 1942.

Leakey, Richard. *The Origin of Humankind.* New York: Basic Books, 1994.

———. "Skull 1470." *National Geographic,* June 1973, 819–29.

Leakey, Richard, and Roger Lewin. *Origins: The Emergence and Evolution of Our Species and Its Possible Future.* New York: E. P. Dutton, 1977.

Le Gros Clark, W. E. *The Antecedents of Man.* Chicago: Quadrangle Books, 1960.

———. *The Fossil Evidence for Human Evolution.* Chicago: University of Chicago Press, 1955, 1964.

Lehrman. "From Race to DNA." *Scientific American,* 23–24.

Lemonick, Michael D., and Andrea Dorfman. "Up from the Apes." *Time,* August 23, 1999, 51–58.

Lewin, Roger. *In the Age of Mankind.* Washington, D.C.: Smithsonian Books, 1988.

———. *The Origin of Modern Humans.* New York: Scientific American Library, 1993.

Linden, Eugene. "A Curious Kinship: Apes and Humans." *National Geographic,* March 1992, 26–27.

Lovtrup, Soren. *Darwinism: The Refutation of a Myth.* New York: Croom Helm, 1987.

Lubenow, Marvin L. *Bones of Contention: A Creationist Assessment of Human Fossils.* Grand Rapids, Mich.: Baker Books, 2005.

Macbeth, Norman. *Darwin Retried.* New York: A Delta Book, 1973.

Mack, Bill. "Mexico's Little People." *Fate,* August 1984, 38–40.

MacLeod, Mairi, and Duncan Graham-Rowe. "Every Primate's Guide to Schmoozing." *New Scientist,* September 3, 2005, 10.

Marsh, R. O. *White Indians of Darien.* New York: G.P. Putnam's Sons, 1934.

Marshack, Alexander. "Use of Symbols Antedates Neanderthal Man." *Science Digest* 73 (March 1973).

Martinez, Susan B. *The Lost History of the Little People: Their Spiritually Advanced Traditions around the World.* Rochester, Vt.: Bear & Co., 2013.

———. *The Time of the Quickening: Prophecies for the Coming Utopian Age.* Rochester, Vt.: Bear & Co., 2011.

Mayell, Hillary. "Hobbit-Like Human Ancestor Found in Asia." *National Geographic News,* October 27, 2004. http://news.nationalgeographic .com/news/2004/10/1027_041027_homo_floresiensis.html.

Mayr, Ernst. *What Evolution Is.* New York: Basic Books, 2001.

McAuliffe, Kathleen. "The Incredible Shrinking Brain." *Discover,* September 2010, 54–56.

Mellars, Paul. *The Neanderthal Legacy.* Princeton, N.J.: Princeton University Press, 1996.

Milton, Richard. *Shattering the Myths of Darwinism.* Rochester, Vt.: Park Street Press, 1997.

Minkel, J. R. "Food for Symbolic Thought." *Scientific American,* January 2008, 18.

Mithen, Steven. *After the Ice.* Cambridge, Mass.: Harvard University Press, 2003.

Montagu, Ashley. *Man: His First Two Million Years; A Brief Introduction to Anthropology.* New York: Columbia University Press, 1969.

Moore, James F. *The Post-Darwinian Controversies.* Cambridge, UK: Cambridge University Press, 1979.

Morley, George. *A Path of Light.* Salt Lake City: Universal Faithists of Kosmon, 1962.

Morris, Henry M. *Scientific Creationism.* San Diego: CLP Publishers, 1981.

Nardo, Don, ed. *Evolution.* Farmington Hills, Mich.: Greenhaven Press, 2005.

Neimark, Jill. "Meet the New Human Family." *Discover,* May 2011, 48–55, 76.

———. "Pinkie Pokes Holes in Human Evolution." *Discover,* December 16, 2010. http://discovermagazine.com/2011/jan-feb/45#.Ub8sidfm_3A.

———. "Stone-Age Romeos and Juliets." *Discover*, January/February 2011, 67.

Norman, Eric. *Gods and Devils from Outer Space*. New York: Lancer Books, 1973.

Nowak, Rachel. "Symbiosis." *New Scientist*, April 9, 2005, 35.

Oahspe: A New Bible in the Words of Jehovih and His Angel Embassadors. New York: Oahspe Publishing, 1960. First published New York and London: Oahspe Publishing, 1882.

Oppenheimer, Stephen. *Eden of the East*. London: Weidenfeld & Nicolson, 1998.

———. *The Real Eve*. New York: Carroll & Graf, 2003.

Orwant, Robin. "Lessons from Our Closest Cousin." *New Scientist*, September 3, 2005, 6.

Oxnard, Charles E. *The Order of Man*. New Haven, Conn.: Yale University Press, 1984.

Palmer, Douglas. *Origins: Human Evolution Revealed*. London: Octopus Publishing Group, 2010.

Pike, Albert. *Morals and Dogma of the Ancient and Accepted Scottish Rite of Free Masonry*. Kila, Mont.: Kessinger Publishing, 1946. First published 1872.

Radin, Paul. *Indians of South America*. New York: Greenwood Press, 1969.

Reader, John. *Missing Links*. Boston: Little, Brown, 1981.

Reader's Digest Association, eds. *Man and Beast*. In *Quest for the Unknown* series. Karl P. N. Shuker, consultant. Pleasantville, N.Y.: Reader's Digest, 1993.

Renfrew, Colin. *Prehistory*. New York: Modern Library, 2008.

Ridley, Mark, ed. *Evolution*. New York: Oxford University Press, 2004.

Roach, John. "Neandertals, Modern Humans May Have Interbred, Skull Study Suggests." *National Geographic News*, January 16, 2008. http://news.nationalgeographic.com/news/2007/01/070116-neanderthals.html.

Robbins, Stephen E. "How Old Is the Earth, Really?" *Atlantis Rising*, July/August, 2008, 42–43, 69–70.

Rosenberg, Alfred. *The Myth of the Twentieth Century*. Costa Mesa, Calif.: Noontide Press, 1982. First published 1930.

Roughgarden, Joan. *Evolution's Rainbow*. Berkeley: University of California Press, 2004.

Rudwick, Martin J. S. *The Meaning of Fossils: Episodes in the History of Palaeontology*. Chicago: University of Chicago, 1985.

Sagan, Carl. *The Dragons of Eden*. New York: Ballantine Books, 1977.

Sarfati, Jonathan D. *Refuting Compromise*. Green Forest, Ark.: Master Books, 2004.

———. *Refuting Evolution*. Green Forest, Ark.: Master Books, 1999.

Saunders, William T., and Joseph Marino. *New World Prehistory*. Englewood Cliffs, N.J.: Prentice Hall, 1970.

Schoch, Robert. *Forgotten Civilization*. Rochester, Vt.: Inner Traditions, 2012.

Schulter, Dolph. *The Ecology of Adaptive Radiation*. Oxford: Oxford University Press, 2000.

Schwartz, Jean-Michel. *The Mysteries of Easter Island*. New York: Avon Books, 1973.

Schwartz, Jeffrey H. *Sudden Origins*. New York: John Wiley & Sons, 1999.

Seiglie, Mario. *Creation or Evolution?* Cincinnati, Ohio: United Church of God, 2000.

Shapiro, Robert. *Origins: A Skeptic's Guide*. New York: Bantam Books, 1987.

Shermer, Michael. *In Darwin's Shadow*. New York: Oxford University Press, 2002.

———. "Unweaving the Heart." *Scientific American,* Sept. 2005, 38.

Sherratt, Andrew, ed. *The Cambridge Encyclopedia of Archaeology*. New York: Crown Publishers, 1980.

Shreeve, James. *The Neandertal Enigma*. New York: William Morrow & Co., 1995.

Simpson, George Gaylord. *The Major Features of Evolution*. New York: Columbia University Press, 1953.

Spence, Lewis. *The Myths of Mexico and Peru*. New York: Dover Publications, 1994.

Steen-McIntyre, Virginia. "The Enigmatic Ostrander Skull." *Pleistocene Coalition News* 2, no. 5 (September/October 2010): 17

Steiger, Brad. *Monsters among Us*. New York: Berkley Books, 1989.

———. *Mysteries of Time and Space*. New York: Dell Publishing, 1974.

Stix, Gary. "Traces of a Distant Past." *Scientific American,* July 2008, 56–60.

Stuart, George. *Discovering Man's Past in the Americas*. Washington, D.C.: National Geographic Society, 1973.

Sullivan, Charles and Cameron McPherson Smith. "Getting the Monkey off Darwin's Back: Four Common Myths about Evolution." *Skeptical Inquirer*

29, no. 3 (May/June 2005). www.csicop.org/si/show/getting_the_monkey_off_darwins_back.

Switek, Brian. *Written in Stone.* New York: Bellevue Literary Press, 2010.

Sykes, Bryan. *The Seven Daughters of Eve.* New York: W. W. Norton, 2001.

Tattersall, Ian. *The Fossil Trail.* New York: Oxford University Press, 1995.

Tattersall, Ian, and Jeffrey Schwartz. *Extinct Humans.* New York: Westview Press, 2000.

Taylor, Gordon R. *The Great Evolution Mystery.* New York: Harper & Row, 1983.

Tenodi, Vesna "Problems in Australian Art and Archaeology." *Pleistocene Coalition News* 5, no. 1 (March/April 2013): 15–17.

Teresi, Dick. "Q and A." *Discover,* April 2009, 69.

Than, Ker. "Neanderthals Didn't Mate with Modern Humans, Study Says." *National Geographic News,* August 12, 2008. http://news.nationalgeographic.com/news/2008/08/080812-neandertal-dna.html.

Thomson, Keith. *Before Darwin.* New Haven, Conn.: Yale University Press, 2005.

———. *The Watch on the Heath.* London: Harper Collins, 2005.

Thorndike, Joseph, Jr. *Mysteries of the Past.* New York: Scribner, 1977.

Tiffany, John. "Editor's Comments." *The Barnes Review* XVI, no. 3 (May/June 2010): 5–14.

Time-Life, eds. *The Human Dawn.* Alexandria, Va.: Time-Life Books, 1990.

Tomas, Andrew. *We Are Not the First.* New York: Bantam Books, 1973.

Trench, Brinsley Le Poer. *Temple of the Stars.* New York: Ballantine Books, 1962.

Trinkaus, Erik, and Pat Shipman. *The Neandertals.* New York: Alfred Knopf, 1993.

Turnbull, Colin. *The Forest People.* New York: Simon & Schuster, 1968.

Tylor, Edward B. *Researches into the Early History of Mankind and the Development of Civilization.* Chicago: University of Chicago Press/Phoenix Books, 1964.

Urban, Sylvanus. "Antiquarian Researches." *Gentleman's Magazine* 3, no. 8 (1837): 182.

VanLandingham, Sam L. "Blocking Data, Part 2." *Pleistocene Coalition News,* September 2010, 2–3.

Van Sertima, Ivan. *They Came before Columbus.* New York: Random House, 1976.

Velikovsky, Immanuel. *Worlds in Collision*. New York: Dell Publishing, 1965.

Verrill, A. Hyatt. *The American Indian*. New York: New Home Library, 1927.

———. *Old Civilizations of the New World*. New York: New Home Library, 1943.

Von Ward, Paul. *We've Never Been Alone*. Charlottesville, Va.: Hampton Roads Publishing, 2011.

Wade, Nicholas. "Signs of Neanderthals Mating with Humans." *New York Times,* May 6, 2010. www.nytimes.com/2010/05/07/science/07neanderthal.html.

Watson, Lyall. *Supernature*. New York: Bantam Books, 1974.

Weikart, Richard. "Dehumanizing." *Atlantis Rising* 93 (May/June 2012): 34–36.

Wells, Spencer. "Dipping into the Gene Pool." *National Geographic* 207, no. 6 (June 2005): xxviii.

"What Really Killed the Dinosaurs." *Atlantis Rising* 63 (May/June 2007): 10.

"What's So Heavenly about the God Particle?" *Newsweek,* January 2, 2012, 50.

White, John W. "Evolution: An Enlightened View" in "Enlightenment 101: A Guide to God-Realization and Higher Human Culture." Unpublished manuscript.

———. *Pole Shift*. Garden City, N.Y.: Doubleday, 1980.

Wilford, John Noble. *The Riddle of the Dinosaur*. New York: Vintage Books, 1985.

Wilkins, Harold T. *Mysteries of Ancient South America*. Kempton, Ill.: Adventures Unlimited Press, 2000. First published 1947.

Wills, Christopher. *The Darwinian Tourist*. New York: Oxford University Press, 2010.

Wilson, Clifford. *The Chariots Still Crash*. New York: Signet Book, 1975.

Winchester, Simon. *Krakatoa: The Day the World Exploded: August 27, 1883*. New York: HarperCollins, 2003.

Winters, Jeffrey. "Hobbit Wars Heat Up." *Discover,* January 2007. http://discovermagazine.com/2007/jan/paleontology#.UctZCNfm_3A.

Wollaston, A. F. R. *Pygmies and Papuans*. New York: Sturgis and Walton, 1912.

Wolpoff, Milford, and Rachel Caspari. *Race and Human Evolution*. New York: Simon & Schuster, 1997.

Wong, Kate. "Hobbit Hullabaloo." *Scientific American,* June 2008, 22–24.

Woolls, Daniel. "Oldest Known Human Fossil Found in Europe." Associated Press, Madrid, March 26, 2008.

Wright, Frederick G. *Origin and Antiquity of Man.* Oberlin, Ohio: Bibliotheca Sacra, 1912.

Zimmer, Carl. "What Is a Species?" *Scientific American,* June 2008, 72–79.

INDEX

BOOKS OF RELATED INTEREST

The Lost History of the Little People
Their Spiritually Advanced Civilizations around the World
by Susan B. Martinez, Ph.D.

Time of the Quickening
Prophecies for the Coming Utopian Age
by Susan B. Martinez, Ph.D.

Before Atlantis
20 Million Years of Human and Pre-Human Cultures
by Frank Joseph

Lost Race of the Giants
The Mystery of Their Culture, Influence, and
Decline throughout the World
by Patrick Chouinard

Forgotten Worlds
From Atlantis to the X-Woman of Siberia and the Hobbits of Flores
by Patrick Chouinard

The Ancient Giants Who Ruled America
The Missing Skeletons and the Great Smithsonian Cover-Up
by Richard J. Dewhurst

DNA of the Gods
The Anunnaki Creation of Eve and the Alien Battle for Humanity
by Chris H. Hardy, Ph.D.

Göbekli Tepe: Genesis of the Gods
The Temple of the Watchers and the Discovery of Eden
by Andrew Collins

INNER TRADITIONS • BEAR & COMPANY
P.O. Box 388
Rochester, VT 05767
1-800-246-8648
www.InnerTraditions.com

Or contact your local bookseller